W9-BWU-855

recent advances in phytochemistry

volume 39

Chemical Ecology and Phytochemistry of Forest Ecosystems

RECENT ADVANCES IN PHYTOCHEMISTRY

Proceedings of the Phytochemical Society of North America
General Editor: John T. Romeo, *University of South Florida, Tampa, Florida*

Recent Volumes in the Series:

Cover design: Image of an odor plume taken using planar, laser-induced fluorescence
Concentration of the odor in the plume indicated by color. From Crimaldi *et al.*, *J. Turbul.* 3:
1–24, 2002. Reproduced with the permission of the authors and Taylor and Francis.
(Chapter 8)

recent advances in phytochemistry

volume 39

Chemical Ecology and Phytochemistry of Forest Ecosystems

Edited by

John T. Romeo
University of South Florida
Tampa, Florida, USA

2005

ELSEVIER

Amsterdam – Boston – Heidelberg – London – New York – Oxford
Paris – San Diego – San Francisco – Singapore – Sydney – Tokyo

ELSEVIER B.V.
Radarweg 29
P.O. Box 211, 1000 AE Amsterdam
The Netherlands

ELSEVIER Inc.
525 B Street, Suite 1900
San Diego, CA 92101-4495
USA

**ELSEVIER Ltd
The Boulevard, Langford Lane
Kidlington, Oxford OX5 1GB
UK**

ELSEVIER Ltd
84 Theobalds Road
London WC1X 8RR
UK

First edition 2005

Library of Congress Cataloging in Publication Data
A catalog record is available from the Library of Congress.

British Library Cataloguing in Publication Data
A catalogue record is available from the British Library.

ISBN: 0-08-044712-0

♾ The paper used in this publication meets the requirements of ANSI/NISO Z39.48-1992 (Permanence of Paper).
Printed in The Netherlands.

PREFACE

The Phytochemical Society of North America held its forty-fourth annual meeting in Ottawa, Ontario, Canada from July 24-28, 2004. The meeting was hosted by the University of Ottawa and the Canadian Forest Service, Great Lakes Forestry Centre and was held jointly with the International Society of Chemical Ecology. All of the chapters in this volume are based on papers presented in the symposium entitled "Chemical Ecology and Phytochemistry of Forest Ecosystems". The Symposium Committee was Mamdouh Abou-Zaid, John T. Arnason, Vincenzo deLuca, Constance Nozzolillo, and Bernard Philogene. They assembled an international group of phytochemists and chemical ecologists working primarily in northern forest ecosystems. While most of these scientists defy the traditional labels we are accustomed to, they brought to the symposium expertise in phytochemistry, insect biochemistry, molecular biology, genomics and proteomics, botany, entomology, microbiology, mathematics, and ecological modeling. It was a unique interdisciplinary forum of scientists working on the cutting edge in their respective fields and collectively creating a higher level of understanding for those present.

The chapter by Schmidt *et al.* focuses on induced defenses in conifers. They use biochemical and molecular approaches to study changes in terpenes, phenolics, and chitinases in response to herbivore and pathogen damage. The work emphasizes the importance of multiple defense systems in woody plants, an already well-know phenomenon in herbaceous plants. Inducible terpene resins are a special feature of conifers, and their long induction time sets them apart from other inducible defenses. They appear to act against both pathogens and herbivores. The work suggests that the application of jasmonates and other signaling compounds is particularly suited for manipulating plant defense systems *in vivo*, thus leading to the identification of genes regulating the formation of a single or a class of defensive substances. Martin and Bohlmann focus their chapter on transgenic resin duct development in direct defense and induced volatile emission in tritrophic defense that have both developed in species of spruce. They discuss the biochemical regulation of the diversity of this defense as well as its evolution. Particularly interesting is the multigenic nature of conifer terpene defense, the phylogeny of terpene synthase genes, and the phenotypic plasticity in long-lived plants. They provide, additionally, an outlook towards new genomics research in a forest health context.

Tittiger *et al.* shift the focus from the trees to the insects. The bark beetle *Ips pinii* has long been known to produce monoterpenes. Here, by applying functional genomics tools, it is shown to be useful as an unusual model tissue to study both digestive and pheromone-biosynthetic functions. Little is yet known about the biochemical tools bark beetles may use to tolerate the hostile environment of the host tree, but detoxification of phloem and resin by their midgets may be controlled by two novel genes that are highly transcribed there.   Additionally, multiple developmental, environmental, and endocrine signals appear to act on numerous genes in the pheromone-biosynthetic pathway. The regulatory schemes can be active post-transciptionally as well.   The midguts of bark beetles can, thus, be viewed as superb multi-taskers!   Raffa *et al.*, take the story up a notch.   They discuss interactions among conifer terpenoids and bark beetles across multiple levels of scale. This is an attempt to understand the links between physiological processes and population dynamics.   Each level, be it conifer, beetle, or microorganism, is characterized by a threshold whose *qualitative* outcome is determined by *quantitative* factors.   An individual compound can affect interactions across multiple levels of scale - from molecular through landscape-, and both mechanistic and landscape methodologies are needed to understand these.

Constabel and Major bring the topic to Angiosperms. *Populus* is often an ecological keystone species in forest ecosystems, and has complex interactions with symbionts, pathogens, and pests. They study the molecular biology and biochemistry of induced defense in this model system in regard to several phytochemicals including flavonoids, tannins, and terpenoids, and focuse on Kunitz trypsin inhibitors, polyphenol oxidase, chitinases, and several other induced genes and proteins. The impact of genomics on poplar defense is discussed with emphasis on EST libraries and transcript profiling.   Rapid progress in our understanding of defense signaling, such as identifying transcription factors that regulate sets of induced genes and signal transduction cascades, is the result.   In moving to the tropics, Isman reminds us that although tropical forests have produced a wealth of phytochemicals with bioactivity, only neem has been exploited to any degree. He emphasizes that the same factors that prevent the introduction of phytochemicals in industrialized societies should enhance their adoption in developing ones where many of the species grow.   Yoshida *et al.* focus their attention on polyphenols from medicinal plants, edible fruits, and beverages. They discuss their potential, in light of recent claims, for use in treatment of chronic diseases that are related to oxidative stress and microbial infections. Polyphenol roles as antioxidants, cancer chemopreventatives, and antimicrobial activities are becoming better understood.

The last three chapters focus on insects.   These will be particularly enlightening to the majority of the phytochemists who will read this volume. Examining things from the insect perspective is a place most of us have visited less often. Insects are challenged with a large diversity of volatiles emitted from plants. Monoterpenes predominate in gymnosperms, and sesquiterpenes in angiosperms, and both concentration and blend are an additional part of the mix that creates a plant's unique chemical signature.   Wright and Thomson show us how the study of the *statistic*s of naturally occurring odor scenes can yield information about the way organisms produce and use scent in their interactions. By relating olfactory physiology to the statistics of natural scenes, as has been previously done for visual and auditory sciences, we gain insight into the way in which odors are presented to the nervous system, not only at the level of the olfactory periphery but also higher up. Higher order coding contributes to the way the brain organizes information about odors, and this may provide a means of reliably producing a neural representation in the face of variation in natural stimuli.

Insects meet their phytochemical challenge with a large number of receptor protein types, and Honson *et al.* review in depth our current understanding of both the odorant and pheromone-binding proteins present in insects as well as their chemosensory-specific proteins. These insect proteins belong to superfamilies, each with a characteristic alpha-helical fold reinforced with disulfide bridges. Ligand binding has been studied, and a number of insights have arisen regarding functional significance. Dose and blend effects are being unraveled giving insight into roles as both semio- and allelo-chemicals. Mustaparta and Stranden discuss olfaction and learning in moths and weevils that live on both angiosperm and gymnosperm hosts. Comparisons between receptor neuron specificities among insects using different hosts have shown that these are narrowly tuned and can be classified according to one compound having the strongest effect -- the primary odorant-- and to a few related compounds with weak effects -- secondary odorants. Major differences between species that use gymnosperm and angiosperm hosts appear in the number of neurons tuned to mono- and sesquiterpenes, respectively. Future studies on receptor neuron specificity are needed if we are to understand how odorants influence the behavior of the insects, both as innate and as learned responses.

Certainly, the Symposium with its multi dimensions changed the way most us will view forest ecosystems in the future. The "out of symposium" interactions between the phytochemists and the chemical ecologists, who had not met together since 1990 in Quebec City, were both stimulating and fruitful. The setting of the Ottawa campus with the smell of conifer terpenes ever-lingering in the air seemed an appropriate venue. We thank the local organizing committee for their warmth and helpfulness, the University of Ottawa, and Natural Resources Canada, Canadian Forest Service for making it all possible. JTR, once again thanks Darrin T. King, who lessens the stress of putting this annual volume together, and all the contributing authors for their cooperation and good will.

John T. Romeo
University of South Florida

CONTENTS

CONTENTS

Chapter One

INDUCED CHEMICAL DEFENSES IN CONIFERS: BIOCHEMICAL AND MOLECULAR APPROACHES TO STUDYING THEIR FUNCTION

Axel Schmidt,[1]* Gazmend Zeneli,[1] Ari M. Hietala,[2] Carl G. Fossdal,[2]
Paal Krokene,[2] Erik Christiansen[2] and Jonathan Gershenzon[1]

[1]*Max Planck Institute for Chemical Ecology*
Hans-Knöll Strasse 8
D-07745 Jena, Germany

[2]*Norwegian Forest Research Institute (Skogforsk)*
Høgskoleveien 12
N-1432 Ås, Norway

*Author for correspondence, e-mail: aschmidt@ice.mpg.de

INTRODUCTION

Although our understanding of plant defense mechanisms has grown rapidly in recent years, most of the new knowledge has been obtained through studies on herbaceous species, especially the model plants *Arabidopsis thaliana, Medicago truncatula*, tomato, potato, maize, and rice.[1] Much less is known about the types of defenses employed by woody plants. Consequently, it is not clear if the deployment of chemical defense in woody taxa is fundamentally the same as that in herbs. Woody plants usually have a much greater size and longer lifetime than herbaceous plants, as well as a different life history,[2] and thus may be subject to different patterns of herbivore and pathogen pressure. In addition, woody plants have unique tissues, such as those resulting from secondary growth of the stem, and so may require different modes of protection.

To gain a complete picture of defense in the plant kingdom, it is essential to know more about the defenses of a variety of woody plants, both angiosperms and gymnosperms. Conifers are a distinctive and widespread group of woody gymnosperms whose 500-600 species include some of the largest and longest lived representatives of the plant kingdom.[3,4] They are significant climax species, dominating most of the major forest ecosystems of Europe, Asia, and North America. Of the 8-9 recognized families, the largest and geographically most widespread is the Pinaceae which includes *Pinus, Abies, Larix, Pseudotsuga,* and *Picea*. As a model species for studying conifer defense, we chose *Picea abies* (Norway spruce), the most abundant and economically-important conifer species in northern and central Europe.[5] In addition, much is already known about the herbivore and pest problems of *P. abies,*[6-11] which will be valuable in studying its defense mechanisms.

In this review, we examine the induced chemical defenses of *P. abies,* defenses whose levels increase following herbivore or pathogen attack. Induced defenses have attracted much attention in recent years because of their widespread occurrence in plants and their usefulness as subjects for study.[12] Here, we cover the induction of several different classes of induced defenses in *P. abies,* including terpene-containing resins, phenolic compounds, and chitinases. Our focus is not only on their defensive roles, but also on how the levels of these compounds may be manipulated by biochemical and molecular methods while minimizing other phenotypic changes. Manipulation of defense compounds in intact plants is a valuable approach to assessing their value to the plant.

TERPENES

The best studied chemical defense of *P. abies* and other conifers is the oleoresin found in foliage, stems, and other organs, a defense system that has existed for at least 50 million years.[13] Oleoresin is composed largely of terpenes, the largest class of plant secondary compounds.[14] Terpenes are formed by the fusion of C_5 isopentenoid units and classified by the number of such units present in their basic skeletons. Conifer resin is composed chiefly of monoterpenes (C_{10}) and diterpenes (C_{20}), with small amounts of sesquiterpenes (C_{15}) and other types of compounds. Oleoresin has long been believed to play a crucial role in conifer defense because of its physical properties (viscosity) and repellency to many herbivores and pathogens. In addition, oleoresin exudes under pressure from the tree following rupture of the ducts or blisters in which it is stored, often expelling or trapping invaders.[15-18] After rupture, the monoterpenes volatilize upon exposure to the air, while the diterpenes polymerize sealing the wound. However, it is still not clear to what extent the defensive properties of terpene resins are based on the repellency and toxicity of individual components or on the physical properties of the total resin (see chapter by Raffa, *et al.*, in this book).

In *P. abies*, oleoresin is found constitutively, but may also be induced by herbivore or pathogen attack.[19,20] We are focusing on the induced resin because of the potential of altering its production to test its protective role. In preliminary studies, we tried wounding and fungal inoculation in an attempt to induce terpene formation in large trees, but this gave variable and inconsistent results (Martin, D., Krokene, P., Gershenzon, J., and Christiansen, E., unpublished data). Since wounding itself can cause the loss of resin, especially the volatile components, we explored the utility of a non-invasive procedure for resin induction involving the application of methyl jasmonate, an elicitor of plant defense responses in many species.[21,22]

Methyl Jasmonate Application to Saplings in the Laboratory

When methyl jasmonate was sprayed on the foliage of 1-2 year-old *P. abies* saplings from a uniform genetic background, this treatment triggered a dramatic increase in terpene levels.[23] There was a more than 10-fold increase in monoterpenes and a nearly 40-fold increase in diterpenes in wood tissue. In contrast, in the bark there was a much smaller increase in monoterpenes and no significant change in diterpene levels. Curiously, the response to methyl jasmonate took much longer than previously-observed inductions of plant defenses with this elicitor. Significant increases were not seen until 15 days after application.[23] Examination of the anatomy of the treated saplings revealed that methyl jasmonate had stimulated the formation of a ring of new resin ducts (traumatic resin ducts) in the newly-formed xylem (Fig. 1.1). Franceschi and co-workers had previously shown that wounding of *P. abies* or infection with *Ceratocystis polonica*, a blue-stain fungus vectored by the bark beetle *Ips typographus,* could induce the appearance of traumatic resin ducts over a 36 day period.[19] Apparently, this response also occurred with methyl jasmonate.[24] A change is triggered in the developmental program of the cambium whereby some of the xylem mother cells become resin duct cells rather than tracheids. The fact that traumatic duct formation requires the differentiation of entirely new cells explains why terpene induction requires such a long interval after methyl jasmonate application. Formation of traumatic resin ducts represents a major investment for *P. abies* and puts heavy demands on limited resources. Careful anatomical studies have shown that both height growth and stem growth is reduced by about 50% in 2-year-old plants after traumatic ducts are induced by application of 100 mM methyl jasmonate externally to the stem bark (Krokene P. *et al.*, unpublished results).

Methyl jasmonate treatment not only triggers a dramatic change in terpene quantity, but also causes changes in terpene composition.[23] For example, of the two major monoterpenes in the wood, α-pinene and β-pinene, the proportion of α-pinene to β-pinene changed from about 1:1 in control saplings to 1:2 after methyl jasmonate treatment, with increases in the relative amounts of the (-)-enantiomers in relation to the (+)-enantiomers of both compounds. Among the diterpenes, levopimaric acid increased over 5-fold after methyl jasmonate treatment in comparison to a 2.5-fold increase in most of the other major diterpene acids.

Methyl jasmonate spraying also induced some increases in monoterpene and sesquiterpene levels in needles, but these were only 2-fold.[25] More significant was that methyl jasmonate application led to a 5-fold increase in the emission of terpenes from the foliage, and emission had a pronounced diurnal rhythm, with the maximum amount released during the light period. The composition of the emitted volatiles also shifted dramatically from a blend dominated by monoterpene olefins, such as α-pinene and β-pinene, to one in which the major compounds were sesquiterpenes, principally (*E*)-β-farnesene and (*E*)-α-bisabolene, as well as the oxygenated monoterpene, linalool. These compounds are of particular ecological interest, as they have been reported to attract natural enemies of herbivores or repel herbivores directly in other plant species.[25] Recent work has shown that methyl jasmonate treatment of large Norway spruce trees reduces both the attack rate and colonization success of the spruce bark beetle in the field (Krokene, P. and Christiansen E., unpublished results).

The dramatic increase in terpene formation, accumulation, and emission in *P. abies* in response to methyl jasmonate is consistent with the effect of methyl jasmonate or jasmonic acid on many defense compounds in angiosperms.[22] In conifers, jasmonates had been previously shown to promote the formation of heat shock[26] and defense signaling proteins,[27] to enhance resistance to pathogenic fungi,[28] and to promote colonization with ectomycorrhizae.[29] In relation to terpenes, jasmonates had been shown to promote formation of an oxygenated sesquiterpene, todomatuic acid, and an oxygenated diterpene, paclitaxel (taxol), in cell cultures,[30,31] but had never before been reported to enhance terpene accumulation in intact plants.

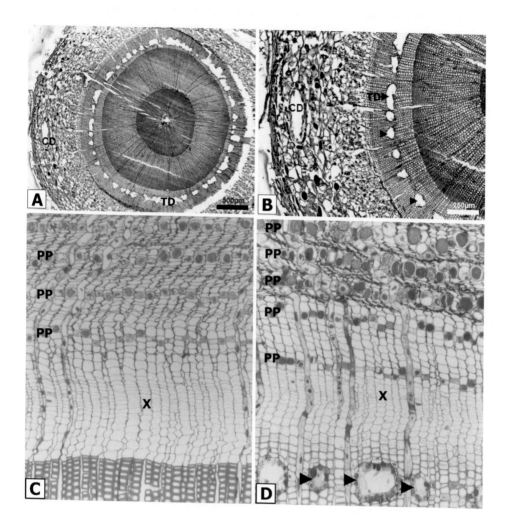

Fig. 1.1: Induced anatomical defense responses in Norway spruce. (A, B) Formation of a ring of new, traumatic resin ducts (TD, arrowheads) in the xylem of 2-year-old Norway spruce saplings after application of methyl jasmonate. A large cortical resin duct (CD) can be observed in the phloem, but these ducts do not appear to respond to methyl jasmonate treatment. (C) Normal phloem and sapwood anatomy of an older tree, with concentric rings of polyphenolic parenchyma cells (PP) in the phloem above the cambium (X) and normal wood below. (D) After treatment with methyl jasmonate or fungal infection the PP cells increase greatly in size and traumatic resin ducts (arrowheads) forms in the wood.

The finding has been corroborated by a study including small plants and larger trees of Norway spruce, where methyl jasmonate induced increased resin flow and other defense reactions when applied externally to intact bark.[32]

The concentrations of methyl jasmonate found to be effective in spraying *P. abies* saplings in our work (maximum effect at 10 mM) were relatively high compared to those typically used on angiosperm foliage: 10 μM – 1 mM.[33-35] This may only be a consequence of the need for higher concentrations to penetrate the thick cuticle of conifer needles. More recently, we have shown that a 100 μM spray of methyl jasmonate is effective in inducing terpene accumulation in *P. abies* saplings when formulated as a 0.5 % solution in Tween 20 detergent (Schmidt, A., unpublished results).

Methyl Jasmonate Application to Mature Trees in the Field

We were curious to see what effect methyl jasmonate treatment would have on the terpene oleoresin content of established trees in the field. If we could manipulate the amount of terpenes in mature trees, this might allow us to learn something about the effects of resin on many serious pest insects, such as bark beetles. Several clones of 30 year-old *P. abies* were treated with methyl jasmonate by painting an aqueous solution of 100 mM methyl jasmonate in 0.1 % Tween 20 on the bark with a paint roller. Samples taken from the methyl jasmonate-treated zones after four weeks showed that wood tissue from these areas had five times as many traumatic resin ducts as samples from untreated control trees (Fig. 1.1C, 1.1D) and a content of monoterpenes that was 2 – 2.5 times greater than that of control trees. Although the increase of terpenes in wood was much less than that observed in 1-2 year-old saplings after methyl jasmonate spraying in the laboratory, the mature trees in this study showed considerable flow of resin on the outer bark surface (Fig. 1.2). Presumably, this resin originates in the traumatic ducts, flowing to the outer bark via rays,[19,32] and so should be added to that found in the wood to get a true measure of terpene production triggered by methyl jasmonate. Significant variation was observed, however, among the clones used in this study, with one of the six clones showing no significant differences from control trees after the 100 mM methyl jasmonate treatment.

Fig. 1.2: External resin flow on a ca. 30-year-old Norway spruce clone after application of methyl jasmonate (100 mM in 0.1% Tween 20) and subsequent inoculation with the blue stain fungus *Ceratocystis polonica*.

To determine if the increased terpene content of methyl jasmonate-treated trees might be associated with increased resistance to enemies, we inoculated treated trees with the blue-stain fungus, *Ceratocystis polonica*, four weeks after methyl jasmonate application. Treatment with 100 mM methyl jasmonate dramatically reduced fungal growth in sapwood (2 % of control) and cambium necrosis caused by the fungus (19 % of control) (Fig. 1.3). It was satisfying to see how significantly the defensive potential of Norway spruce could be manipulated by methyl jasmonate in the field. However, jasmonates trigger a variety of induced defense systems in angiosperms,[22,36] and so resistance to *C. polonica* cannot be attributed to the increased terpene level without further experiments.

Fig. 1.3: Symptoms of fungal infection in Norway spruce after massive inoculation with the blue-stain fungus *Ceratocystis polonica*. In areas that have been successfully colonized by the fungus the phloem and cambial areas are necrotic and the sapwood is blue-stained (arrowheads).

The Search for Genes Encoding Short-Chain Isoprenyl Diphosphate Synthases-Branch-point Enzymes of Terpene Biosynthesis

To study more precisely the function of induced terpene resins in defense, it is necessary to develop a method to manipulate terpene formation without affecting other possible defenses. For this purpose, we initiated a long-term study of the molecular biology of induced terpene biosynthesis in *P. abies*. The basic outline of terpene formation is well understood[14] (Fig. 1.4). The C_5 building blocks of all terpenes are synthesized via the mevalonate pathway (localized in the cytosol) from acetyl-CoA or via the methylerythritol phosphate pathway (localized in the plastids) from pyruvate and glyceraldehyde-3-phosphate. These pathways produce isopententyl diphosphate (IPP, C_5) and its isomer dimethylallyl diphosphate (DMAPP, C_5). IPP and DMAPP then condense in a series of reactions to form isoprenyl diphosphates of 10, 15, 20 or more carbon atoms. Next, these different length diphosphate intermediates undergo reactions catalyzed by terpene synthases (principally cyclizations) to form the parent carbon skeletons of the major terpene classes. The parent carbon skeletons in turn undergo a series of secondary reactions, largely oxidations or reductions, to form the enormous variety of terpenes found in plants.

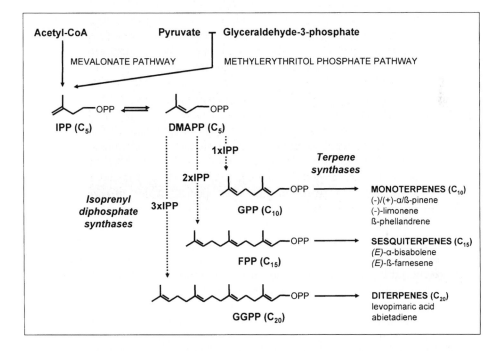

Fig. 1.4: Outline of terpenoid biosynthesis from isopentenyl diphosphate (IPP) via dimethylallyl diphosphate (DMAPP), geranyl diphosphate (GPP), farnesyl diphosphate (FPP) and geranylgeranyl diphosphate (GGPP). These reactions are catalyzed by isoprenyl diphosphate synthases and terpene synthases. The major products of the monoterpene, sesquiterpene, and diterpene pathways that constitute the oleoresin of *Picea abies* are listed. The general precursor IPP is derived either from the plastidial methylerythritol phosphate (MEP) pathway or the cytosolic mevalonate pathway.

While many of the molecular investigations of terpene biosynthesis have concentrated on terpene synthases (see chapter by Martin and Bohlmann, this book), we chose to study the isoprenyl diphosphate synthases, the enzymes that catalyze the assembly of the C_5 units into different chain length intermediates. By determining whether the final product will have 10, 15, 20 or more carbon atoms in its parent skeleton, isoprenyl diphosphate synthases specify the class of terpene to be formed. A series of three isoprenyl diphosphate synthase-catalyzed reactions occurs in all plants[14] (Fig. 1.5). First, DMAPP (C_5) and IPP (C_5) condense to form geranyl diphosphate (GPP, C_{10}). Addition of another molecule of IPP to GPP gives farnesyl diphosphate (FPP, C_{15}), while reaction of FPP with IPP gives geranylgeranyl diphosphate (GGPP, C_{20}). GPP, FPP, and GGPP are the branch-point intermediates leading to the different major classes of terpenes. These three compounds are produced by product-specific isoprenyl diphosphate synthases, catalyzing one, two, or three successive condensations between an allylic diphosphate and a unit of IPP (Fig. 1.5). For example, GPP synthase catalyzes a single condensation of IPP and DMAPP, while FPP synthase catalyzes the sequential condensation of DMAPP with two molecules of IPP (without releasing the C_{10} intermediate) to form a C_{15} final product. In a similar way, GGPP synthase catalyzes three successive condensations with IPP to form the C_{20} prenyl diphosphate. Other plant isoprenyl diphosphate synthases catalyze the formation of intermediates longer than 20 carbons that are involved in rubber, dolichol, ubiquinone, and plastoquinone biosynthesis.

Considerable research has been carried out on the short chain isoprenyl diphosphate synthases of plants,[14] though our knowledge of these enzymes in conifers is restricted to work on just two species, *Abies grandis* and *Taxus canadensis*.[37-39] All short-chain isoprenyl diphosphate synthases share some basic properties, including an absolute catalytic requirement for a divalent metal ion (usually Mg^{2+}), a pH optimum near neutrality, K_m values for both substrates in the 1-100 μM range and a homodimeric architecture with subunits of 30-50 kDa (except for a few GPP synthases that are heterodimeric).[39-42] Short-chain isoprenyl diphosphate synthases function at the metabolic branch-points to the major terpene classes, and so may be important in controlling the relative rates of formation of different terpene types. When *P. abies* saplings were treated with methyl jasmonate, GGPP synthase in wood tissue increased many-fold compared to activity in untreated saplings, although GPP synthase and FPP synthase activities did not show significant changes.[23] The rise in GGPP synthase activity paralleled the increase in diterpenes observed in the wood, suggesting that this enzyme might closely regulate the production of the terpene resin constituents.

We searched for isoprenyl diphosphate synthase gene sequences in *P. abies* using a homology-based approach. First, RNA was isolated from bark and wood of methyl jasmonate-treated spruce saplings from a single clone. This was then used as a template for reverse transcriptase PCR carried out with degenerate primers

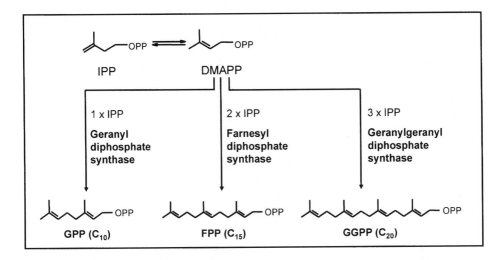

Fig. 1.5: Reactions catalyzed by short chain isoprenyl diphosphate synthases in plants. All use the same precursors, IPP and DMAPP, but the enzymes may reside in different subcellular compartments. The reactions catalyzed by geranyl diphosphate synthase and geranylgeranyl diphosphate synthase are thought to occur in the plastid, and the reaction catalyzed by farnesyl diphosphate in the cytosol.

designed to conserved sequence regions of known GPP, FPP, and GGPP synthases. PCR products were cloned, sequenced and used to screen a cDNA library. Six different short-chain isoprenyl diphosphate synthase-like sequences (*PaIDS 1-6*) were isolated. All share similarities to other *IDS*-encoded proteins including two major aspartate-rich motifs that are thought to be responsible for substrate binding.[43] These sequences can be provisionally classified as GPP synthases (*PaIDS1-3*), FPP synthases (*PaIDS4*), and GGPP synthases (*PaIDS5-6*) based on their similarities to sequences already on deposit in public databases. Each sequence was actually represented by multiple versions in the cDNA library with 96-99% similarity in the coding region and larger differences in the 3'-untranslated region. These variants may represent different genes or different alleles of a single gene.

When the *P. abies IDS* sequences are portrayed in a phylogenetic tree with other plant isoprenyl diphosphate synthases, some interesting patterns are evident (Fig. 1.6). First, there is a major division between the FPP synthases including *PaIDS4*, on the one hand, and the large group of GPP and GGPP synthases, on the other. Among the latter group, several clusters of GPP synthases can be

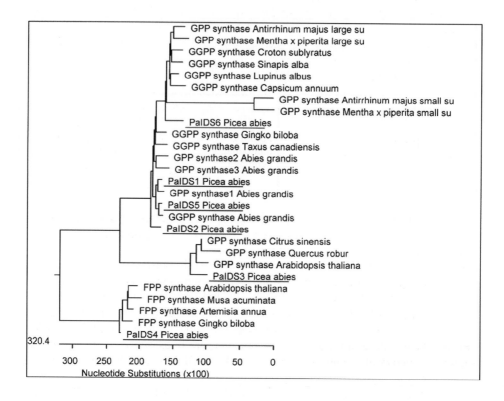

Fig. 1.6: Phylogenetic tree of gymnosperm and angiosperm isoprenyl diphosphate synthase sequences. The isolated *Picea abies* sequences are marked; other isoprenyl diphosphate synthases sequences are listed according to the major reaction product of the recombinant protein. Abbreviations: GPP synthase, geranyl diphosphate synthase; FPP synthase, farnesyl diphosphate synthase; and GGPP synthase, geranylgeranyl diphosphate synthase; large su, large subunit of protein; small su, small subunit of protein.

distinguished. One is a group of homodimeric GPP synthases (sequences from *Citrus sinensis, Quercus robur, Arabidopsis thaliana,* and *PaIDS3*).[44] Another group of sequences represents the small subunit of the heterodimeric GPP synthases from *Antirrhinum majus* and *Mentha* x *piperita*.[42,45] The corresponding large subunits of these proteins nest separately within the main group of GGPP synthases, an appropriate position, since these are reported to have GGPP synthase activity when heterologously expressed along without their small subunit partners.[39,42] A final group of GPP synthases is part of conifer homodimeric GPP and GGPP synthases, including *PaIDS1, 2* and *5*. It appears that *P. abies* has a great number of different types of GPP and GGPP synthases, perhaps appropriate for a plant that makes such a variety of terpene metabolites. However, further speculation is unwarranted until the catalytic function of these genes has been determined. For this purpose, it is necessary to express them heterologously and assay the enzymatic activity of the encoded proteins.

To this point, we have tested the expression of four of the six *P. abies* IDS genes in *E. coli* by cloning them into expression vectors which produce proteins with a fused His-tag to facilitate purification. Sequences for *PaIPS1, PaIPS5,* and *PaIPS6* were first truncated to remove putative transit peptides. The recombinant proteins were extracted, purified on a Ni^{2+}-agarose column (Fig. 1.7) and assayed with IPP and DMAPP (Fig. 1.8). The PaIDS4 protein was shown to make FPP with a small quantity of GPP. The proteins designated PaIDS5 and PaIDS6 make solely GGPP, as might be expected from their sequences, while curiously PaIDS1 makes GPP and GGPP in an approximate 2:1 ratio. There is as yet no precedent for an isoprenyl diphosphate synthase that makes both GPP and GGPP in substantial amounts, but does not produce any FPP. However, since *P. abies* terpene resin contains about equal amounts of GPP products (monoterpenes) and GGPP products (diterpenes), the existence of an isoprenyl diphosphate synthase that makes both *in vivo* is an intriguing possibility. Additional study is underway to see if PaIDS1 makes both products *in vivo*. To learn more about the role of these genes in the plant, we are also investigating their expression pattern in various organs and tissues in relation to terpene formation. In addition, we are developing a transformation system for *P. abies* and hope to use these genes to try to manipulate terpene formation in transgenic saplings. Plants with altered terpene profiles would supply ideal material for experiments to test the roles of terpene resins against herbivores and pathogens.

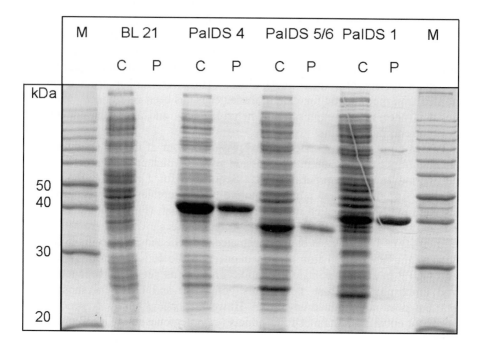

Fig. 1.7: Heterologous expression of *P abies* isoprenyl diphosphate synthases in *E. coli* after Coomassie stain of an SDS-polyacrylamide gel with extracts from bacteria expressing *PaIDS1, PaIDS4, PaIDS5,* and *PaIDS6. Lane M,* molecular mass markers; *Lane BL21,* extracts from bacteria containing only the expression vector without an isoprenyl diphosphate synthase sequence; *C lanes,* bacterial crude extracts; *P lanes,* purified recombinant proteins.

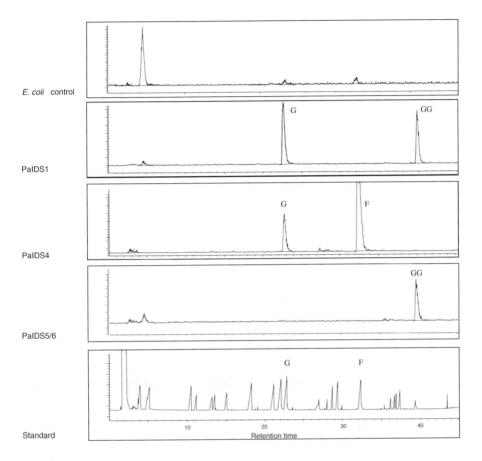

Fig. 1.8: Catalytic activities of the isoprenyl diphosphate synthases PaIDS1, PaIDS4, PaIDS5, and PaIDS6 after heterologous expression in *E. coli*. Products were measured by radio-gas chromatography (plotted in Bequerel, upper four panels) and identified by co-injection of non-radioactive terpene standards, detected via a thermal conductivity detector (plotted as detector response, bottom panel). The main products after acid hydrolysis are listed: G, geraniol; F, farnesol; and GG, geranylgeraniol. Bacteria containing only the expression vector without an isoprenyl diphosphate synthase sequence showed no enzyme activity (top panel).

PHENOLICS

Another large group of secondary metabolites that often has defensive roles is phenolics.[46] In extensive anatomical studies of *P. abies* stems, Franceschi, Krekling and coworkers showed that wounding or fungal infection not only induced the formation of traumatic ducts with terpene resin, but also changes in certain cells believed to produce phenolics. They described a cell type found in secondary phloem referred to as polyphenolic parenchyma (PP) cells[24,47-50] (Fig. 1.1). These occur in concentric rings, 1-2 cells thick, surrounded by sieve cells. One ring of PP cells is formed per year.[48] The vacuoles of these cells harbor a material that appears to be phenolic, based on its intense fluorescence under 450-490 nm light[47] and strong staining with the periodic acid-Schiff procedure.[24] In addition, phenylalanine ammonia lyase, a major enzyme in plant phenolic formation has been localized to the PP cells by immunolocalization.[47] Upon wounding or fungal infection, the PP cells increase in size with a strong increase in periodic acid-Schiff's staining, and the phenolic material appears to be released to the wall of surrounding cells[24,49] (Fig. 1.1). Are these changes associated with alterations in phenolic chemistry?

In recent years, phenolic compounds have been identified in spruce bark, including stilbenes, flavonoids, and tannins.[51-54] Studies have looked for changes in phenolic quantity and composition after wounding or fungal infection.[55-58] However, the changes observed were unremarkable (increase or decrease of 2-fold or less) or poorly replicated. Thus, it is still unclear what changes in phenolic chemistry are associated with the dramatic changes observed in the anatomy of the PP cells.

We looked for changes in the soluble phenolic content of *P. abies* saplings and mature trees after methyl jasmonate spraying by HPLC analysis of methanol bark extracts (Fig. 1.9). However, as in previous work, none of the major stilbenes or flavonoids showed any substantial changes over 4 weeks after treatment (Fig. 1.10), although the PP cells showed similar anatomical changes as after fungal infection. Perhaps other phenolics that have not yet been measured in *P. abies* bark are the ones associated with the anatomical changes in the PP cells, such as high molecular weight condensed tannins[59,60] or cell wall-bound substances.[61] These other phenolics may also be responsible for conifer protection against pathogens and bark beetles since the defensive role of stilbenes and flavonoids and simple phenylpropanoids described is ambiguous. *In vitro* tests showed that these compounds have antifungal properties against certain pathogens,[62,63] but not others,[57] and they did not affect bark beetle feeding.[64] Moreover, crude methanol extracts (which would be expected to contain nearly all stilbenes, flavonoids, and simple phenolic conjugates present) exhibited little or no inhibition of fungal growth.[57] Perhaps the situation *in vitro* is different due to the presence of activating enzymes or other factors. To clarify their importance in conifer defense, attempts must be made to carry out more extensive phenolic analyses and to manipulate their levels *in vivo*.

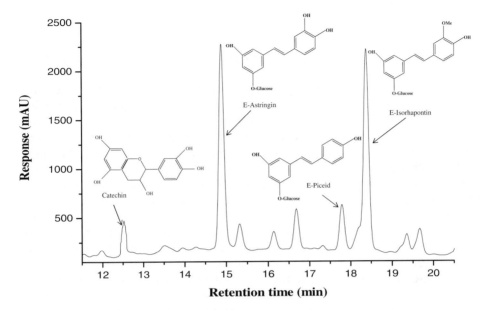

Fig. 1.9: HPLC chromatogram of methanol extract of two year-old Norway spruce saplings showing the presence of soluble phenolic compounds. The chemical structures of the major compounds are shown.

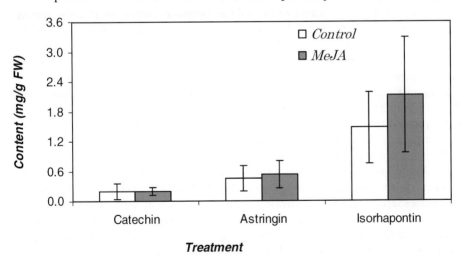

Fig. 1.10: Effect of methyl jasmonate (MeJA) treatment on the levels of the major soluble phenolic compounds in two year-old *P. abies* saplings. No significant differences were observed between treated and control plants four weeks after treatment.

CHITINASES

Among the proteinaceous plant defenses are chitinases, a group of enzymes that hydrolyze the 1,4-*N*-acetyl-D-glucosamine (GlcNAc) linkages of chitin, a component of cell walls of higher fungi. Hydrolysis of chitin results in the swelling and lysis of the hyphal tips,[65,66] and the chitinolytic breakdown products generated can act as elicitors of further defense reactions in plants.[67] Proof of the role of chitinases in plant defense comes from studies in which chitinases were constitutively overexpressed in transgenic plants leading to increased resistance against pathogens *in vivo*.[68,69] However, chitinases can also hydrolyze other substrates, such as arabinogalactan proteins, rhizobial Nod factors, and other lipo-chitooligosaccharides,[70,71] and so may have roles in plants other than defensive ones. Chitinases have been divided into two families (18 and 19) of glycosylhydrolases (E.C. 3.2.1.14) on the basis of their hydrolytic mechanisms, and into seven classes (class I-VII) based on their primary structure.[67,72] Within an individual plant, chitinases are present as multiple isoforms that differ in their size, isoelectric point, primary structure, cellular localization, and pattern of regulation[73-75] and function (*e.g.*[70]).

In conifers, chitinases have been reported to be induced by pathogen attack and wounding in both *P. abies* and *Pinus elliottii*.[75-79] Induction occurs at the level of the transcript, the protein, and the active enzyme. Based on EST sequences from a cDNA library prepared from the bark of methyl jasmonate-sprayed *P. abies* sapling stems, we cloned genes for class I, II, and IV chitinases whose expression changed after mature trees were infected with *Heterobasidion annosum*.[75] The role of these chitinases in resistance was demonstrated by showing that a *P. abies* clone resistant to *H. annosum* had more rapid temporal and spatial accumulation of class II and IV chitinase gene transcripts than a susceptible clone in areas immediately adjacent to inoculation.[75] Data for the class IV gene is given in Fig. 1.11. Similarly, *Pinus elliottii* seedlings resistant to the fungal pathogen *Fusarium subglutinans* f. sp. *pini* accumulated chitinase class II transcripts faster than susceptible seedlings.[78] Resistant plants may perceive the pathogen faster and have more efficient signaling mechanisms than susceptible plants.

The presence of local and systemic signaling cascades involved in inducing defenses in *P. abies* has been suggested based on the systemic expression of peroxidases and chitinases following pathogen infection along with associated anatomical changes.[50,80,81] The radial and vertical rates of signal movement in different tissues of conifers as well as the components of signal transduction pathways are not well known. In stems, terpene resin-containing-traumatic ducts are induced by fungal inoculation with a signal that moves away from the inoculation point at 2.5 cm per day in the axial direction.[49] In needles of seedlings, chitinolytic activity increased within 2 to 4 days after inoculation with the root pathogen *Rhizoctonia*, sp.[79] For chitinase expression, as for terpene induction as discussed above, jasmonates are clearly a component of the signal transduction cascade in both pine and spruce, as is ethylene in other Pinaceae.[82]

Many aspects of the defensive function of conifer chitinases remain to be clarified. For example, the multiple chitinases induced by infection probably are not redundant defense enzymes based on data from angiosperms, but instead are complementary hydrolases with synergistic action on N-acetylglucosamine-containing substrates.[70] Of the class IV chitinases, at least one member has been proposed to release chito-oligosaccharides from an endogenous substrate in *P. abies*, promoting programmed cell death needed for proper embryo development.[83] Notably, a practically identical class IV chitinase of *P. abies* shows differential spatiotemporal expression in bark among clones that display variation in resistance to *H. annosum*.[75] This raises the question of whether certain conifer chitinases might have an indirect role in host defense by eliciting programmed cell death through the release of elicitors from an endogenous substrate. In contrast, the class III chitinases are hypothesized to promote symbiotic interaction with ectomycorrhizal fungi by fragmenting chitin derived elicitors and thereby preventing induction of host defense responses.[84] Kinetic studies of chitin hydrolysis by purified conifer chitinases, genetic manipulation of chitinase levels in intact plants, and direct application of purified chitinases to plant tissue should help to elucidate their roles in both defense and developmental processes of conifers.

CONCLUSIONS

As this survey has shown, *P. abies* produces three distinct types of chemical defenses: terpenes, phenolics, and chitinases, in response to herbivore damage or pathogen infection. Defenses induced by herbivores and pathogens are widespread in the plant kingdom, and are thought to be favored if the incidence of attack is uncertain and the costs of producing and storing defenses are high.[12] For *P. abies* and other forest trees, the pattern of insect attack can be variable from year to year as evidenced by the occurrence of *I. typographus* outbreaks.[6,85,86] In addition, the metabolic cost of producing and storing defenses, such as terpene resins are high in the concentrations present in *P. abies*.[87] However, to be effective, induced defenses must be made rapidly enough to significantly reduce herbivore or pathogen damage. While changes in chitinases and phenolics appear to occur within 6 days,[49,75] the increase in terpene resins requires 15 days. This may not be fast enough to protect *P. abies* from attack by the bark beetle, *Ips typographus*, which can mass attack and successfully colonize a mature tree within a week.[6,11,24] However, during a typical

DAY 3

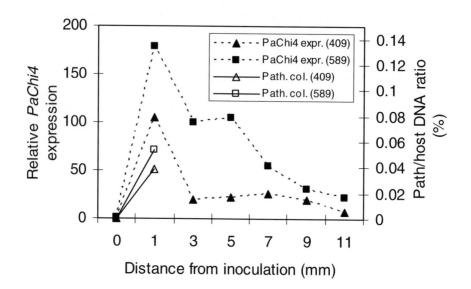

DAY 14

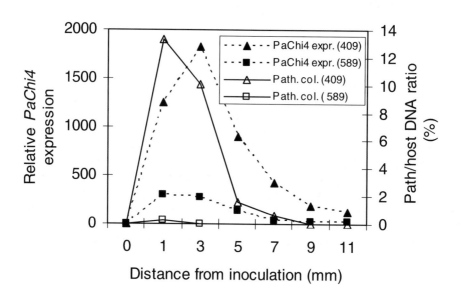

Figure 1.11: Relative gene expression profiles of *PaChi4*, a class IV chitinase, and pathogen colonization levels in bark of two Norway spruce clones following inoculation with *Heterobasidion annosum*. The bark around the inoculation site was spatially sampled 3 and 14 days after inoculation. Clone 409 is highly susceptible and clone 589 is moderately resistant to this pathogen. The transcript levels of the chitinase in clone 409 at the time of inoculation (shown at 0 mm distance from inoculation) were used as a reference transcript level and defined as the 1x expression level, and the transcript levels of all the other samples are expressed as the fold change over this reference level. Pathogen colonization (Path. col.) was measured as the ratio of pathogen to host DNA. Each data point represents the mean of two ramets. For further details, see Hietala *et al.* (2004).

Scandinavian summer, favorable weather conditions may not last long enough to allow for these rapid attacks, and so induced resin may form an effective defense against *I. typographus*.

Based on our current knowledge, the chemistry of induced defenses in *P. abies* and other woody plants is not materially different from that of herbaceous plants. Terpenes, phenolics, and chitinases are all common metabolites in herbs, and in many cases are inducible upon herbivory or pathogen infection. However, the presence of inducible terpene resins is a special feature of conifers. In the rest of the plant kingdom, mixtures of terpenes accumulate in resin ducts, cavities or glandular hairs of many taxa, but are usually not reported to be inducible.[88] Moreover, the long induction time of conifer resin also sets it apart from other induced defenses.

The occurrence of multiple defense systems in a single plant species is also not unique to gymnosperms or any other plant group. It has been suggested that different defense systems target different types of pests. On the one hand, chitinases, because of their specific activity on fungal cell walls, are expected to serve as defenses chiefly against fungi. However, terpene resins, based on their bioassay results and physical properties, could act as barriers against both herbivores or pathogens. Indeed, attack on *P. abies* by an herbivore such as *I. typographus*, is often accompanied by infestation of fungi dispersed by the herbivore. Thus, the possession of defenses active against multiple enemies is easy to explain. Further studies are needed to determine the intended targets of *P. abies* defenses.

Further research is also necessary to demonstrate the actual defensive roles of these metabolites. Terpenes, phenolics, and chitinases have all been suggested to function in defense in conifers based on their toxicity and repellency *in vitro*, their induction around the point of attack and, in the case of terpene resins, their physical properties and mode of storage. However, the results of *in vitro* bioassays have sometimes been equivocal and inducibility by enemy attack *per se* does not prove the roles of these substances in defense. Clearly, it would be best to test defensive roles *in vivo* because it is difficult to simulate the physical arrangements of the living plant (*e.g.*, the exudation of resin) *in vitro*. *In vivo* experiments require a method of manipulating defense level that has only minimal effects on other aspects of plant phenotypes. Application of jasmonates and other signaling compounds may be suitable for this purpose, but the specificity of these substances to one or a few classes of defenses has not yet been demonstrated in conifers. Increasing knowledge of the biochemistry and molecular biology of defense metabolism is now making it easier to identify genes regulating the formation of a single defensive substance or class of defensive compounds in conifers. Such genes can be used to prepare transgenic conifers with altered defense profiles, which should provide ideal material to test defensive roles. Even without transformation, the monitoring of defense gene expression around the site of damage can readily implicate certain metabolites in a defense response.

ACKNOWLEDGMENTS

We thank Andrea Bergner, Beate Rothe, Marion Staeger, and Katja Witzel for outstanding technical assistance, Vince Franceschi, Trygve Krekling, and Nina Nagy for stimulating discussions, and Diane Martin and Jörg Bohlmann for getting our methyl jasmonate studies on spruce off to a flying start. The Max Planck Society, the Norwegian Forestry Research Institute, the Norwegian Research Council and The Academy of Finland provided funds to support the research described. G.Z. was the recipient of a fellowship from the German Academic Exchange Service.

REFERENCES

1. HEATH, M.C., BOLLER, T., eds, Biotic interactions., *Curr. Opinion Plant Biol.*, 2002, **5**, 265-361.
2. BRYANT, J.P., KUROPAT, P.J., REICHARDT, P.B., CLAUSEN, T.P., Controls over the allocation of resources by woody plants to chemical antiherbivore defense, *in*: Plant Defenses Against Mammalian Herbivory (R.T. Palo and C.T. Robbins, eds,), CRC Press, Boca Raton. 1991, pp. 83-102.
3. KUBITZKI, K., The Families and Genera of Vascular Plants. Springer-Verlag, Berlin, 1990, 404 p.
4. DOYLE, J.A., Phylogeny of vascular plants., *Annu. Rev. Ecol. Syst.*, 1998, **29**, 567-599.
5. SCHMIDT-VOGT, H., Die Fichte, Band I. Taxonomie-Verbreitung-Morphologie-Ökologie-Waldgesellschaften. (Spruce, Volume 1, Taxonomy, Distribution,

Morphology, Ecology, Forest Communities). Verlag Paul Parey, Hamburg, 1987, 647 p.

6. CHRISTIANSEN, E., BAKKE, A., The spruce bark beetle of Eurasia, *in:* Dynamics of Forest Insect Populations (A.A. Berryman, ed.,), Plenum, New York. 1988, pp. 479-503.

7. CHRISTIANSEN, E., SOLHEIM, H., The bark beetle-associated blue-stain fungus *Ophiostoma polonicum* can kill various spruces and Douglas-fir., *Eur. J. For. Pathol.,* 1990, **20,** 436-446.

8. OKSANEN, J., HOLOPAINEN, J.K., NERG, A., HOLOPAINEN, T., Levels of damage of Scots pine and Norway spruce caused by needle miners along a SO$_2$ gradient., *Ecography,* 1996, **19,** 229-236.

9. STADLER, B., MICHALZIK, B., The impact of spruce aphids on nutrient flows in the canopy of Norway spruce., *Agr. For. Entomol.,* 1999, **1,** 3-9.

10. LIEUTIER, F., DAY, K.R., BATTISTI, A., GREGOIRE, J.-C., EVANS, H., Bark and Wood Boring Insects in Living Trees in Europe, a Synthesis, Kluwer, Dordrecht, 2004, 569 p.

11. WERMELINGER, B., Ecology and management of the spruce bark beetle *Ips typographus*- a review of recent research., *For. Ecol. Manage.,* 2004, **202,** 67-82.

12. KARBAN, R., BALDWIN, I.T., Induced Responses to Herbivory. University of Chicago Press, Chicago, 1997, 319 p.

13. LABANDEIRA, C., LEPAGE, B., JOHNSON, A., A *Dendroctonus* bark engraving (Coleoptera: Scolytidae) from a middle Eocene *Larix* (Coniferales: Pinaceae): early or delayed colonization?, *Amer. J. Bot.,* 2001, **88,** 2026-2039.

14. GERSHENZON, J., KREIS, W., Biosynthesis of monoterpenes, sesquiterpenes, diterpenes, sterols, cardiac glycosides and steroid saponins, *in:* Biochemistry of Plant Secondary Metabolism, Annual Plant Reviews, Vol. 2 (Wink, M., ed,), Sheffield Academic Press, Sheffield. 1992, pp. 222-299.

15. SCHWERDTFEGER, F., Pathogenese der Borkenkäfer-Epidemie 1946–1950 in Nordwestdeutschland., *Schr. Reihe forstl. Fak. Univ. Göttingen,* 1955, **13/14,** 1-135.

16. RAFFA, K.F., BERRYMAN, A.A., The role of host plant resistance in the colonization behavior and ecology of bark beetles (Coleoptera: Scolytidae)., *Ecol. Monogr.* 1983, **53,** 27-49.

17. BYERS, J.A., Host-tree chemistry affecting colonization of bark beetles, *in:* Chemical Ecology of Insects 2 (R.T. Cardé and W.J. Bell, eds,), Chapman and Hall, New York. 1995, pp. 154-213.

18. TRAPP, S., CROTEAU, R., Defensive resin biosynthesis in conifers., *Annu. Rev. Plant Physiol. Plant Mol. Biol.,* 2001, **52,** 689-724.

19. NAGY, N.E., FRANCESCHI, V.R., SOLHEIM, H., KREKLING, T., CHRISTIANSEN, E., Wound-induced traumatic resin duct development in stems of Norway spruce (Pinaceae): anatomy and cytochemical traits., *Amer. J. Bot.,* 2000, **87,** 302-313.

20. BAIER, P., FUHRER, E., KIRISITS, T., ROSNER, S., Defense reaction of Norway spruce against bark beetles and the associated fungus *Ceratocystis polonica* in secondary pure and mixed species stands., *For. Ecol. Manage.,* 2002, **159,** 73-86.

21. BEALE, M.H., WARD. W.L., Jasmonates: key players in plant defense., *Nat. Prod. Rep.*, 1998, **15**, 533-548.
22. WASTERNACK, C., HAUSE B., Jasmonates and octadecanoids: Signals in plant stress responses and development., *Prog. Nucleic Acid Res. Mol. Biol.*, 2002, **72**, 165-221.
23. MARTIN, D., THOLL, D., GERSHENZON, J., BOHLMANN, J., Methyl jasmonate induces traumatic resin ducts, terpenoid resin biosynthesis, and terpenoid accumulation in developing xylem of Norway spruce stems., *Plant Physiol.*, 2002, **129**, 1003-1018.
24. FRANCESCHI, V.R., KROKENE, P., KREKLING, T., CHRISTIANSEN, E., Phloem parenchyma cells are involved in local and distant defense responses to fungal inoculation or bark beetle attack in Norway spruce (Pinaceae)., *Amer. J. Bot.*, 2000, **87**, 314-326.
25. MARTIN, D.M., GERSHENZON, J., BOHLMANN, J., Induction of volatile terpene biosynthesis and diurnal emission by methyl jasmonate in foliage of Norway spruce., *Plant Physiol.*, 2003, **132**, 1586-1599.
26. KAUKINEN, K.H., TRANBARGER, T.J., MISRA, S., Post-termination-induced and hormonally dependent expression of low-molecular-weight heat shock protein genes in Douglas fir., *Plant Mol. Biol.*, 1996, **30**, 1115-1128.
27. LAPOINTE, G., LUCKEVICH, M.D., SEGUIN, A., Investigation on the induction of 14-3-3 in white spruce, *Plant Cell Rep.*, 2001, **20**, 79-84.
28. KOZLOWSKI, G., BUCHALA, A., METRAUX, J.-P., Methyl jasmonate protects Norway spruce [*Picea abies* (L.) Karst.] seedlings against *Pythium ultimum* Trow, *Physiol. Mol. Plant Pathol.*, 1999, **55**, 53-58,
29. REGVAR, M., GOGALA, N., ZNIDARSIC, N., Jasmonic acid effects mycorrhization of spruce seedlings with *Laccaria laccata.*, *Trees*, 1997, **11**, 511-514.
30. BOHLMANN J., CROCK, J., JETTER, R., CROTEAU, R., Terpenoid-based defenses in conifers: cDNA cloning, characterization, and functional expression of wound-inducible (*E*)-α-bisabolene synthase from grand fir (*Abies grandis*), *Proc. Natl. Acad. Sci. USA*, 1998, **95**, 6756-6761.
31. KETCHUM, R.E., GIBSON, D.M., CROTEAU, R.B., SHULER, M.L., The kinetics of taxoid accumulation in cell suspension cultures of *Taxus* following elicitation with methyl jasmonate., *Biotechnol. Bioeng.*, 1999, **62**, 97-105.
32. FRANCESCHI, V.R., KREKLING, T., CHRISTIANSEN, E., Application of methyl jasmonate on *Picea abies* (Pinaceae) stems induces defense-related responses in phloem and xylem., *Amer. J. Bot.*, 2002, **89**, 578-586.
33. DICKE, M., GOLS, R., LUDEKING, D., POSTHUMUS, M.A., Jasmonic acid and herbivory differentially induce carnivore-attracting plant volatiles in lima bean plants., *J. Chem. Ecol.*, 1999, **25**, 1907-1922.
34. THALER, J.S., Jasmonate-inducible plant defenses cause increased parasitism of herbivores., *Nature.* 1999, **3996**, 686-688.
35. THALER, J.S., KARBAN, R., ULLMAN, D.E., BOEGE, K., BOSTOCK, R.M., Cross–talk between jasmonate and salicylate plant defense pathways: effects on several plant parasites, *Oecologia,*. 2002, **131**, 227-235.

36. CREELMAN, R.A., MULLET, J.E., Biosynthesis and action of jasmonates in plants., *Annu. Rev. Plant Physiol. Plant Mol. Biol.*, 1997, **48**, 355-381.
37. HEFNER, J., KETCHUM, R.E.B., CROTEAU, R., Cloning and functional expression of a cDNA encoding geranylgeranyl diphosphate synthase from *Taxus canadensis* and assessment of the role of this prenyltransferase in cells induced for Taxol production., *Arch. Biochem. Biophys.*, 1998, **360**, 62-74.
38. THOLL, D., CROTEAU, R., GERSHENZON, J., Partial purification and characterization of the short-chain prenyltransferases, geranyl diphosphate synthase and farnesyl diphosphate synthase, from *Abies grandis* (Grand fir)., *Arch. Biochem. Biophys.*, 2001, **386**, 233-242.
39. BURKE, C., CROTEAU, R., Geranyl diphosphate synthase from *Abies grandis*: cDNA isolation, functional expression, and characterization., *Arch. Biochem. Biophys.*, 2002, **405**, 130-136.
40. POULTER, C.D., RILLING, H.C., Prenyl transferases and isomerase, *in*: Biosynthesis of Isoprenoid Compounds, Vol. 1 (J.W. Porter and S.L. Spurgeon, eds,), John Wiley and Sons, New York. 1981, pp. 162-224.
41. OGURA, K., KOYAMA, T., Enzymatic aspects of isoprenoid chain elongation., *Chem. Rev.*, 1998, **98**, 1263-1276.
42. THOLL, D., KISH, C.M., ORLOVA, I., SHERMAN, D., GERSHENZON, J., PICHERSKY, E., DUDAREVA, N., Formation of monoterpenes in *Antirrhinum majus* and *Clarkia breweri* flowers involves heterodimeric geranyl diphosphate synthases., *Plant Cell*, 2004, **16**, 977-992.
43. CHEN, A., KROON, P.A., POULTER, C.D., Isoprenyl diphosphate synthases: protein sequence comparisons, a phylogenetic tree, and predictions of secondary structure., *Protein Sci.*, 1994, **3**, 600-607.
44. BOUVIER, F., SUIRE, C., D'HARLINGUE, A., BACKHAUS, R.A., CAMARA, B., Molecular cloning of geranyl diphosphate synthase and compartmentation of monoterpene synthesis in plant cells., *Plant J.*, 2000, **24**, 241-252.
45. BURKE, C.C., WILDUNG, M.R., CROTEAU, R. Geranyl diphosphate synthase: Cloning, expression, and characterization of this prenyltransferase as a heterodimer., *Proc. Natl. Acad. Sci. USA*, 1999, **96**, 13062-13067.
46. NICHOLSON, R.L., HAMMERSCHMIDT, R., Phenolic compounds and their role in disease resistance., *Annu. Rev. Phytopathol.*, 1992, **30**, 369-389.
47. FRANCESCHI, V.R., KREKLING, T., BERRYMAN, A.A., CHRISTIANSEN, E., Specialized phloem parenchyma cells in Norway spruce (Pinaceae) bark are an important site of defense reactions., *Amer. J. Bot.*, 1998, **85**, 601-615.
48. KREKLING, T., FRANCESCHI, V.R., BERRYMAN, A.A., CHRISTIANSEN, E., The structure and development of polyphenolic parenchyma cells in Norway spruce (*Picea abies*) bark., *Flora*, 2000, **195**, 354-369.
49. KREKLING, T., FRANCESCHI, V.R., KROKENE, P., SOLHEIM, H., Differential anatomical response of Norway spruce stem tissues to sterile and fungus infected inoculations., *Trees*, 2004, **18**, 1-9.
50. NAGY, N.E., FOSSDAL, C.G., KROKENE, P., KREKLING, T., LÖNNEBORG, A., SOLHEIM, H., Induced responses to pathogen infection in Norway spruce

phloem: changes in polyphenolic parenchyma cells, chalcone synthase transcript levels and peroxidase activity., *Tree Physiol.*, 2004, **24**, 505-515.

51. PAN, H.F., LUNDGREN, L. N., Phenolic extractives from root bark of *Picea abies.*, *Phytochemistry*, 1985, **39**, 1423-1428.

52. SOLHAUG, K.A., Stilbene glucosides in bark and needles from *Picea* species., *Scan. J. For. Res.*, 1990, **5**, 59-67.

53. TOSCANO UNDERWOOD, C.D., PEARCE, R.B., Astringin and isorhapontin distribution in Sitka spruce trees., *Phytochemistry,* 1991, **30,** 2183-2189.

54. TOSCANO UNDERWOOD, C.D., PEARCE, R.B., Variation in the levels of the antifungal stilbene glucosides astringin and isorhapontin in the bark of Sitka spruce [*Picea sitchensis* (Bong.) Carr.]., *Eur. J. For. Pathol.,* 1991, **21,** 279-289.

55. BRIGNOLAS, F., LACROIX, B., LIEUTIER, F., SAUVARD, D., DROUET, A., CLAUDOT, A.-C., YART, A., BERRYMAN, A.A., CHRISTIANSEN, E., Induced responses in phenolic metabolism in two Norway spruce clones after wounding and inoculations with *Ophiostoma polonicum*, a bark beetle-associated fungus., *Plant Physiol.*, 1995, **109**, 821-827.

56. BRIGNOLAS, F., LIEUTIER, F., SAUVARD, D., CHRISTIANSEN, E., BERRYMAN, A.A., Phenolic predictors for Norway spruce resistance to the bark beetle *Ips typographus* (Coleoptera: Scolytidae) and an associated fungus, *Ceratocystis polonica.*, *Can. J. For. Res.*, 1998, **28**, 720-728.

57. EVENSEN, P.C., SOLHEIM, H., HØILAND, K., STENERSEN, J., Induced resistance of Norway spruce, variation of phenolic compounds and their effects on fungal pathogens., *For. Pathol.*, 2000, **30**, 97-108.

58. LIEUTIER, F., BRIGNOLAS, F., SAUVARD, D., YART, A., GALET, C., BURNET, M., VAN DE SYPE, H., Intra- and inter-provenance variability in phloem phenols of *Picea abies* and relationship to a bark beetle-associated fungus., *Tree Physiol.*, 2003, **23**, 247-256.

59. BEHRENS, A., MAIE, N., KNICKER, H., KÖGEL-KNABNER, I., MALDI-TOF mass spectrometry and PSD fragmentation as means for the analysis of condensed tannins in plant leaves and needles., *Phytochemistry*, 2003, **62**, 1159-1170.

60. MAIE, N., BEHRENS, A., KNICKER, H., KÖGEL-KNABNER, I., Changes in the structure and protein binding ability of condensed tannins during decomposition of fresh needles and leaves., *Soil Biol. Biochem.,* 2003, **35**, 577-589.

61. STRACK, D., HEILEMANN, J., KLINKOTT, E.-S., Cell wall-bound phenolics from Norway spruce (*Picea abies*) needles., *Z. Naturforsch.*, 1988, **43c**, 37-41.

62. SHAIN, L., The response of sapwood of Norway spruce to infection by *Fomes annosus.*, *Phytopathology*, 1971, **61**, 301-307.

63. WOODWARD, S. PEARCE, R.B., The role of stilbenes in resistance of Sitka spruce (*Picea sitchensis*) to entry of fungal pathogens., *Physiol. Mol. Plant Pathol.*, 1988, **33**, 127-149.

64. McNEE, W.R., BONELLO, P., STORER, A.J., WOOD, D.L., GORDON, T.R., Feeding response of *Ips paraconfusus* to phloem and phloem metabolites of *Heterobasidion annonsum*-inoculated Ponderosa pine., *J. Chem. Ecol.*, 2003, **29**, 1183-1202.

65. SCHLUMBAUM, A., MAUCH, F., VÖGELI, U., BOLLER, T., Plant chitinases are potent inhibitors of fungal growth., *Nature*, 1986, **324**, 365-367.
66. MAUCH, F., MAUCH-MANI, B., BOLLER, T., Antifungal hydrolases in pea tissue., *Plant Physiol.*, 1988, **88**, 936-942.
67. COLLINGE, D.B., KRAGH, K.M., MIKKELSEN, J.D., NIELSEN, K.K., RASMUSSEN, U., VAD, K., Plant chitinases., *Plant J.*, 1993, **3**, 31-40.
68. BROGLIE, K., CHET, I., HOLLIDAY, M., CRESSMAN, R., BIDDLE, P., KNOWLTON, S., MAUVAIS, C.J., BROGLIE, R., Transgenic plants with enhanced resistance to the fungal pathogen *Rhizoctonia solani.*, *Science*, 1991, **254**, 1194-1197.
69. GRISON, R., GREZES-BESSET, B., SCHNEIDER, M., LUCANTE, N., OLSEN, L., LEGUAY, J., TOPPAN, A., Field tolerance to fungal pathogens of *Brassica napus* constitutively expressing a chimeric chitinase gene., *Nat. Biotechnol.*, 1996, **14**, 643-646.
70. BRUNNER, F., STINTZI, A., FRITIG, B., LEGRAND, M., Substrate specificities of tobacco chitinases., *Plant J.*, 1998, **14**, 225-234.
71. DYACHOK, J.V., WIWEGER, M., KENNE, L., VON ARNOLD, S., Endogenous nod-factor-like signal molecules promote early embryo development in Norway spruce., *Plant Physiol.*, 2002, **128**, 523-533.
72. GOMEZ, L., ALLONA, I., CASADO, R., ARAGONCILLO, C., Seed chitinases., *Seed Sci. Res.*, 2002, **12**, 217-230.
73. PETRUZZELLI, L., KUNZ, C., WALDVOGEL, R., MEINS, F., JR., LEUBNER-METZGER, G., Distinct ethylene- and tissue-specific regulation of beta-1,3-glucanases and chitinases during pea seed germination., *Planta*, 1999, **209**, 195-201.
74. SALZER, P., BONAMONI, A., BEYER, K., VÖGELI-LANGE, R., AESCHBACHER, R.A., LANGE, J., WIEMKEN, A., KIM, D., COOK, D.R., BOLLER, T., Differential expression of eight chitinase genes in *Medicago truncatula* roots during mycorrhiza formation, nodulation, and pathogen infection., *Mol. Plant Microbe Interact.*, 2000, **13**, 763-777.
75. HIETALA, A.M., KVAALEN, H., SCHMIDT, A., JØHNK, N., SOLHEIM, H., FOSSDAL, C.G., Temporal and spatial profiles of chitinase expression by Norway spruce in response to bark colonization by *Heterobasidion annosum.*, *Appl. Environ. Microbiol.*, 2004, **70**, 3948-3953.
76. SHARMA, P., BØRJA, D., STOUGAARD, P., LÖNNEBORG, A., PR-proteins accumulating in spruce roots infected with a pathogenic *Pythium sp.* isolate include chitinases, chitosanases and beta-1-3-glucanases., *Physiol. Mol. Plant Pathol.*, 1993, **43**, 57-67.
77. KOZLOWSKI, G., METRAUX, J.-P., Infection of Norway spruce (*Picea abies* (L.) Karst.) seedlings with *Pythium irregulare* Buism. and *Pythium ultimum* Trow.: histological and biochemical responses., *Eur. J. Plant Pathol.*, 1998, **104**, 225-234.
78. DAVIS, J.M., WU, H., COOKE, J.E., REED, J.M., LUCE, K.S., MICHLER, C.H., Pathogen challenge, salicylic acid, and jasmonic acid regulate expression of chitinase gene homologs in pine., *Mol. Plant Microbe Interact.*, 2002, **15**, 380-387.
79. NAGY, N., FOSSDAL, C.G., DALEN, L.S., LÖNNEBORG, A., HELDAL, I., JOHNSEN, Ø., Effects of *Rhizoctonia* infection and drought on peroxidase and

chitinase activity in Norway spruce (*Picea abies*)., *Physiol. Plant.*, 2004, **120**, 465-473.

80. CHRISTIANSEN, E., KROKENE, P., BERRYMAN, A.A., FRANCESCHI, V.R., KREKLING, T., LIEUTIER, F., LÖNNEBORG, A., SOLHEIM, H., Mechanical injury and fungal infection induce acquired resistance in Norway spruce., *Tree Physiol.*, 1999, **19**, 399-403.

81. FOSSDAL, C.G., SHARMA, P., LÖNNEBORG, A., Isolation of the first putative peroxidase cDNA from a conifer and the local and systemic accumulation of related proteins upon pathogen infection., *Plant Mol. Biol.*, 2001, **47**, 423-435.

82. HUDGINS, J.W., FRANCESCHI, V.R., Methyl jasmonate-induced ethylene production is responsible for conifer phloem defense responses and reprogramming of stem cambial zone for traumatic resin duct formation., *Plant Physiol.*, 2004, **135**, 2134-2149.

83. WIWEGER, M., FARBOS, I., INGOUFF, M., LAGERCRANTZ, U., VON ARNOLD, S., Expression of Chia4-Pa chitinase genes during somatic and zygotic embryo development in Norway spruce (*Picea abies*): similarities and differences between gymnosperm and angiosperm class IV chitinases., *J. Exp. Bot.*, 2003, **54**, 2691-2699.

84. SALZER, P., HÜBNER, B., SIRRENBERG, A., HAGER, A., Differential effect of purified spruce chitinases and ß-1,3-glucanases on the activity of elicitors from ectomycorrhizal fungi., *Plant Physiol.*, 1997, **114**, 957-968.

85. WALLNER, W.E., Factors affecting insect population dynamics: Differences between outbreak and non-outbreak species., *Annu. Rev. Entomol.*, 1987, **32**, 317-340.

86. ØKLAND, B., BERRYMAN, A.A., Resource dynamic plays a key role in regional fluctuations of the spruce bark beetles *Ips typographus.*, *Agric. For. Entomol.*, 2004, **6**, 141-146.

87. GERSHENZON, J., The metabolic costs of terpenoid accumulation in higher plants., *J. Chem. Ecol.*, 1994, **20**, 1281-1328.

88. GERSHENZON, J., CROTEAU, R., Terpenoids, *in*: Herbivores: Their Interaction with Secondary Metabolites, 2nd edition, Vol. 1 (G.A. Rosenthal and M.R. Berenbaum, eds.), Academic Press, New York. 1991, pp. 165-219.

Chapter Two

MOLECULAR BIOCHEMISTRY AND GENOMICS OF TERPENOID DEFENSES IN CONIFERS

Diane Martin and Jörg Bohlmann*

Michael Smith Laboratories, and Departments of Botany and Forest Science
University of British Columbia, Vancouver, B.C., Canada

*Author for correspondence, email: bohlmann@interchange.ubc.ca

INTRODUCTION

Successful chemical defense of long-lived conifers against herbivores and pathogens is largely dependent on the formation, accumulation, and release of oleoresin monoterpenoids, sesquiterpenoids, and diterpenoids. In addition, conifers also produce a large array of phenolic and other defense compounds. The topic of terpenoid defenses in conifers has previously been reviewed.[1-7] Oleoresin terpenoids are stored in large quantities in resin canals, resin blisters, or resin cells in stems, roots, or foliage of many conifer species. The development of these specialized anatomical structures can be induced with insect or fungal attack, mechanical wounding, and chemical elicitation. Terpenoids may also be released as constitutive or induced volatiles from the foliage of conifers. These volatiles can act as chemical signals by attracting natural enemies of herbivores. The terpene synthases (TPS) play a central role in the formation of terpenoid chemical diversity, in maintaining phenotypic plasticity in conifer defense, and they are a major part of the genomic hardwiring of conifer resistance.[7,8] Recent work has revealed much of the multigenic nature of this successful conifer terpenoid defense system. Application of methyl jasmonate (MeJA) has enabled a detailed characterization of inducible terpenoid defenses in several conifer species, including Norway spruce (*Picea abies*), Sitka spruce (*P. sitchensis*), and Douglas fir (*Pseudotsuga menziesii*).[9-15] For example, species of spruce produce copious amounts of oleoresin terpenoids, which are stored in constitutive resin ducts, mainly in the bark, or in inducible traumatic resin ducts (TD) in the xylem. Terpenoid accumulation in traumatic resin ducts is regulated, at least in part, by *TPS* gene expression and by TPS enzyme activities, which are induced upon MeJA treatment or in response to insect attack. *TPS* gene expression and TPS enzyme activities are also elevated in foliage following MeJA elicitation. The cDNA cloning, functional characterization, and gene expression analysis of a large family of *TPS* genes from Norway spruce and Sitka spruce enabled an association of the biochemical function of these *TPS* genes with the accumulation of oleoresin terpenoids in stems or the release of terpenoid volatiles from needles.[8,12,16] Recent work in Sitka spruce compared the MeJA- and insect-induced terpenoid defenses at the molecular, biochemical, and anatomical levels.[15] The initial targeted gene characterization of the complex insect- and MeJA-induced terpenoid defense system in spruce has laid the foundation for new genome-scale characterization of insect-induced defenses in conifers (www.treenomix.com).

In this chapter, we emphasize recent research, published and unpublished, of the last five years that has established spruce as one of the best characterized systems for molecular, biochemical, and genomic research of conifer defense against insect pests.[8,10-12,15,16] The chapter touches on molecular and biochemical regulation of chemical diversity of conifer defense and provides an outlook towards new genomics research in a forest health context, specifically the genomics of trees interacting with insect pests and insect-associated fungal pathogens. Most of the chapter covers our recent research with two species of spruce, Norway spruce and Sitka spruce, including terpenoid defense responses induced by MeJA and the white pine weevil (*Pissodes strobi*). Results from our research with spruce terpenoid defenses are relevant to the larger field of secondary metabolite structural diversity in plant defense and the evolution of such defenses and their phenotypic plasticity in long-lived, sessile organisms.[7]

TERPENOID BIOCHEMISTRY AND MOLECULAR GENETICS IN CONIFERS

The terpenoids represent the largest group of known plant secondary metabolites.[17] Three classes, monoterpenoids (10 carbon atoms, C10), sesquiterpenoids (15 carbon atoms, C15), and diterpenoids (20 carbon atoms, C20) are especially prominent among the conifer defenses, as they constitute much of the conifer oleoresin (Fig. 2.1).[7,10,15] In contrast to the mono-, sesqui-, and diterpenoids other terpenoids, such as hemiterpenoids and triterpenoids, have not yet been adequately characterized at the biochemical and molecular levels in conifers. Terpenoid biosynthesis begins with the formation of the five carbon building blocks, isopentenyl diphosphate (IDP) and its isomer, dimethylallyl diphosphate (DMADP) (Fig. 2.2). Two pathways exist for the formation of these precursors.[17] The mevalonate (MEV) pathway is found in the cytosol and endoplasmic reticulum, and the 2-C-methyl erythritol-4-phosphate (MEP) pathway, which proceeds via 1-deoxyxylulose-5-phosphate, occurs in plastids. Most genes of the MEV and MEP pathways have been identified in the spruce expressed sequence tag (EST) database and full-length cDNA clone collections (www.treenomix.com) and are now being used for gene expression analysis in conifer defense.

Upon the formation of IDP and DMADP, prenyltransferases (PT) perform 1'-4 condensation reactions coupling IDP with an allylic prenyl diphosphate. Geranyl diphosphate (GDP) synthase forms the C10 precursor of monoterpenes, farnesyl diphosphate (FDP) synthase forms the C15 precursor of sesquiterpenes, and geranylgeranyl diphosphate (GGDP) synthase produces the precursor for diterpene biosynthesis. A molecular characterization of conifer PTs was first initiated in grand fir (*Abies grandis*).[18] Recent work with Norway spruce PTs is described elsewhere in this volume (see chapter by Schmidt, *et al.*, this volume). Several Sitka spruce and white spruce PTs have also been discovered in the spruce EST database (www.treenomix.com). PTs of the three classes, GDP synthase, FDP synthase, and GGDP synthase, share many common features including sequence similarities, a divalent metal ion requirement for catalysis, and the catalytically active sequence motif DDXXD. GDP synthase and GGDP synthase are localized to the

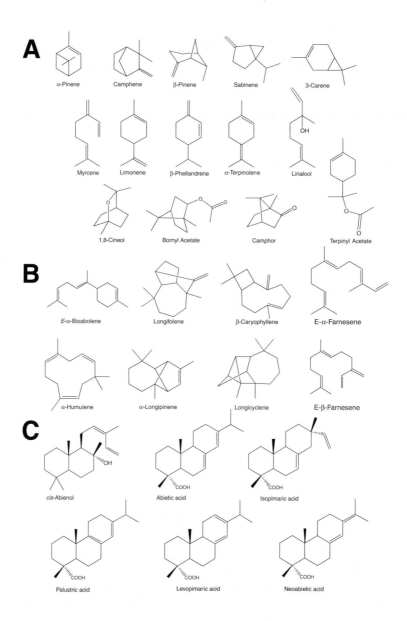

A α-Pinene, Camphene, β-Pinene, Sabinene, 3-Carene, Myrcene, Limonene, β-Phellandrene, α-Terpinolene, Linalool, 1,8-Cineol, Bornyl Acetate, Camphor, Terpinyl Acetate

B E-α-Bisabolene, Longifolene, β-Caryophyllene, E-α-Farnesene, α-Humulene, α-Longipinene, Longicyclene, E-β-Farnesene

C cis-Abienol, Abietic acid, Isopimaric acid, Palustric acid, Levopimaric acid, Neoabietic acid

Fig. 2.1: Characteristic monoterpenes (A), sesquiterpenes (B) and (C) diterpenes present in conifer oleoresin and conifer volatiles emissions.

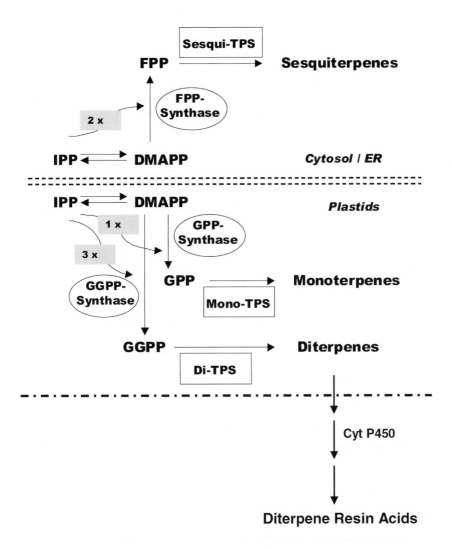

Fig. 2.2: Scheme of the terpenoid biosynthetic pathway. Formation of monoterpenes and diterpenes occurs in plastids. Formation of sesquiterpenoids occurs in the cytsol. Secondary transformation of diterpenes to diterpene resin acids involves membrane- associated cytochrome P450 dependent monooxygenases. IPP is isopentenyl disphosphate; DMAPP is dimethylallyl diphosphate; GPP is geranyl disphosphate; FPP is farnesyl disphosphate; GGPP is geranylgeranyl disphosphate; ER is endoplasmatic reticulum; Cyt P450 is cytochrome P450 dependent monooxygenase; TPS is terpene synthase.

plastid, and FDP synthase is localized to the cytosol / endoplasmic reticulum. Flux of pathway intermediates between these two compartments has been reported.[19] Although PTs are critical components of terpenoid biosynthesis, relatively little is known about the regulation of PTs in induced defense responses in conifers.[10]

The three prenyl diphosphates GDP, FDP, and GGDP are the substrates of a large family of TPS enzymes that catalyze the formation of an amazing structural diversity of monoterpenoids, sesquiterpenoids, and diterpenoids.[20] The TPS utilize an electrophilic reaction mechanism, assisted by divalent metal ion cofactors.[21-23] The prenyl diphosphate substrates are ionized or protonated by TPS to produce reactive carbocation intermediates, which can be rearranged and eventually quenched to yield a wide variety of cyclic and acyclic terpenoid products.[21-26] TPS enzymes exist as single or multiple product enzymes.[27] Class specific TPS, mono-TPS, sesqui-TPS, and di-TPS, are responsible for the formation of the many simple (acyclic or single ring structure) and intricate (two or more ring structures) terpene skeletons of conifer mono-, sesqui-, and diterpenoids.[8] While many terpenoids are further modified by an array of secondary transformations, the basic terpenoid skeletal structures are the products of unique catalytic activities of TPS enzymes. TPS also tend to exert tight control over the stereochemistry of products formed, and usually one enantiomer dominates any TPS product profile, both for single- and multiple-product TPS.[28] A number of conifer TPS have been cloned and characterized from grand fir (*Abies grandis*),[27,29-32] loblolly pine (*Pinus taeda*),[33] Sitka spruce,[16] Norway spruce,[8,12] and Douglas fir[35] (Huber *et al.*, unpublished results), and the enantiomeric specificity and product profiles of these enzymes has been investigated.

The recent cloning and characterization of ten functionally different *TPS* in Norway spruce provided valuable information concerning the biochemistry of conifer defense and established the necessary tools for *TPS* gene expression analysis and for TPS enzyme structure-function analyses in this system.[8,12,15] Additionally, the characterization of more than 30 different conifer *TPS*, from closely and distantly related members of the pine family (Pinaceae), including Norway spruce,[8,12] Sitka spruce,[16] loblolly pine,[33] grand fir (*Abies grandis*),[27,29-32] and Douglas fir (Huber *et al.*, unpublished results), enabled a robust phylogenetic reconstruction of the conifer *TPS* gene family, thereby shedding light on the evolution of the larger plant *TPS* family, including members from angiosperms and gymnosperms.[8,20] Phylogenetic analyses positioned conifer di-TPS closest to the putative ancestor of plant TPS.[8,20,34] Phylogenetic analyses also showed that specific functions and subfamilies of gymnosperm and angiosperm TPS evolved largely independently in these major plant lineages.[8,20] Specific biochemical functions of some TPS also evolved independently and repeatedly in multiple subfamilies of the TPS gene family.[8,20] In contrast to repeated gene duplication and evolution of new functions of TPS in angiosperm and conifer secondary metabolism, TPS involved in gibberellic acid phytohormone formation seem to be more conserved and have undergone less radiation.[8,20,35] Gymnosperm *di-TPS* genes of gibberellic acid formation have so far not been identified and are, therefore, missing in any reconstruction of the phylogeny of the plant TPS family. Such di-TPS would likely be near the base of the TPS family tree.

Recent work on terpenoid modifying enzymes in conifer defense resulted in the first cDNA cloning and functional characterization of a cytochrome P450 dependent monoxygenase involved in diterpene resin acid biosynthesis in loblolly pine (Fig. 2.3) (Ro *et al.*, unpublished results). The discovery and biochemical identification of this P450 gene, which encodes an unusual multi-substrate and multi-functional diterpene oxidase, *PtAO* abietadienol/abietadienal oxidase, was possible by combining a phylogenetic genomics approach of EST mining with a functional biochemical approach of P450 screening in transformed yeast cells. The newly discovered P450 gene is responsible for at least two of three consecutive oxidation steps in the formation of conifer diterpene resin acids. The PtAO P450 enzyme efficiently uses several different diterpene alcohols and diterpene aldehydes as substrates, leading to the formation of a suite of diterpene resin acids found in loblolly pine oleoresin.[38] Expression of this P450 gene is inducible in stems of loblolly pine by MeJA treatment, which mimics insect attack. Closely related P450 genes have also been identified in the spruce EST databases and are now being tested for biochemical functions and their role in spruce defense.

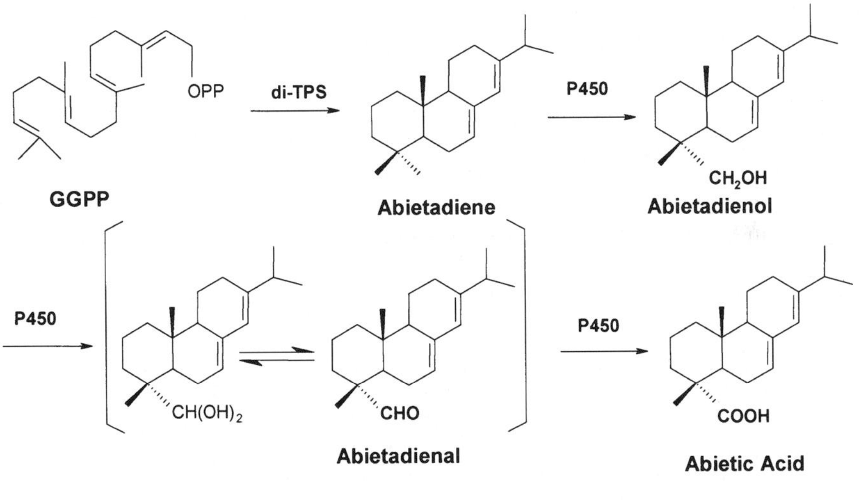

Fig. 2.3: Biosynthetic pathway scheme for diterpene resin acids. Di-TPS is diterpene synthase; P450 is cytochrome P450 dependent monooxygenase.

PHYLOGENY OF THE CONIFER TPS-d FAMILY IN COMPARISON WITH ANGIOSPERM TPS

The known plant TPS constitute a large family of enzymes containing members from angiosperms and gymnosperms (Fig. 2.4). Based on sequence similarities, catalytic mechanisms, and exon and intron patterns, all plant TPS are believed to have arisen from a common ancestor.[8,20,34,35] Such an ancestral TPS may have resembled the known conifer di-TPS and may have been involved in gibberellic acid metabolism, as known TPS of these functions all share an ancestral sequence motif of approximately 200 amino acids.[8,20,34] The evolution of *TPS* genes occurred to some large extent separately in angiosperms and gymnosperms, and appears to have involved the discrete loss of exons and introns that gave rise to the gene structures of known *TPS*.[21,34] Based on amino acid similarity, the *TPS* gene family has been divided into seven subfamilies designated *TPS-a* through *TPS-g*.[8,20,36] TPS subfamilies are designated by distinct branching patterns visualized by phylogenetic analyses, and each subfamily usually consists of TPS from the same biochemical class. Members of the *TPS-a* group include angiosperm sesqui-TPS and one known di-TPS with a similar catalytic mechanism to that of sesqui-TPS. The *TPS-b* group is composed of mono-TPS and one hemi-TPS, isoprene synthase,[37] all of angiosperm origin. The TPS-g family was recently identified and contains members of mono-TPS from snapdragon (*Antirrhinum majus*)[36] and *Arabidopsis thaliana*[38] that form acyclic products. The *TPS-f* branch contains the monotype linalool synthase from *Clarkia brewerii*.[39] This mono-TPS is the only characterized mono-TPS to contain the ancestral 200 amino acid motif. Also coding for this motif are the angiosperm di-TPS producing precursors of gibberellic acid, which are grouped into the *TPS-c* subfamily (copalyl diphosphate synthases) and into the *TPS-e* subfamily (kaurene synthases). All known TPS of conifer origin, regardless of the biochemical class and whether or not the ancestral 200 amino acid motif is present, cluster into the *TPS-d* subfamily (Fig. 2.5). TPS of plant secondary metabolism are believed to have evolved from TPS involved in primary metabolism. Through gene duplication events followed by functional diversification, TPS could have evolved to fill new niches and gained functions to create the wealth of terpene products known.[8,20,35] Within the TPS family, there are highly related enzymes that make different products, as there are highly divergent enzymes which make the same products.[8,20] Therefore, accurate *a priori* predictions of enzyme function based on sequence relatedness cannot be made. Each *TPS* gene must be functionally characterized to ascertain its biochemical role.

Fig. 2.4: The plant TPS (terpene synthase) family is divided into seven subfamilies, TPS-a through TPS-g, based on biochemical function and sequence similarity. Details of the phylogeny are described in Martin *et al.*[8]

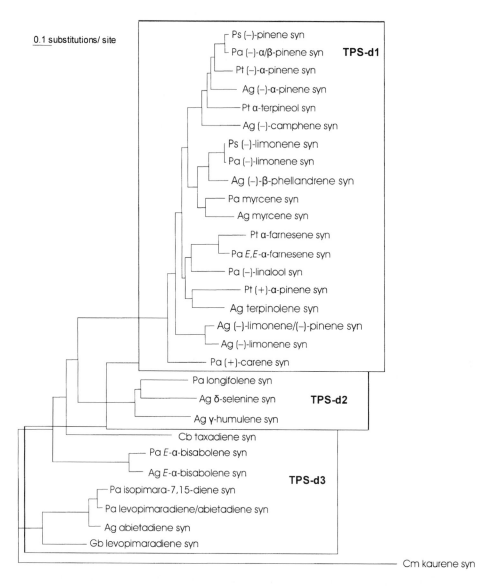

Fig. 2.5: The known conifer TPS are members of the TPS-d subfamily. The TPS-d subfamily is divided into three groups, TPS-d1 through TPS-d3, based on sequence similarity and biochemical function. Details of the phylogeny are described in Martin *et al.*[8]

Until recently, relatively few gymnosperm *TPS* genes had been cloned and characterized, compared to a larger number of well-characterized angiosperm *TPS* genes. Furthermore, the conifer *TPS-d* subfamily contained mainly TPS from grand fir and only few TPS from other gymnosperms.[3,20] For these reasons, the phylogeny of the gymnosperm *TPS-d* subfamily needed to be re-addressed with the inclusion of new genes functionally characterized in other conifer species. The cloning and characterization of ten *TPS* genes from Norway spruce enabled a reconstruction of the *TPS-d* subfamily and a re-examination of the larger TPS family (Figs. 2.4 and 2.5).[8] The inclusion of these ten TPS greatly expanded the *TPS-d* subfamily and provided evidence that distinct branching patterns exist within the *TPS-d* group based on sequence similarity and function in this subfamily.[8] The *TPS-d* subfamily was, therefore, divided into *TPS-d1* (mostly mono-TPS), *TPS-d2* (mostly sesqui-TPS), and *TPS-d3* (mostly di-TPS). This recently expanded phylogeny of the conifer *TPS-d* subfamily showed some similarities with regard to multiple events of independent evolution of sesqui-TPS in both the angiosperm and gymnosperm lineages.[8]

ASPECTS OF THE CHEMICAL ECOLOGY OF CONIFER TERPENOID DEFENSES

Oleoresin terpenoid biosynthesis and accumulation is a critical component of conifer chemical ecology. As some of the most ancient living organisms, long-lived conifers have developed both constitutive and inducible terpenoid defenses enabling them to combat herbivores and pathogens.[7,40] The resin, or pitch, of conifer oleoresin is a complex mixture composed of approximately equal molar concentrations of monoterpenes and diterpene resin acids present with a smaller fraction of sesquiterpenes. Recent reviews have highlighted the complexity and plasticity of terpene-based conifer defenses.[1-4,7] These chemicals can protect the tree through toxicity, by providing mechanical barriers, by interrupting essential processes in insect biology, or by signaling predators and parasites to the attacking herbivore. As toxicants, many terpenes, mono-, sesqui-, and diterpenes can have antibiotic activities against insects and insect-associated fungal pathogens. Some terpenes can interrupt the normal development and maturation of insects by interfering with insect endocrine systems. Examples of chemicals with this activity are the sesquiterpenes juvabione, originally isolated from *Abies balsamea*,[41] farnesol, and farnesal.[42] The precursor to juvabione is *E*-α-bisabolene, the synthesis of which is wound-inducible and MeJA-inducible in *Abies grandis*.[31] The mechanical component of oleoresin defenses enables the tree to physically seal wounds, and insect invaders are often pushed out of the tree in this process.[43] When an opening is created by a stem boring insect, volatile monoterpene constituents act as solvents to facilitate the movement of the diterpene resin acids to the wound. Subsequently, the monoterpenes evaporate, and the crystallization of the resin acids forms a protective barrier sealing the point of injury.

Terpenes are known to play important ecological roles not only in the defense of the tree, but also in bark beetle chemical ecology. Two chapters in this volume (see chapters by Raffa, *et al.*, and Tittiger, *et al.*) review aspects of terpenoids as chemicals in conifer-bark beetle interactions and isoprenoid formation in bark beetle chemical ecology. While some species of bark beetles may be able to produce monoterpene pheromones *de novo*,[44,45] bark beetles can also use host-derived monoterpenes as precursors to their own sex and aggregation pheromones.[46] Bark beetle monoterpene pheromones are often used in a stereospecific manner such that often only one enantiomer is able to serve as a sex or aggregation pheromone. Since both enantiomeric host-derived monoterpene precursors are accepted and chemically modified by the beetle, the exact monoterpene enantiomeric mixture of conifer resin is important to both the beetle's ability to produce the correct pheromone and to the trees' ability to escape an attack.[47,48]

Recently, new attention has been drawn to the activity of conifer terpenes in tritrophic or indirect defense interactions. Volatile terpenes may function in indirect or tritrophic defense interactions by signaling predators or parasites of the herbivore. For example, indirect defense interactions involving Norway spruce, the European spruce bark beetle (*Dendroctonus micans*), and the predatory beetle (*Rhizophagus grandis*) are mediated through oxygenated monoterpene volatiles where these cues also stimulate oviposition behavior by the predator.[49,50] An investigation of predator responses to Ips spp. pheromones revealed that pine monoterpenes act synergistically on the attractiveness of beetle pheromones to the predators, *Thanasimus dubius*, *Platysoma cylindrica*, and *Corticeus parallelus*.[51,52] Volatile terpenes emitted from Scotts pine (*Pinus sylvestris*) in response to pine sawfly egg deposition were found to attract the egg parasitoid *Chrysonotomyia ruforum*.[53] The biological activities of terpenes in either direct defense or attraction of predators or parasites can be specific to a given monoterpene or its enantiomer or to a blend of terpenes and other volatiles. Other important factors governing indirect defense in conifers are only now becoming apparent, including the molecular and biochemical control of active monoterpene and sesquiterpene volatile emissions in conifers.[11,15] Feeding by the white pine weevil on Sitka spruces induces the emission of a blend of monoterpenes and sesquiterpenes.[15] Many of these volatiles are passively released by evaporation of oleoresin terpenoids at the wound site. However, molecular and biochemical studies along with selective monitoring of induced volatile emissions demonstrated that new volatiles, including the monoterpene alcohol (−)-linalool, are formed *de novo* in spruce upon attack by weevils or after MeJA-treatment.[11,15] Linalool was released with diurnal emission profiles, and apparently originated from the foliage of induced spruce saplings. Thus conifers, which constitutively sequester large amounts of monoterpenes in resin ducts or blisters, can use stored resin terpenes as a pool for passive volatile release upon wounding, and can also use additional active mechanisms for induced volatile emissions that are controlled at the levels of insect-induced TPS gene expression and TPS enzyme activities.[11,15] The actively controlled, herbivore-induced volatiles in Sitka spruce and Norway spruce, specifically that of (–)-linalool, are not sequestered in substantial amounts in constitutive or traumatic oleoresins. A more complete understanding of the mechanisms involved in conifer tritrophic interactions may provide biological means to control insect pest outbreaks in conifer forests.

Conifer chemical ecology includes many other herbivores, predators and pathogens, aside from the bark beetles, weevils, sawflies, or budworms, and in these diverse interactions, terpenoid volatile emissions as well as accumulated oleoresin terpenes are likely to play many important roles in signaling and defense.

SPECIALIZED STRUCTURES FOR RESIN ACCUMULATION, CONSTITUTIVE AND TRAUMATIC RESINOSIS

Conifers have evolved specialized anatomical structures for the sequestration and accumulation of large amounts of hydrophobic oleoresin terpenoids, which would otherwise interfere with essential processes of metabolisms and membrane integrity in the terpenoid producing cells. A classical taxonomy of the Coniferales is based on Bannan's classification of these structures.[54] Terpenoid sequestration structures can be as simple and short-lived as the resin blisters found in true firs (*Abies*), cedars (*Cedrus*), hemlock (*Tsuga*), and golden larches (*Pseudolarix*), or they can be a complex long lasting highly organized system of interconnected canals found in species of spruce (*Picea*), larch (*Larix*), pine (*Pinus*), and Douglas-fir (*Pseudotsuga*). The complexity and the longevity of these structures are defining characteristics for members of this order.[43] Species producing high constitutive levels of terpenes have a well developed system of resin canals, while those with little constitutive resin lack extensive pre-existing ducts, but retain the ability to induce terpenoid defenses when challenged.[13,55]

Spruces represent an interesting system displaying both complex constitutive resin canals along with the inducible response of traumatic resin duct development.[9,10,56] Spruce bark tissue has numerous axial resin canals formed by epithelial cells capable of producing terpenes for many years. However, under normal conditions, there is little axial resin duct development in spruce xylem. The stress-induced formation of traumatic resin canals in spruces occurs within the outermost layers of xylem cells and can extend over great distances longitudinally. Recent work on Norway spruce has described the histology of induced formation of traumatic resin ducts and indicated that xylem mother cells in the cambium zone are the primary source of traumatic resin duct formation.[56] The development of traumatic resin ducts is due to a change in cambial activity that initiates resin duct epithelial cells in lieu of tracheids. Nagy *et al.* also found lumenal continuity between the traumatic resin ducts and the radial duct rays and direct contact between the epithelial cells and the ray parenchyma cells.[56] Together these features enable enhanced resin flow and communication between xylem and bark tissues.

Conifer traumatic resin canals can be induced by a wide variety of both external and internal stimuli. These include, but are not limited to, insect attack,[15,16,57] pathogens,[58,59] mechanical wounding,[58] elicitor preparations made from fungal cell walls,[60] and treatment of trees with MeJA.[9,10,13,15] However, under normal growth conditions, these ducts rarely appear, or if they do, they are few in number and scattered, as opposed to the nearly complete concentric rings of ducts that occur in response to pathogens or insects. The formation of traumatic resin ducts as an induced anatomical defense that supports accumulation of terpenoid chemical defenses is not restricted to conifer stems, but is also induced in roots. For example, in Douglas fir seedlings, treatment with MeJA induces development of traumatic resin ducts in root xylem accompanied by massive accumulation of resin terpenoids in these tissues.[61] Studies with white spruce (*Picea glauca*) found a positive correlation of traumatic resin duct formation with resistance to white pine weevil attack.[5,62,63] The rather long time of two to three weeks for traumatic resin duct development upon elicitation in species of spruce corresponds well with the lag phase of larval development of the white pine weevil in the phloem and outer xylem tissues of the host tree. Traumatic resin ducts, by spilling their terpenoid contents into the cavities created by egg deposition and feeding larvae, can effectively increase the rate of weevil brood mortality.[57] Furthermore, the formation of traumatic resin ducts can provide a long-lasting, induced defense system. The formation of traumatic resin ducts in conifer stems and roots is associated with induced terpenoid resin accumulation and induced terpenoid biosynthesis.[10,11,15,61] Recent work with species of spruce, Norway spruce, and Sitka spruce, in which trees were treated with MeJA or attacked by white pine weevil, revealed that resin duct development, accumulation of resin terpenoids, increased enzyme activities of mono-TPS, sesqui-TPS and di-TPS, and increased expression of the corresponding *TPS* genes are all closely coordinated events in the induced traumatic resin defense response in spruce stem tissues.[10-12,15,16]

Prior to these studies of induced terpenoid defenses in spruce, there was a considerable amount of research on induced terpene biosynthesis within the stem tissue of grand fir, as previously reviewed.[1-4] As in the spruce system, mono-TPS and di-TPS activities and *TPS* gene expression are wound-inducible in grand fir. In both systems, the increases in TPS activities and *TPS* gene expression were reflected in the quality and quantity of oleoresin accumulation. Characterization of the genes and enzymes involved in induced conifer terpenoid biosynthesis has been critical to the understanding of these defenses and the processes governing them. In particular, the use of MeJA as a non-destructive means of defense elicitation of traumatic resinosis allowed a detailed analysis of induced biochemical and molecular processes of insect- or pathogen-inducible conifer defenses.[10-12,15,16]

METHYL JASMONATE AS A BIOLOGICAL ELICITOR OF CONIFER DEFENSE

Plants control and coordinate metabolic processes, development, and defense through a system of phytohormones and other regulatory compounds. Signal molecules such as jasmonates or other octadecanoids and oxylipins, ethylene, and salicylate are critical to the orchestration of defenses, and they enable plants to respond to pathogen or herbivore challenges. The octadecanoid or jasmonate pathway is one of the best characterized plant defense signaling pathways.[64-66] MeJA has been widely used to induce plant defense responses in a number of angiosperm species. Jasmonic acid and MeJA are also known to induce the emission of volatiles of indirect defense in several angiosperm species.[67-75]

In conifers, the use of MeJA to induce plant defenses has been paramount in the characterization of terpenoid defense responses recently characterized in Norway spruce and Sitka spruce.[9-13,15] Treatment of seedlings, saplings, and mature spruce trees with solutions of MeJA induced terpenoid responses that are largely similar to those induced by real insect attack or induced by fungal inoculation. Ectopic treatment with MeJA induces the complex reprogramming of traumatic resin duct development and induces terpenoid accumulation as well as terpenoid volatile emissions. MeJA provided a means by which to induce conifer defense responses without mechanically injuring the tree. This is especially vital to the quantification of induced terpene defenses where any injury not only limits the biological capacity of the plant tissue to respond, but also, enhances the loss of terpenes through the wound site. MeJA treatment of Norway spruce and Sitka spruce allowed us to characterize the induced TPS activities and *TPS* gene expression in bark, wood and foliage, enabled the quantitative and qualitative comparison of terpene constituents present in control and MeJA treated trees, and enabled the characterization of a distinct volatile emission profile induced by MeJA treatment.[10,11,15] Furthermore, the accumulation of resin components as well as the increased activity of TPS enzymes and *TPS* gene expression is correlated with the development of TD in MeJA treated saplings.[10,15]

MeJA has also been used successfully in other gymnosperm systems to induce secondary metabolism and defense responses. In species of *Taxus*, enzymes involved in the synthesis of the diterpene taxol are induced by MeJA in suspension cultures.[76-78] Studies in white spruce seedlings and cell cultures have shown that MeJA treatment induces the expression of a 14-3-3 protein and chalcone synthase.[79,80] MeJA application assists ectomycorrhizal colonization of Norway spruce roots,[81] and MeJA application to Norway spruce seedlings also enhances their survival rate when they were challenged by *Pythium ultimun*.[82] In recent work with Sitka spruce, we have established comprehensive gene expression profiles of thousands of transcript species in response to MeJA and insect attack monitored with microarrays.

BIOCHEMICAL FUNCTIONS OF NORWAY SPRUCE TPS GENES AND THEIR ROLES IN CONSTITUTIVE AND INDUCED TERPENOID DEFENSES

Recent studies have described the complex, constitutive, and MeJA- and insect-induced terpenoid defenses in Norway spruce and Sitka spruce at anatomical, chemical, and biochemical levels.[10,11,15] Miller *et al.* compared the effect of MeJA treatment and white pine weevil attack on induced terpenoid defenses in Sitka spruce.[15] They found that MeJA and real insects induce very similar terpenoid responses at the biochemical and molecular levels. The cloning and functional characterization of a family of ten, functionally diverse Norway spruce *TPS* genes not only allowed an improved phylogeny of the conifer *TPS-d* gene family but also allowed a detailed molecular and biochemical characterization of induced changes in terpenoid metabolite profiles in spruce defense. Many of the products of the recently identified Norway spruce *TPS* genes[8,12] correspond with components of terpenoid metabolite profiles after MeJA treatment or insect attack in species of spruce. Our work on Norway spruce terpenoid responses to MeJA treatment[10,11] is discussed here with an emphasis on those compounds that are products of the newly characterized spruce *TPS* genes.[8,12]

While nearly all monoterpenoids produced by the cloned Norway spruce *TPS* are found constitutively in Norway spruce, increases in many monoterpenoids occur after MeJA treatment in wood and bark, but to a much lesser extent in needles.[10,11] The most prominent monoterpenes in Norway spruce are β-pinene and α-pinene. Accumulation of these compounds is substantially induced following MeJA treatment. Analysis of enantiomers showed that following MeJA treatment both (–)-α-pinene and (–)-β-pinene increased relative to the (+)-α-pinene and (+)-β-pinene enantiomers.[10] This finding indicates the presence of enantiomer-specific pinene synthases in Norway spruce. Indeed, we identified a stereospecific (–)-α/β-pinene synthase (*PaTPS-Pin*).[8] Another highly abundant compound in bark, wood, and needles of Norway spruce is the acyclic monoterpene myrcene. Concentrations of myrcene increased following MeJA treatment, suggesting a role for the cloned myrcene synthase gene (*PaTPS-Myr*) in this response. Limonene is the fifth most abundant monoterpene in Norway spruce tissues tested in our studies. Following MeJA treatment, concentrations of limonene increased dramatically in all three tissues, bark, wood, and foliage. These findings indicate a role for the (–)-limonene synthase gene (*PaTPS-Lim*) in the induced defense response. Accumulation of (+)-3-carene increased dramatically in MeJA induced wood tissue, a function that can be attributed to the activity of the *PaTPS-Car* gene.[12] 3-Carene was also found constitutively in bark, foliage, and in volatile emissions. When Norway spruce mono-TPS probes were employed for gene expression analysis in weevil- or MeJA-induced Sitka spruce stems we found strong increases of mono-TPS transcript levels with all available *mono-TPS* gene probes.[15]

Volatile terpenoid emissions in response to insect feeding or MeJA treatment were recently described for Sitka spruce[15] and Norway spruce[11] and have also been shown in Scotts pine in response to insect oviposition.[53,83] Following MeJA treatment, monoterpenoids released from Norway spruce trees changed in composition from mainly monoterpene hydrocarbons to mainly oxygenated monoterpenes. The most prominent monoterpene volatile released was linalool.[11] A Norway spruce (–)-linalool synthase (*PaTPS-Lin*) was subsequently cloned, and the recombinant enzyme was functionally characterized.[8] The inducible gene expression and enzyme activity of this TPS may be important in signaling predators or parasites of herbivores. Using the *PaTPS-Lin* probe, we found strong MeJA-induced transcript accumulation in needles of Sitka spruce, corresponding with an induced, active emission of (–)-linalool in response to MeJA-treatment and weevil attack in this system.[15]

Although amounts of sesquiterpenoids in Norway spruce bark, wood, and needles are only a fraction of those of monoterpenoids and diterpenoids, there are profound increases in some sesquiterpenoids after MeJA treatment.[10,11] The same is true for weevil- and MeJA-induced sesquiterpenoid accumulations in stems of Sitka spruce.[15] The release of the sesquiterpene volatile E-α-bisabolene was increased after MeJA treatment in Norway spruce, and E-α-bisabolene also increased in bark with MeJA treatment.[10,11] Activities of the E-α-bisabolene synthase gene (*PaTPS-Bis*) could account for these responses.[8] The *PaTPS-Bis* probe revealed strongly up-regulated transcript levels in needles of MeJA-treated Sitka spruce.[15] The sesquiterpenes longifolene and longipinene decreased in Norway spruce xylem tissue following MeJA treatment, but increased in bark tissue.[10] It is likely that the cloned longifolene synthase (*PaTPS-Lon*) controls the levels of these two compounds.[8] E,E-α-Farnesene, the product of the cloned *PaTPS-Far* gene,[8] was found only as a volatile emission in Norway spruce, but was not increased in response to elicitation.[11]

Diterpenes occur in Norway spruce and in other members of the pine family, mainly in the form of diterpene resin acids. The precursor in the formation of diterpenes is GGDP formed by GGDP synthase. Activities of GGDP synthase were increased in MeJA-treated wood.[10] This suggests that a level of control of diterpene resin acids may reside partly in the flux of GGDP and the regulation of GGDP synthase, followed by regulation of di-TPS. We cloned and functionally characterized two different Norway spruce *di-TPS* genes, levopimaradiene/abietadiene synthase *PaTPS-LAS* and isopimara-7,15-diene synthase *PaTPS-Iso*.[8] The four most abundant diterpene products of these *di-TPS* genes are precursors for the major diterpene resin acids in Norway spruce bark and wood.[10] Induced levels of the following ditepene resin acids, abietate, levopimarate, and neoabietate, increased in both bark and wood of Norway spruce following MeJA treatment. This indicates a role of the *PaTPS-LAS* gene in Norway spruce induced defense. Isopimarate, the diterpene resin acid of isopimara-7,15-diene, was found in bark and wood, but was not increased after MeJA treatment. The finding of two different *di-TPS* genes and identification of their product profiles suggests that coordinated up-regulation of diterpene resin acids abietate, levopimarate, and neoabietate but lack of increase of isopimarate is regulated by differential expression of these two *di-TPS*. Transcripts hybridizing with Norway spruce *di-TPS* probes were strongly increased in stems of Sitka spruce treated with MeJA or attacked by white pine weevil.[15] In contrast, MeJA treatment did not cause an induction of *di-TPS* gene expression in needles of Sitka spruce, which matches the lack of induced diterpenoid accumulation in needles.[15]

This correlative analysis of terpenoid accumulation, TPS enzyme activities, and functional characterization of a family of cloned *TPS* genes in Norway spruce and Sitka spruce in response to MeJA treatment and insect attack will guide future studies regarding differential expression of *TPS* genes in conifer defense, including *in situ* localization of gene expression and TPS enzymes.

NEW DIRECTIONS IN CONIFER DEFENSE RESEARCH

Despite progress of the last five years, we are only beginning to understand the full complexity of constitutive and inducible terpenoid defenses in conifers. To date, terpenoid defenses have only been studied in a few conifer species at the molecular biochemical levels, namely in Norway spruce, Sitka spruce, and in grand fir, and more work is required in these systems. Recently, new research on conifer terpenoid defenses has also been initiated in Douglas fir and in loblolly pine.

Signaling of Induced Defenses

The use and investigation of endogenous conifer defense signal molecules as well as the study of elicitors present in herbivore oral secretions and oviposition fluids offers exciting avenues for future research into conifer defense. Despite the effective use of MeJA as an elicitor of conifer defense, the role of endogenous octadecanoid signaling is still largely unknown. Miller *et al.* recently identified several candidate genes of the octadecanoid pathway in the spruce EST-database.[15] Transcripts resembling allene oxide synthase and allene oxide cyclase, two enzymes in the octadecanoid / oxylipin pathway, were induced in stems of Sitka spruce in response to weevil feeding and in response to MeJA treatment.[15] The induction of this signaling pathway seems well coordinated with subsequent induced terpenoid defenses. Hudgins and Franceschi also provided convincing evidence for a function of the defense signal molecule ethylene in wound- and MeJA-induced traumatic resin duct formation in Douglas fir and in giant redwood (*Sequoiadendrum giganteum*).[84] The recent development and application of spruce cDNA microarray chips with more than 15,000 cDNA elements spotted on glass slides (www.treenomix.com) has revealed many additional candidate genes for defense signaling, including inducible transcription factors, in the weevil- and spruce budworm-induced defense response in Sitka spruce (Ralph and Bohlmann, unpublished results).

Constitutive and Traumatic Resinosis and Volatile Emissions and Resistance

By using an integrated molecular and biochemical approach, we have established an understanding of some of the fundamental mechanisms of constitutive and inducible, terpenoid defenses in bark, wood, and needles of a few conifer species. However, more work is needed to prove the effectiveness of these defenses against herbivores and pathogens, and their contributions to resistance against some specialist conifer herbivores. New research in this direction is being developed primarily along two lines. One, in our laboratory, is testing the effect of MeJA pretreatment of conifers on the success of subsequent attack by insect herbivores and fungal infections. If, indeed, induced conifer defenses provide better protection against herbivores or pathogens, prior induction of defenses should increase the survival of conifer hosts in interactions with herbivores or pathogens. A second line of research is the use of known resistant and susceptible trees from established long-term Sitka spruce breeding programs[91] and trying to establish associations of defense traits and genes with resistance and susceptibility. This work is targeting selected terpenoid defenses, and is also using a population genomics approach. While the former builds primarily on established molecular and biochemical characterization of constitutive and induced defenses in Sitka spruce, the latter is building on additional population and quantitative genetics methods developed in the Treenomix (www.treenomix.com) conifer forest health genomics program funded by Genome Canada and the Province of British Columbia.[86]

Structure-Function Analysis and New Applications of TPS

Ongoing research with the newly characterized *TPS* genes from Norway spruce[8] and other conifers is aiming at an understanding of the structure-function relationships of TPS enzymes in conifer defense. For example, we are targeting specific amino acids in the active sites of closely related Norway spruce di-TPS, levopimaradiene/abietadiene synthase PaTPS-LAS and isopimara-7,15-diene synthase PaTPS-Iso, based in part on comparative active site modeling for site directed mutagenesis. By identifying amino acids responsible for the differences in the product profiles of these newly divergent *di-TPS* genes, we will be able to better understand reaction mechanisms of these enzymes. This information is particularly relevant since there is currently no crystal structure known for any di-TPS or for any conifer TPS. A detailed knowledge of the processes determining product profiles of TPS may also lead to tailor-made TPS enzymes with new functions and improved biological functions in the formation of terpenoid natural products following lead structures such as those of the anticancer diterpene drug taxol or the antimalaria sesquiterpene compound artemisinin. Along the same line of research, genome wide screening for new *TPS* and cytochrome *P450* genes in conifers can lead to the discovery of new enzymes with potential application in biotechnology, which continues to be an important niche area in the non-timber forest products industry.

Tissue and Cell-Specific Localization of Terpenoid Pathways in Defense

Our recent research suggests organ-, tissue-, and cell-specific localization of constitutive and induced terpenoid defense pathways in conifers. For example, linalool synthase (PaTPS-Lin) seems to be preferentially expressed in needles of Norway spruce and Sitka spruce with little or no expression in stems.[11,15] It is also likely that expression of PaTPS-Lin in spruce needles is not associated with resin ducts but could reside in other cells involved with induced terpenoid emission.[15] In contrast, we can speculate that most other mono-TPS and di-TPS are associated with epithelial cells of constitutive and induced resin ducts. The possible localization of conifer sesqui-TPS is difficult to predict. Furthermore, the exact spatial and temporal patterns of terpenoid pathway gene expression associated with traumatic resin duct development in the cambium zone and outer xylem remain to be studied at the tissue and cell level. *In situ* hybridization and immuno-localization of *TPS* will address these open questions. These methods have worked well in identifying cell type specific gene and protein expression of alkaloid formation in opium poppy (*Papaver somniferum*).[87] As the biochemistry of induced terpene defenses and the development of traumatic resin ducts have been well described in spruce, this system is ideal for future studies of tissue- and cell-specific localization of transcripts and proteins associated with oleoresin defense and induced volatile emissions in conifers. In addition, the advent of laser dissection microscopy techniques presents a fascinating means by which to further address RNA and protein analysis in a tissue- and cell-specific manner.[88] These techniques, when applied to the cambium zone, xylem mother cells, and the epithelial cells that surround traumatic resin ducts, and will allow a temporal and spatial analysis of cellular functions occurring in the traumatic resin response.

Genomic and Proteomic Analyses of Conifer Defenses

While the understanding of induced conifer terpenoid defenses has increased dramatically with much information now available from terpenoid metabolite profiling and the study of associated terpenoid pathways, there is still much to be learned about other possible defense and resistance mechanisms. Our ongoing genomic and proteomic analyses of spruce defenses against insects and pathogens are already providing new insights into conifer defense systems, including the discovery of herbivore induced transcription factors and other defense pathways that involve a group of weevil-induced dirigent proteins and other genes for phenolic defense compounds (Ralph *et al.*, unpublished results). These genomics and proteomics approaches will result in a better understanding of which genes are involved in defense responses and will provide a plethora of candidate genes for further characterization and for marker development. Most importantly, both genomic and proteomic analyses will quite possibly offer information about genes and proteins not previously known or regarded to participate in conifer defense. The synthesis of these upcoming results will enable us to conceptualize conifer defenses on many different levels, and will eventually lead to a more complete understanding of conifer defense systems.

SUMMARY

In the last five years, much progress has been made in the characterization of conifer defenses using an integrated biochemical, molecular, and genomics approach. With a focus on the fascinating processes of traumatic resin duct development in direct defense and on induced volatile emissions in possible tritrophic defense, we have developed species of spruce, Sitka spruce, and Norway spruce, as two of the best characterized systems of conifer defense research at the molecular and biochemical levels. The economic importance of spruce for the forest industry and the enormous economic, environmental, and social impacts of recent, unprecedented outbreaks of conifer insect pests has mandated substantial new investment into conifer genomics research with an emphasis on the genomics of defense and resistance of conifers against insect pests.

ACKNOWLEDGEMENTS

Research in our laboratory has been generously supported by grants to JB from the Natural Sciences and Engineering Research Council of Canada (NSERC), the Human Frontiers Science Program (HFSP), Genome Canada, Genome British Columbia, and the Province of British Columbia. The Canadian Foundation for Innovation (CFI) and the BC Knowledge and Development Funds (BCKDF) provided infrastructure support. JB acknowledges the many contributions of an outstanding group of undergraduate and graduate students, postdoctoral associates, and technicians and the inspiring and supportive collaborations with scientists in the BC Ministry of Forests, the Canadian Forest Service, and academia. JB wishes to thank his mentor, Rodney Croteau, who started biochemical research on conifers terpenoid defenses.

REFERENCES

1. BOHLMANN, J., CROTEAU, R., Diversity and variability of terpenoid defenses in conifers: molecular genetics, biochemistry and evolution of the terpene synthase gene family in grand fir (*Abies grandis*). *in*: Insect Plant Interactions and Induced Plant Defense (D.J. Chadwick and J.A. Goode, eds,), John Wiley and Sons Ltd., West Sussex. 1999, pp. 132-146.
2. PHILLIPS, M.A., CROTEAU, R.B., Resin-based defenses in conifers., *Trends Plant Sci.*, 1999, **4**, 184-190.
3. BOHLMANN, J., GERSHENZON, J., AUBOURG, S., Biochemical, molecular genetic and evolutionary aspects of defense-related terpenoid metabolism in conifers., *Rec. Adv. Phytochem.*, 2000, **34**, 109-149.
4. TRAPP, S.C., CROTEAU, R., Defensive resin biosynthesis in conifers., *Annu. Rev. Plant Physiol. Plant Mol. Biol.*, 2001, **52**, 689-724.
5. ALFARO, R.I., BORDEN, J.H., KING, J.N., TOMLIN, E.S., MCINTOSH, R.L., BOHLMANN, J., Mechanisms of resistance in conifers against shoot infesting insects., *in*: Mechanisms and Deployment of Resistance in Trees to Insects (M.R. Wagner, K.M. Clancy, F. Lieutier, and T.D. Paine, eds,), Kluwer Academic Press, Dordrecht. 2002, pp. 101-126.
6. BOHLMANN, J., MARTIN, D.M., MILLER, B., HUBER, D.P.W., Terpenoid synthases in conifers and poplars., *in*: Plantation Forest Biotechnology for the 21st Century (C. Walter and M. Carson, eds,), Research Signpost, Kerala, 2004, pp. 181-201.
7. HUBER, D.P.W., RALPH, S., BOHLMANN, J., Genomic hardwiring and phenotypic plasticity of terpenoid-based defenses in conifers., *J. Chem. Ecol.*, 2004, **30**, 2399-2418.
8. MARTIN, D.M., FÄLDT, J., BOHLMANN, J., Functional characterization of nine Norway spruce *TPS* genes and evolution of gymnosperm terpene synthases of the *TPS-d* subfamily. *Plant Physiol.*, 2004, **135**, 1908-1927.
9. FRANCESCHI, V.R., KREKLING, T., CHRISTIANSEN, E., Application of methyl jasmonate on *Picea abies* (Pinaceae) stems induces defense-related responses in phloem and xylem. *Am. J. Bot.*, 2002, **89**, 578-586.
10. MARTIN, D., THOLL, D., GERSHENZON, J., BOHLMANN, J., Methyl jasmonate induces traumatic resin ducts, terpenoid resin biosynthesis, and terpenoid accumulation in developing xylem of Norway spruce stems. *Plant Physiol.*, 2002, **129**, 1003-1018.
11. MARTIN, D.M., GERSHENZON, J., BOHLMANN, J., Induction of volatile terpene biosynthesis and diurnal emission by methyl jasmonate in foliage of Norway spruce (Picea abies). *Plant Physiol.*, 2003, **132**, 1586-1599.
12. FÄLDT, J., MARTIN, D., MILLER, B., RAWAT, S., BOHLMANN, J., Traumatic resin defense in Norway spruce (*Picea abies*): Methyl jasmonate-induced terpene

synthase gene expression, and cDNA cloning and functional characterization of (+)-3-carene synthase. *Plant Mol. Biol.*, 2003, **51**, 119-133.

13. HUDGINS, J.W., CHRISTIANSEN, E., FRANCESCHI, V.R., Methyl jasmonate induces changes mimicking anatomical defenses in diverse members of the Pinaceae. *Tree Physiol.*, 2003, **23**, 361-371.

14. HUBER, D.P.W., PHILIPPE, R.N., MADILAO, L.L., STURROCK, R.N., BOHLMANN, J., Changes in anatomy and terpene chemistry in roots of Douglas-fir seedlings following treatment methyl jasmonate. *Tree Physiol*, 2005, in press.

15. MILLER, B., MADILAO, L.L., RALPH, S., BOHLMANN, J., Insect-induced conifer defense. White pine weevil and methyl jasmonate induce traumatic resinosis, *de novo* formed volatile emissions, and accumulation of terpenoid synthase and octadecanoid pathway transcripts in Sitka spruce. Plant Physiol, 2005, **137**, 369-382.

16. BYUN MCKAY, A., HUNTER, W., GODDARD, K., WANG, S., MARTIN, D., BOHLMANN, J., PLANT, A., Insect attack and wounding induce traumatic resin duct development and gene expression of (–)-pinene synthase in Sitka spruce. *Plant Physiol.*, 2003, **133**, 368-378.

17. CROTEAU, R., KUTCHAN, T., LEWIS, N. Natural products (secondary metabolism). *in*: Biochemistry and Molecular Biology of Plants (B.B. Buchanan, W. Gruissem, and R.L. Jones, eds,), American Society of Plant Biologists, Rockville. 2000, pp. 1250-1318.

18. BURKE, C., CROTEAU, R., Geranyl diphosphate synthase from *Abies grandis*: cDNA isolation, functional expression, and characterization. *Arch. Biochem. Biophys.*, 2002, **405**:130-136.

19. BICK, J.A., LANGE, B.M., Metabolic cross talk between cytosolic and plastidial pathways of isoprenoid biosynthesis: unidirectional transport of intermediates across the chloroplast envelope membrane. *Arch. Biochem. Biophys.*, 2003, **415**, 146-154.

20. BOHLMANN, J., MEYER-GAUEN, G., CROTEAU, R., Plant terpenoid synthases: Molecular biology and phylogenetic analysis. *Proc. Natl. Acad. Sci. USA*, 1998, **95**, 4126-4133.

21. CANE, D.E., Sesquiterpene biosynthesis: Cyclization mechanisms. *in:* Comprehensive Natural Products Chemistry: Isoprenoids, Including Carotenoids and Steroids, Vol 2. (D.E. Cane, ed,), Pergamon Press, Oxford. 1999, pp. 155-200.

22. WISE, M.L., CROTEAU, R. (1999) Monoterpene biosynthesis. *in*: Comprehensive Natural Products Chemistry: Isoprenoids, Including Carotenoids and Steroids, Vol 2. (D.E. Cane, ed,), Pergamon Press, Oxford. 1999, pp. 97-154

23. DAVIS, E.M., CROTEAU, R. Cyclization enzymes in the biosynthesis of monoterpenes, sesquiterpenes, and diterpenes. *Top. Curr. Chem.*, 2000, **209**, 53-95.

24. LESBURG, C.A., ZHAI, G., CANE, D.E., CHRISTIANSON, D.W. Crystal structure of pentalenene synthase: mechanistic insights on terpenoid cyclization reactions in biology. *Science*, 1997, **277**, 1820-1824.

25. STARKS, C.M., BACK, K.W., CHAPPELL, J., NOEL, J.P. Structural basis for cyclic terpene biosynthesis by tobacco 5-*epi*-aristolochene synthase. *Science*, 1997, **277**, 1815-1820.

26. WHITTINGTON, D.A., WISE, M.L., URBANSKY, M., COATES, R.M., CROTEAU, R.B., CHRISTIANSON, D.W., Bornyl diphosphate synthase: structure

and strategy for carbocation manipulation by a terpenoid cyclase. *Proc. Natl. Acad. Sci. USA*, 2002, **99**, 15375-15380.

27. STEELE, C.L., CROCK, J., BOHLMANN, J., CROTEAU, R., Sesquiterpene synthases from grand fir (*Abies grandis*): Comparison of constitutive and wound-induced activities, and cDNA isolation, characterization, and bacterial expression of δ-selinene synthase and γ-humulene synthase. *J. Biol. Chem.*, 1998, **273**, 2078-2089.

28. PHILLIPS, M.A., SAVAGE, T.J., CROTEAU, R., Monoterpene synthases of loblolly pine (Pinus taeda) produce pinene isomers and enantiomers. *Arch. Biochem. Biophys.*, 1999, **372**, 197-204.

29. STOFER VOGEL, B., WILDUNG, M.R., VOGEL, G., CROTEAU, R., Abietadiene synthase from grand fir (*Abies grandis*) - cDNA isolation, characterization, and bacterial expression of a bifunctional diterpene cyclase involved in resin acid biosynthesis. *J. Biol. Chem.*, 1996, **271**, 23262-23268.

30. BOHLMANN, J., STEELE, C.L., CROTEAU, R., Monoterpene synthases from grand fir (*Abies grandis*). cDNA isolation, characterization, and functional expression of myrcene synthase, (−)-(4S)-limonene synthase, and (−)-(1S,5S)-pinene synthase. *J. Biol. Chem.*, 1997, **272**, 21784-21792.

31. BOHLMANN, J., CROCK, J., JETTER, R., CROTEAU, R., Terpenoid-based defenses in conifers: cDNA cloning, characterization, and functional expression of wound-inducible (*E*)-α-bisabolene synthase from grand fir (*Abies grandis*). *Proc. Natl. Acad. Sci. USA*, 1998, **95**, 6756-6761.

32. BOHLMANN, J., PHILLIPS, M., RAMACHANDIRAN, V., KATOH, S., CROTEAU, R., cDNA cloning, characterization, and functional expression of four new monoterpene synthase members of the *Tpsd* gene family from grand fir (Abies grandis). *Arch. Biochem. Biophys.*, 1999, **368**, 232-243.

33. PHILLIPS, M.A., WILDUNG, M.R., WILLIAMS, D.C., HYATT, D.C., CROTEAU, R., cDNA isolation, functional expression, and characterization of (+)-α-pinene synthase and (-)-α-pinene synthase from loblolly pine (*Pinus taeda*): Stereocontrol in pinene biosynthesis. *Arch. Biochem. Biophys.*, 2003, **411**, 267-276

34. TRAPP, S.C., CROTEAU, R.B., Genomic organization of plant terpene synthases and molecular evolutionary implications. Genetics, 2001, **158**, 811-832

35. AUBOURG, S., LECHARNY, A., BOHLMANN, J., Genomic analysis of the terpenoid synthase (*AtTPS*) gene family of *Arabidopsis thaliana. Mol. Gen. Genom.*, 2002, **267**, 730-745.

36. DUDAREVA, N., MARTIN, D., KISH, C.M., KOLOSOVA, N., GORENSTEIN, N., FÄLDT, J., MILLER, B., BOHLMANN, J. (*E*)-β-Ocimene and myrcene synthase genes of floral scent biosynthesis in snapdragon: Function and expression of three terpene synthase genes of a new terpene synthase subfamily. *Plant Cell*, 2003, **15**, 1227-1241.

37. MILLER, B., OSCHINSKI, C., ZIMMER, W., First isolation of an isoprene synthase gene from poplar and successful expression of the gene in *Escherichia coli. Planta*, 2001, **213**, 483-487.

38. CHEN, F., THOLL, D., D'AURIA, J.C., FAROOQ, A., PICHERSKY, E., GERSHENZON, J., Biosynthesis and emission of terpenoid volatiles from *Arabidopsis* flowers., *Plant Cell*, 2003, **15**, 481-494.

39. DUDAREAVA, N., CSEKE, L., BLANC, V.M., PICHERSKY, E., Evolution of floral scent in *Clarkia*: Novel patterns of S-linalool synthase gene expression in *C. brewerii* flower. *Plant Cell*, 1996 8, 1137-1148.
40. LANGENHEIM, J.H., Plant Resins: Chemistry, Evolution, Ecology, and Ethnobotany. Timber Press, Inc., Portland, OR, 2003, 586 p.
41. SLAMA, K., WILLIAMS, C.M., Juvenile hormone activity for the bug *Pyrrhocoris apterus.*, *Proc. Natl. Acad. Sci. USA*, 1965, **54**, 411-414.
42. SCHMIALEK, P., Compounds with juvenile hormone action., *Z. Naturforsch.*, 1963, **18**, 516-519.
43. CROTEAU, R., JOHNSON, M.A., Biosynthesis of terpenoid wood extractives. *in*: Biosynthesis and Degradation of Wood Components., Academic Press, Orlando, Fl. 1985, pp. 379-439.
44. SEYBOLD, S.J., TITTIGER, C., Biochemistry and molecular biology of *de novo* isoprenoid pheromone production in the Scolytidae., *Annu. Rev. Entomol.*, 2003, **48**, 425-453.
45. MARTIN, D., BOHLMANN, J., GERSHENZON, J., FRANCKE, W., SEYBOLD, S.J., A novel sex-specific and inducible monoterpene synthase activity associated with a pine bark beetle, the pine engraver, *Ips pini.*, *Naturwissen.*, 2003, **90**, 173-179.
46. SEYBOLD, S.J., BOHLMANN, J., RAFFA, K.F., Biosynthesis of coniferophagous bark beetle pheromones and conifer isoprenoids: Evolutionary perspective and synthesis., *Can. Entomol.*, 2000, **132**, 697-753.
47. BYERS, J.A., Host tree chemistry affecting colonization in bark beetles. *in*: Chemical Ecology of Insects, Vol 2. Chapman and Hall, New York. 1995, pp. 154-213.
48. BORDEN, J.H., Semio-Chemical Mediated Aggregation and Dispersion in the Coleoptera. Academic Press, London, 1984.
49. GREGOIRE, J.C., BAISIER, M., DRUMONT, A., DAHLSTEN, D.L., MEYER, H., FRANCKE, W., Volatile compounds in the larval frass of *Dendroctonus valens* and *Dendroctonus micans* (Coleoptera, Scolytidae) in relation to oviposition by the predator, *Rhizophagus grandis* (Coleoptera, Rhizophagidae). *J. Chem. Ecol.*, 1991, **17**, 2003-2019.
50. GREGOIRE, J.C., COUILLIEN, D., KREBBER, H., KÖNIG, W.A., MEYER, H., FRANCKE, W., Orientation by *Rhizophagus grandis* (Coleoptera: Rhizophagidae) to oxygenated monotepenes in a species-specific predator-prey relationship., *Oecologia*, 1992, **3**, 14-18.
51. RAFFA, K.F., KLEPZIG, K.D., Chiral escape of bark beetles from predators responding to a bark beetle pheromone., *Oecologia*, 1989, **80**, 566-569.
52. ERBILGIN, N., RAFFA, K.F., Modulation of predator attraction to pheromones of two prey species by stereochemistry of plant volatiles., *Oecologia*, 2001, **127**, 444-453.
53. HILKER, M., KOBS, C., VARMA, M., SCHRANK, K., Insect egg deposition induces *Pinus sylvestris* to attract egg parasitoids. *J. Exp. Biol.*, 2002, **205**, 455-461.
54. BANNAN, M.W., Vertical resin ducts in the secondary wood of the abietineae., *New Phytol.*, 1936, **35**, 11-46.

55. LEWINSOHN, E., GIJZEN, M., SAVAGE, T.J., CROTEAU, R., Defense-mechanisms of conifers - Relationship of monoterpene cyclase activity to anatomical specialization and oleoresin monoterpene content., *Plant Physiol.*, 1991, **96**, 38-43.

56. NAGY, N.E., FRANCESCHI, V.R., SOLHEIM, H., KREKLING, T., CHRISTIANSEN, E., Wound-induced traumatic resin duct development in stems of Norway spruce (Pinaceae): Anatomy and cytochemical traits. *Am. J. Bot.*, 2000, **87**, 302-313.

57. ALFARO, R.I., An induced defense reaction in white spruce to attack by the white-pine weevil, *Pissodes strobi*. *Can. J. For. Res.*, 1995, **25**, 1725-1730.

58. CHRISTIANSEN, E., KROKENE, P., BERRYMAN, A.A., FRANCESCHI, V.R., KREKLING, T., LIEUTIER, F., LONNEBORG, A., SOLHEIM, H., Mechanical injury and fungal infection induce acquired resistance in Norway spruce. *Tree Physiol.*, 1999, **19**, 399-403.

59. KREKLING, T., FRANCESCHI, V.R., KROKENE, P., SOLHEIM, H., Differential anatomical response of Norway spruce stem tissues to sterile and fungus infected inoculations. *Trees-Struct. Funct.*, 2004, **18**, 1-9.

60. LIEUTIER, F., BERRYMAN, A.A., Elicitation of Defense Reactions in Conifers. Springer-Verlag, New York, 1988.

61. HUBER, D.P.W., PHILIPPE, R.N., MADILAO, L.L., STURROCK, R.N., BOHLMANN, J., Changes in anatomy and terpene chemistry in roots of Douglas-fir seedlings following treatment methyl jasmonate. *Tree Physiol*, 2005, in press

62. ALFARO, R.I., KISS, G.K., YANCHUK, A., Variation in the induced resin response of white spruce, *Picea glauca*, to attack by *Pissodes strobi*. *Can J For Res.*, 1996, **26**, 967-972.

63. TOMLIN, E.S., ALFARO, R.I., BORDEN, J.H., HE, F.L., Histological response of resistant and susceptible white spruce to simulated white pine weevil damage. *Tree Physiol.*, 1998, **18**, 21-28.

64. FARMER, E.E., RYAN, C.A, Octadecanoid precursors of the jasmonic acid activate the synthesis of wound-inducible proteinase inhibitors. *Plant Cell*, 1992, **4**, 129-134.

65. CREELMAN, R.A., MULLET, J.E., Biosynthesis and action of jasmonates in plants. *Annu. Rev. Plant Phys. Plant Mol. Biol.*, 1997, **48**, 355-381.

66. BALDWIN, I.T., The jasmonate cascade and the complexity of the induced defence against herbivore attack. *Annu. Plant Rev.*, 1999, **3**, 155-186.

67. HOPKE, J., DONATH, J., BLECHERT, S., BOLAND, W., Herbivore-induced volatiles - the emission of acyclic homoterpenes from leaves of *Phaseolus lunatus* and *Zea mays* can be triggered by a beta-glucosidase and jasmonic acid. *FEBS Lett.*, 1994, **352**, 146-150.

68. DICKE, M., GOLS, R., LUDEKING, D., POSTHUMUS, M.A., Jasmonic acid and herbivory differentially induce carnivore- attracting plant volatiles in lima bean plants. *J. Chem. Ecol.*, 1999, **25**, 1907-1922.

69. GOLS, R., POSTHUMUS, M.A., DICKE, M., Jasmonic acid induces the production of gerbera volatiles that attract the biological control agent *Phytoseiulus persimilis*. *Entomol. Exp. Appl.*, 1999, **93**, 77-86.

70. KOCH, T., KRUMM, T., JUNG, V., ENGELBERTH, J., BOLAND, W., Differential induction of plant volatile biosynthesis in the lima bean by early and late intermediates of the octadecanoid-signaling pathway. *Plant Physiol.*, 1999, **121**, 153-162.

71. HALITSCHKE, R., KESSLER, A., KAHL, J., LORENZ, A., BALDWIN, I.T., Ecophysiological comparison of direct and indirect defenses in *Nicotiana attenuata. Oecologia*, 2000, **124**, 408-417.

72. KESSLER, A., BALDWIN, I.T., Defensive function of herbivore-induced plant volatile emissions in nature. *Science*, 2001, **291**, 2141-2144.

73. RODRIGUEZ-SAONA, C., CRAFTS-BRANDNER, S.J., PARÉ, P.W., HENNEBERRY, T.J. Exogenous methyl jasmonate induces volatile emissions in cotton plants. *J. Chem. Ecol.*, 2001, **27**, 679-695.

74. SCHMELZ, E.A., ALBORN, H.T., TUMLINSON, J.H., The influence of intact-plant and excised-leaf bioassay designs on volicitin- and jasmonic acid-induced sesquiterpene volatile release in *Zea mays. Planta*, 2001, **214**, 171-179.

75. ARIMURA, G., HUBER, D.P.W., BOHLMANN, J., Forest tent caterpillars (*Malacosoma disstria*) induce local and systemic diurnal emissions of terpenoid volatiles in hybrid poplar (*Populus trichocarpa x deltoides*): cDNA cloning, functional characterization, and patterns of gene expression of (−)-germacrene D synthase, *PtdTPS1. Plant J.*, 2004, **37**, 603-616.

76. HEFNER, J., KETCHUM, R.E., CROTEAU, R., Cloning and functional expression of a cDNA encoding geranylgeranyl diphosphate synthase from Taxus canadensis and assessment of the role of this prenyltransferase in cells induced for taxol production. *Arch. Biochem. Biophys.*, 1998, **360**, 62-74.

77. YUKIMUNE, Y., TABATA, H., HIGASHI, Y., HARA, Y., Methyl jasmonate-induced overproduction of paclitaxel and baccatin III in Taxus cell suspension cultures. *Nat Biotechnol.*, 1996, **14**, 1129-1132.

78. KETCHUM, R.E., GIBSON, D.M., CROTEAU, R.B., SCHULER, M.L., The kinetics of taxoid accumulation in cell suspension cultures of *Taxus* following elicitation with methyl jasmonate. *Biotechnol Bioeng.*, 1999, **62**, 97-105.

79. RICHARD, S., LAPOINTE, G., RUTLEDGE, R.G., SEGUIN, A., Induction of chalcone synthase expression in white spruce by wounding and jasmonate. *Plant Cell Physiol.*, 2000, **41**, 982-987.

80. LAPOINTE, G., LUCKEVICH, M.D., SEGUIN, A., Investigation on the induction of 14-3-3 in white spruce. *Plant Cell Rep.*, 2001, **20**, 79-84.

81. REGVAR, M., GOGALA, N., ZNIDARSIC, N., Jasmonic acid effects mycorrhization of spruce seedlings with *Laccaria laccata. Trees-Struct. Funct.*, 1997, **11**, 511-514.

82. KOZLOWSKI, G., BUCHALA, A., METRAUX, J., Methyl jasmonate protects Norway spruce [*Picea abies* (L.) Karst.] seedlings against *Pythium ultimum* Trow. *Physiol. Mol. Plant Path.*, 1999, **55**, 53-58.

83. MUMM, R., SCHRANK, K., WEGENER, R., SCHULZ, S., HILKER, M., Chemical analysis of volatiles emitted by *Pinus sylvestris* after induction by insect oviposition. *J. Chem. Ecol.*, 2003, **29**, 1235-1252.

84. HUDGINS, J.W., FRANCESCHI, V.R., Methyl jasmonate-induced ethylene production is responsible for conifer phloem defense responses and reprogramming of stem cambial zone for traumatic resin duct formation. *Plant Physiol.*, 2004, **135**, 2134-2149.

85. KING, J.N., ALFARO, R.I., CARTWRIGHT, C., Genetic resistance of Sitka spruce (*Picea sitchensis*) populations to the white pine weevil (*Pissodes strobi*): distribution of resistance. *Forestry* , 2004, **4**, 269-278.

86. RUNGIS, D., BERUBE, Y., ZHUANG, J., RALPH, S., RITAND, C.E., ELLIS, B.E., DOUGLAS, C., BOHLMANN, J., RITLAND, K., Robust simple sequence repeat (SSR) markers for spruce (*Picea spp.*) from expressed sequence tags (ESTs)., *Theor. Appl. Genet.*, 2004, **109**, 1283-1294.

87. BIRD, D.A., FRANCESCHI, V.R., FACCHINI, P.J., A tale of three cell types: Alkaloid biosynthesis is localized to sieve elements in opium poppy. *Plant Cell*, 2003, **15**, 2626-2635.

88. KEHR, J., Single cell technology. *Curr. Opin. Plant Biol.*, 2003, **6**, 617-621.

Chapter Three

SOME INSIGHTS INTO THE REMARKABLE METABOLISM OF THE BARK BEETLE MIDGUT

Claus Tittiger,[1]* Christopher I. Keeling,[2] Gary J. Blomquist[1]

[1]*Department of Biochemistry and Molecular Biology*
Mail Stop 330
University of Nevada, Reno
Reno, NV, USA 89557

[2]*Michael Smith Laboratories*
The University of British Columbia
#301 - 2185 East Mall
Vancouver,BC V6T 1Z4
Canada

Author for correspondence, email: crt@unr.edu

INTRODUCTION

The midgut has diverse functions for all insects. It is the primary tissue that deals with an incoming meal. It is thought to produce and secrete most of the digestive enzymes and components of the peritrophic matrix and is thus the first site of digestion. Nutrients are also absorbed, making the midgut a kind of "gateway organ" that provides nutrients to the rest of the body.[1] It is an important defense site against microbial pathogens and harmful host chemicals that may be ingested with food, while often harboring prokaryotic symbionts that appear to assist with digestion. Finally, the midgut is both a target of endocrine regulation and a source of enteric hormones. All of these roles combine to make the insect midgut, metabolically, a very busy tissue. For some species of pine bark beetles (Coleoptera: Scolytidae), adult midguts can also produce pheromone components. This is remarkable for two reasons. First, most insect exocrine molecules are produced in specialized cells or glands that are closely associated with the epidermis.[2] Thus, adult bark beetle midguts can function as both a digestive tissue and exocrine organ. Second, the pheromone components synthesized in the midgut are often monoterpenes –an unusual biochemical feat for an animal tissue. This article summarizes our knowledge of these processes in the midgut of the pine engraver, *Ips pini*, and includes recent insights afforded by functional genomics studies.

PHEROMONE PRODUCTION

Background

Bark beetle reproduction requires a pioneer beetle (male or female, depending on the species) to initiate feeding on a host pine tree. The tree is protected by the production of constitutive and induced resin components that serve as physical and toxic barriers to colonization. In order to successfully overcome these defenses, bark beetles rely on a "mass attack" strategy whereby hundreds to thousands of conspecifics aggregate at the same tree. This is coordinated by the production of volatile aggregation pheromone components by the pioneer beetles.

Biochemistry

One of the best understood pheromone systems in bark beetles is from the pine engraver, *Ips pini* (Say) (Coleoptera: Scolytidae). This species is broadly distributed across North America, and populations can be divided into three distinct groups based on the enantiomeric composition of ipsdienol. Ipsdienol is the major pheromone component produced by pioneer males. Eastern populations synthesize and respond to mostly (+) ipsdienol, while western populations rely on the (-)

enantiomer. A third population, loosely associated with the Rocky Mountains at the broad border between eastern and western populations, responds to a mixture of both enantiomers.[3,4]

Production of this monoterpene was originally hypothesized to be due to modification of host tree resin components.[5] *De novo* monoterpene biosynthesis in animals was unprecedented until late in the last century. Until pioneering studies by Seybold *et al.* [6] proved otherwise, it was thought that monoterpenes were produced by plants and bacteria, but not animals. The monoterpene backbone of ipsdienol was thought to be contributed by the plant resin, and only modified into the pheromone components by insect tissue. This made good evolutionary sense, because pine resins contain high amounts of toxic monoterpenes, which the beetles would need to overcome to survive. A common detoxification mechanism is to oxidize hydrophobic chemicals into more hydrophilic and, therefore, easily excretable forms.[7] Furthermore, since many bark beetle pheromones are synergistic with host volatiles,[5] a mixture of oxidized and un-metabolized plant chemicals could serve as a marker for successful tree colonization. Thus, for *I. pini*, it was thought that ingested myrcene was hydroxylated to form ipsdienol, and that what was initially a detoxification mechanism evolved into regulated signaling.[8]

Evidence accumulated for[9-12] and against[13-15] the paradigm that bark beetle pheromone biosynthesis involved direct modification of host precursor monoterpenes. For *I. pini*, the issue was laid to rest with the demonstration that male tissues incorporate radio-labeled acetate into ipsdienol in a manner consistent with pheromone production.[6] Similar experiments proved the *de novo* biosynthesis of frontalin, an important isoprenoid-derived semiochemical produced by male *Dendroctonus jeffreyi.*[16] It is probable that other Coleoptera can also synthesize monoterpenes, either as pheromone components[17-20] or defensive compounds.[21,22] Despite the capacity for *de novo* biosynthesis, plant precursor modification is likely an important source of pheromone components for some species. In these cases, plant chemicals could enter the pheromone biosynthetic pathway at later steps.

In the mevalonate pathway, isoprenoid precursors (C5) produced by early steps are converted to longer chain isoprenoids including farnesyl diphosphate (C15; FPP) and geranylgeranyl diphosphate (C20: GGPP) by corresponding synthases (FPPS and GGPPS, respectively). *De novo* monoterpene biosynthesis in bark beetles requires carbon to be shunted away from the main pathway into pheromone-specific steps at the level of geranyl diphosphate (C10) (rev[5,20]). Preliminary evidence suggests that a specific geranyl diphosphate synthase (GPPS) has evolved in these insects for this purpose (A Gilg-Young, unpublished data).

Regulation and Site of Synthesis

Pheromone production in *I. pini* begins in males when they feed on a host tree. Feeding stimulates production of juvenile hormone (JH) III by the corpora allata, and this hormone in turn stimulates pheromone biosynthesis.[23]   Indeed, starved adults of various bark beetle species can be induced to synthesize pheromone components if they are treated with JH III or a JH III analog.[24-27]   In vertebrates, the major regulatory step in the mevalonate pathway is the first committed step, at the conversion of 3-hydroxy-3-methylglutaryl Coenzyme A to mevalonate.[28]   This reaction is catalyzed by 3-hydroxy-3-methylglutaryl Co-enzyme A reductase (HMG-R).   Surveys of the responses of HMG-R genes (*HMG-R*s) to JH III in *I. pini, I. paraconfusus*, and *Dendroctonus jeffreyi* confirmed that the hormone causes increased expression consistent with pheromone production.[25,29,30]

Determination of the actual site of bark beetle pheromone biosynthesis proceeded with increasingly sophisticated methods. Early behavioral studies used isolated tissues and/or fecal material to implicate the alimentary canal as the anatomical source of aggregation pheromone (ref.[31], rev.[20]). *In vitro* assays showed that the metathorax produces pheromone components in *I. paraconfusus*.[32]   The robust response of *HMG-R*s to JH III provided a molecular tool for these studies. Northern blots prepared from approximate body sections (head, thorax, abdomen) were used to map regions of high HMG-R expression. In all species investigated, the metathorax was the region with highest *HMG-R* expression and thus the likely site of pheromone production.[32-34]   For *D. jeffreyi*, finer mapping was achieved by varying the way in which beetles were sectioned. This localized *HMG-R* expression at the metathorax/abdomen border.[34]   Tissue localization of pheromone production in *I. pini* and *D. jeffreyi* was determined by *in situ* hybridizations of exposed whole mount specimens. Radio-tracer studies confirmed that midguts are the source of monoterpenoid pheromone components in *I. pini* and *D. jeffreyi*.[33,34]

These results are unusual because most insects use epidermally-associated glands or tissues to synthesize pheromones (rev.[2]), and beetles generally follow this trend.[35]   However, there are some exceptions. In the two bark beetle species investigated, pheromones are produced in midguts, and this new paradigm is likely to hold for other Scolytidae. Indeed, parallel studies in the cotton boll weevil, *Anthonomus grandis* (Coleoptera: Curculionidae) strongly suggest that male aggregation pheromone components are also produced in the midgut (H. Nural *et al.*, unpublished), despite early biochemical evidence that suggested that they are synthesized in the fat body.[18,36]   Recently, monoterpenoid pheromone components were described in the Colorado potato beetle, *Leptinotarsa decemlineata*.[17] It would be interesting to determine if they are similarly produced by digestive tissues.[8]

Pheromone-producing midguts have a significant additional metabolic load compared to female midguts. In addition to inducing the mevalonate pathway, pheromone biosynthetic cells likely need to activate associated components including pheromone sequestering and transport proteins, as well as adjustments in basic metabolism. The shift in cellular functions appears reflected in subcellular structure: pheromone-synthesizing midguts undergo a significant sub-cellular reorganization compared to non-producing midguts. Electron micrographs of tissues prepared from male *I. pini* and *D. jeffreyi* that were induced to produce pheromone contain large crystalline arrays of smooth endoplasmic reticulum (SER) that are not present in cells from female or uninduced male midgut cells.[37] The arrays are especially striking in *D. jeffreyi*, and look remarkably like crystalline SER arrays found in mammalian cells that over-produce HMG-R.[38-40] This implies that pheromone-synthesizing midguts similarly have high levels of HMG-R protein, although such a correlation has yet to be confirmed. Since male *I. pini* can produce nearly 1% of their body mass per day in volatile pheromone,[26] metabolic flux through the mevalonate pathway is probably high, consistent with elevated levels of HMG-R protein.

All cells in pheromone-biosynthetic midguts appear to contribute to pheromone production. Evidence for this comes from the even distribution of the *in situ* hybridization signals, and the fact that the SER arrays are found in all midgut cells.[33,34,37] These results also confirm that the pheromones are produced from insect tissues, and not by prokaryotic symbionts. Prokaryotic HMG-Rs are structurally and mechanistically different from their eukaryotic homologs,[41] so the gene expression, activity, and subcellular changes accompanying the induction of pheromone production are clearly all due to bark beetle cells.

PHLOEM DETOXIFICATION

Bark beetle-host tree associations are ancient, which may explain why different beetle species "prefer" or are even restricted to certain hosts (rev.[5,42]). Beetles seem to be able to tolerate the resin components of host trees, but not those from non-host trees,[43] suggesting that detoxification mechanisms for different beetle species are tuned to their host trees. Indeed, it has been hypothesized that bark beetle feeding may contribute to compositional variation of resin within different pine species over time.[44]

Pine resins may be divided into volatile (usually mono- and sesqui-terpenoid) and non-volatile (diterpenoid) and phenolic components (*e.g.*, lignin and its precursors).[45] The two components are thought to act differently on invading beetles: the volatiles are chemical toxins, whereas the non-volatiles are more of a physical barrier.[45,46] Phloem and oleo-resin compositions differ among tree species,[47-49] and some components can be induced by beetle attacks.[50,51] Predominant resin monoterpenes in various hosts include: Δ3-carene, limonene, and β-pinene in ponderosa pine; Δ3-carene, limonene, and β-phellandrene in Jeffrey pine; and α-pinene and Δ3-carene in pinyon pine (*P. edulis*).[47,48,52,53] Sesqui-, and di-terpenes are also found.[51,54]

There is extensive literature documenting biochemical/molecular adaptations of herbivorous insects to host toxins,[55-57] though almost none of it deals with bark beetles. Beyond noting the toxicity of various components to different beetles,[58-61] little is known about the biochemical adaptations that bark beetles have evolved to deal with their meal. Lipophilic terpenes can be ingested with the meal, enter through spiracles, and may possibly penetrate the cuticle. Detoxification mechanisms may include any combination of oxidation to more readily excreted derivatives, inactivation by conjugation to transport proteins, or evolution of target proteins to less susceptible forms. Since terpenes are lipophilic solvents that can disrupt membranes, it is reasonable to suggest that oxidation and/or conjugation are the two most likely detoxification mechanisms.

It is probably worth noting that the distinction between resin detoxification and pheromone production sometimes can be blurred. For example, *D. ponderosae* pupae contain hydroxylated α-pinene,[62] presumably as a detoxification product, but hydroxylated α-pinene is *trans*-verbenol, an important semiochemical for adults. Similarly, adult *I. pini* exposed to myrcene vapors will hydroxylate the monoterpene to a racemic mixture of (+) and (-)-ipsdienol, which is readily excreted.[14] Thus, there may be two systems to hydroxylate myrcene in *I. pini*: one that is stereo-nonspecific for detoxification, and a second that is stereospecific for pheromone production.[63] Myrcene may also be hydroxylated to form myrcenol and/or linalool; both compounds have been found in volatile extracts of fed *I. pini*, though they are also present in host phloem so the contribution of beetle tissues to their production is unclear.[26] Interestingly, the P450 activities of *D. ponderosae* populations infesting different host species (*P. ponderosae, P. controta,* and *P. flexilis*; each with different relative monoterpene abundances) are essentially identical.[64] This suggests that "detoxification P450s," if they exist, have broad substrate preferences. Alternatively, monoterpene detoxification may be accomplished by other proteins (see below, transcript cataloging).

RECENT ADVANCES – FUNCTIONAL GENOMICS

Transcript Cataloging (EST Project)

The motivation for functional genomics work in *I. pini* stemmed from the unusual activities of adult midguts. The fact that adult male *I. pini* midguts respond strongly to JH III by elevating *HMG-R* expression suggested that the system could be used to study one of entomology's longest-standing mysteries: the mode of action of JH.[65] Since traditional approaches to this question such as testing models based on precedents had not been successful, it made sense to approach the question with as little bias as possible. Functional genomics accomplishes this by allowing the tissue to tell us which genes have prominent roles in different functions. In addition, this approach can shed light on several other midgut functions, including regulation of the entire pheromone-biosynthetic pathway, identification of genes involved in the conversion of geranyl diphosphate to ipsdienol, and non-pheromone-related roles, including phloem detoxification. Finally, some of the genes identified through functional genomics could be expected to be bark beetle-specific, and they could be targets for future control strategies if ways to disrupt them (*e.g.*, RNAi or specific inhibitors) are developed. Future pest management strategies may also involve developing strains of trees resistant to beetle attack through an understanding of these molecular targets.

A directionally-cloned cDNA library of mRNA isolated from pheromone-biosynthetic midguts was constructed, arrayed, and sequenced to yield 1671 useful expressed sequence tags (ESTs).[66] Sequencing was stopped after only 30 plates because of unacceptably high redundancy values, despite attempts to reduce redundancy by pre-screening. The ESTs were sorted into clusters or singlets, resulting in 574 "tentative unique genes" (TUGs). Most of the redundancy was contributed by two clusters, representing genes or gene families provisionally named *IPG001B01* and *IPG001D12*. Other highly-represented sequences include catabolic enzymes, especially glycosidases, and ribosomal RNAs.[66]

The two highly redundant clusters represented by IPG001B01 and IPG001D12 are of particular interest. Altogether, 35% of all recovered ESTs sort into these two clusters, suggesting they are abundant transcripts.[66] Preliminary expression analyses suggest that they are not significantly regulated by JH III, and thus are not likely involved in pheromone biosynthesis (J. Bearfield, unpublished data). Their predicted translation products have only 25% identity, despite a highly conserved N-terminal putative signal peptide. BLAST searches of GenBank nr and dbEST consistently return no significant hits. PCR-based surveys identify orthologs in other bark beetle species, but not cotton boll weevils (S. Woydziak, unpublished data). Thus, genes in these clusters so far appear limited to the Scolytidae. Given their uniqueness and high transcript abundance, we hypothesize that they are involved in phloem detoxification, perhaps by metabolizing or sequestering and solubilizing lignin or terpenoid resin components.[66] The possible role(s) of IPG001B01 and IPG001D12 proteins are still under investigation.

This project has had good success in identifying genes involved in pheromone biosynthesis and JH regulation. ESTs of significance to pheromone production include seven of nine mevalonate pathway enzymes (thiolase, HMG-S, HMG-R, mevalonate kinase, mevalonatediphosphate decarboxylase, isopentenyldiphosphate isomerase, and farnesyldiphosphate synthase). One TUG corresponded to a geranyldiphosphate synthase, which is the likely step at which carbon is shunted from the mevalonate pathway into the ipsdienol-biosynthetic pathway (A. Gilg-Young, unpublished data). The project also identified two putative JH-metabolizing enzymes –JH esterase and JH epoxide hydrolase- which may clear the hormone from the tissue.[66] Some of the "unknown" ESTs may represent enzymes catalyzing uncharacterized late steps in ipsdienol biosynthesis,[63] or may be involved in JH signaling. Indeed, one of at least five cytochrome P450 genes has an expression pattern consistent with a pheromone biosynthetic role, and we are in the process of determining whether this enzyme hydroxylates myrcene to form ipsdienol (P. Sandstrom, unpublished data).

Comparative Transcript Cataloging

Inter- and intra-specific comparisons of different EST data sets can reveal different transcription patterns and thus inform about differing roles of the tissues from which they were isolated.[67-69] We applied this principle to the bark beetles by comparing our ESTs with those generated from larval silkworm (*Bombyx mori*) midguts. We expected to see similarities because both tissues share digestive functions, and indeed, when ESTs from both databases were sorted into functional categories as per Adams *et al.*,[70] most of the 94 commonly expressed orthologs sorted into the metabolism category. They represent predominantly glycosidases, cellulases, and other carbohydrate metabolizing enzymes, as would be expected from tissue digesting a plant diet.[66,71]

We also expected differences in expression patterns, because the *B. mori* library was constructed from actively feeding larvae, whereas the *I. pini* guts were isolated from starved adults. Interestingly, peritrophin/mucin-like ESTs, which may contribute to the protective peritrophic matrix in the midgut lumen, were identified in the *B. mori* ESTs, but not the *I. pini*. Furthermore, male *I. pini* midguts are receptive to JH signaling, and synthesize large amounts of monoterpenoid pheromone while larval *B. mori* midguts do not. In support of this, the fraction of metabolism ESTs was almost three times larger in the *I. pini* ESTs than in *B. mori*, and no mevalonate pathway enzymes were identified in the *B. mori* ESTs.[66] EST representation also appeared to reflect dietary differences between the species. Although total lipid content is similar (as a percent of total mass), the ratio of neutral lipids to free fatty acids is much higher in mulberry leaves compared to pine phloem.[47,72] Similarly, the fraction of lipase ESTs in *B. mori* is also larger than in *I. pini* (Fig. 3.1).[66]

While predicting biological functions from EST sets may be unfeasible, different expression profiles should reflect different biological roles. ESTs were recently produced from the larval midgut of the cowpea weevil (*Callosobruchus maculatus*),[73] allowing further comparison among these three phytophagous species. *C. maculatus* is a significant stored-crop pest because the larvae develop inside the pods of mature cowpeas (*Vigna unguiculata*), thus destroying them. Cowpeas are rich in sugars and proteins, and in addition produce protease inhibitors to defend against herbivores.[74] Cowpea weevils, in turn, evidently produce a battery of digestive enzymes and utilize various strategies to overcome the plant defenses.[75] Indeed, the EST study identified twelve glycoside hydrolases and seven peptide hydrolases from a pool of 296 cDNAs.[73]

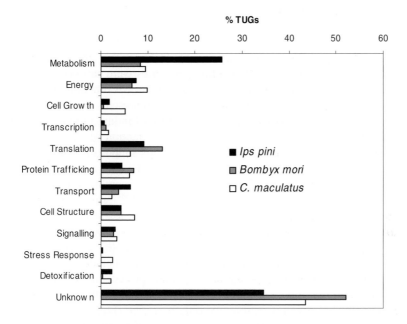

Fig. 3.1: EST representation in functional categories. BlastX and TeraBlastX searches were used to assign putative identities to tentative unique genes (TUGs) from *I. pini, Bombyx mori,* and *Callosobruchus maculatus.* Putatitve identities were based on the best hit, or, if the best hit was an un-annotated sequence (*e.g.,* from a genome project), then a lower, but still confident hit was used to assign function. The cut-off E value was set at 10^{-7}.

Although cowpea weevils are phylogenetically more closely related to bark beetles, we expected the expression profile to be more similar to that of *B. mori* larvae since the silkworm and pea weevil libraries were constructed from "fed" samples, and also because neither of these insects synthesize monoterpenoid pheromone components. We downloaded 502 *C. maculatus* ESTs, each representing a TUG, from dbEST (accession numbers CB377223 – CB377725), assigned putative identities using tera-BLAST (M. Gollery, UNR, unpublished), and manually sorted them into functional groups according to Adams *et al.*[70] Our *C. maculatus* assignments differ slightly from Pedra *et al.*[73] because we searched a later nr release.

Indeed, CB377224 had a strong BLAST hit to a lambda phage protein (E = 4.4 x 10^{-110}) and was omitted from our analysis. Detoxification TUGs included radical scavengers and cytochromes P450, while "Stress" TUGs include serpins and heat shock proteins.

Among all three insects, the fractions of TUGs within the transcription, protein trafficking, signaling, and detoxification categories were similar within ~2% between species (Fig. 3.1). Most of the TUGs have unknown functions in all three species. Interestingly, most *C. maculatus* TUGs represent energy-producing proteins, and the fraction representing cell growth (DNA-binding proteins) is much higher than in *B. mori* and *I. pini*. The "metabolism" fraction of *C. maculatus* is similar to *B. mori* (much smaller than in *I. pini*), perhaps supporting the hypothesis that more *I. pini* ESTs sort into this category because of the dual nature of the male bark beetle midgut. Similarly, the absence of peritrophic membrane ESTs from *I. pini* may reflect the starved status of the insects assayed,[66] since corresponding ESTs were recovered from fed *B. mori* and *C. maculatus* larvae.

To further investigate different expression patterns in these three tissues, we subdivided the "metabolism" TUGs based on substrate classes. Thus, "carbohydrate" includes all glycoside hydrolases, "lipid" includes lipases, beta-oxidation enzymes, JH-metabolizing enzymes, etc., and "protein" includes proteases. TUGs for all subcategories are contributed by all three insects, with the exception of the mevalonate pathway, which is represented only by *I. pini* (Fig. 3.2). This further supports the observation that their presence in *I. pini* midguts reflects the pheromone-biosynthetic activity of that tissue.

Despite this general correlation, the utility of TUG distribution patterns to provide information about biological function appears limited. Most *I. pini* metabolism TUGs are involved in carbohydrate metabolism, whereas larger fractions of *B. mori* and *C. maculatus* TUGs are devoted to lipid and protein substrates. To suggest that this correlates with relatively high sugar content in pine phloem compared to mulberry leaves and cowpeas may be misleading. If such a correlation were true, the distribution of *B. mori* and *C. maculatus* "carbohydrate" and "protein" ESTs should also reflect the relative amounts of these substrates in mulberry leaves and cowpeas. In fact, while the fraction of "carbohydrate" TUGs is similar between the two species (Fig. 3.2), the sugar content of cowpeas[76] is approximately seven-fold higher than that of mulberry leaves.[77] Similarly, *B. mori* has a higher fraction of protease TUGs than *C. maculatus*, despite the fact that mulberry leaves and cowpeas have similar protein contents per gram total weight (~18 – 28% total),[76,77] Thus, the presence or absence of cDNAs can indicate metabolic characteristics, however, the relative fractions contributed by different enzyme types probably does not reliably indicate biological function.

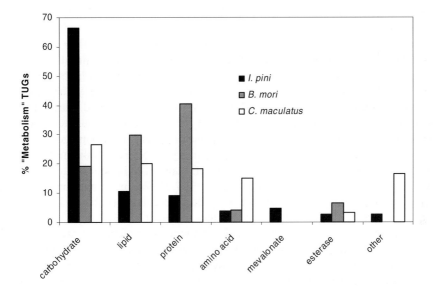

Fig. 3.2: Classification of "metabolism" ESTs. Tentative unique genes (TUGs) from each species that clustered into the "metabolism" category (see Fig. 3.1) were further subdivided. Here, "carbohydrate" incorporates all glycoside hydrolases, "lipid" includes lipases, B-oxidation and steroid metabolizing enzymes, "protein" includes predominantly proteases and their precursors, and "amino acid" includes all enzymes involved in the interconversion of amino acids. "Esterase" includes those esterase-like enzymes with unknown substrates, and "other" contains those TUGs that do not sort into other categories. These include predominantly oxidoreductases and purine/pyrimidine metabolizing enzymes.

Transcript Profiling

To examine global changes in gene expression in the midgut upon phloem feeding, we prepared cDNA microarrays representing 384 TUGs chosen from the first 1,900 clones sequenced from the EST project.[66] Complete details of these microarrays, the experimental materials and methods used, and the data obtained are available on-line at NCBI GEO,[78] accession numbers GPL551 and GSM14812-14815. Briefly, beetles and un-infested Jeffrey pine (*Pinus jeffreyi* Grev. & Balf.) bolts were obtained from the Sierra Nevada in California and Nevada, USA. Male beetles were allowed to feed for 24 hours on host phloem before they were removed from the bolts and their midguts excised. Control beetles were simultaneously held in the dark in plastic cups. For each of four replicates, 30 midguts were pooled. RNA was extracted, converted to fluorescently labeled cDNA (Cy3 and Cy5 dyes), and hybridized to the microarray. Microarrays were then scanned, normalized using online SNOMAD tools,[79] and then analyzed for significantly regulated genes using the SAM method of Tusher *et al.*[80] with a false discovery rate of one gene.

As would be expected in a tissue actively involved in phloem digestion and pheromone biosynthesis, the expression of many TUGs in the anterior midgut changed with feeding (not shown). Of the 384 TUGs on the microarray, 40 were significantly up- and 62 were significantly down-regulated (Table 31), with most of the up-regulated TUGs belonging to the metabolism (23) and unknown (10) functional categories. Many are putatively involved in digestion and carbohydrate metabolism, not surprising considering their phloem meal, and also include putative calreticulin, carboxypeptidase a, and β-glucan binding protein genes. Contrary to the data from the more sensitive quantitative real-time PCR analysis (see below), only one mevalonate pathway gene, isopentenyldiphosphate isomerase was significantly regulated at 24 hours on the microarray. The pheromone-biosynthetic capacity of the midgut cells is thus not strongly indicated in this analysis. However, a cytochrome P450 thought to catalyze one of the last steps in the conversion of geranyldiphosphate to ipsdienol (P. Amos, unpublished data) was the most strongly up-regulated by feeding. Most of the down-regulated TUGs belong to the unknown (23) and translation (23) functional categories. Most of the translation TUGs are ribosomal, suggesting that feeding is accompanied by a decrease in *de novo* protein synthesis in midgut cells, though this has not been confirmed.

Many genes in the male *I. pini* midgut are regulated upon phloem-feeding. Consistent with other insects,[71,75,81] these include digestive enzymes that are induced by the presence of nutrients. Other regulated genes might include pheromone biosynthetic genes that are regulated by the feeding-induced JH III signal. To examine gene expression regulated by JH III and potentially involved in pheromone biosynthesis directly, an analysis of juvenile hormone-regulated changes in gene expression in the midgut using a more comprehensive microarray platform (NCBI GEO accession GPL575) has recently been completed (Keeling, *et al.*, unpublished).

Table 3.1:   **Genes with statistically significant changes in expression upon feeding**

Putative Function/BlastX I.D.	Fold change $(\log_2)^a$	std. error	# TUGs
Upregulated			
cytochrome P450	1.56	0.45	1
calreticulin	1.39	0.19	1
β-glucuronidase	1.28	0.04	1
β-glucosidase	1.27	0.23	1
α-Esterase	1.11	0.29	1
carboxylesterase	1.08 - 1.13	0.19 - 0.24	2
poly-U binding splicing factor	1.07	0.20	1
ATPase V-type subunit	0.98	0.16	1
putative rhamnogalacturonase	0.97 - 1.55	0.18 - 0.28	5
ribosomal protein L29	0.95	0.25	1
protein disulphide isomerase	0.87	0.25	1
acyl-Coenzyme A oxidase	0.86	0.17	1
β-galactosidase	0.83 - 1.08	0.04 - 0.2	3
α-glucosidase 2	0.82	0.14	1
β-mannosidase	0.73	0.11	1
cellulase	0.73	0.02	1
alcohl dehydrogenase like	0.72	0.20	1
isopentenyldiphosphate isomerase	0.68	0.14	1
glucan binding protein	0.57	0.16	1
NADH dehydrogenase subunit	0.55	0.03	1
unknown	0.49 - 1.14	0.36 - 0.08	10
endopolygalacturonase	0.48	0.05	1
aldose reductase	0.43	0.07	1
carboxypeptidase A	0.41	0.06	1
Total			40
Downregulated			
actin	-0.31	0.04	1
peroxiredoxin	-0.37	0.06	1
putative glucose translocator	-0.40	0.09	1
heat shock cognate protein	-0.46	0.06	1
translation elongation factors	(-0.46) - (-1.19)	0.13 - 0.18	2
ribosomal proteins	(-0.47) - (-0.60)	0.11 - 0.29	20
MIP like protein	-0.51	0.13	1
Glutathione S-transferase	-0.57	0.10	1
malate dehydrogenase	-0.60	0.21	1
proferredoxin	-0.62	0.11	1
moesin homolog (fruitfly)	-0.68	0.14	1
ferritin	-0.69	0.08	1
unknown	(-0.77) - (-1.76)	0.07 - 0.33	23
translation initiation factor 5A	-0.85	0.07	1
polygalacturonase	-0.90	0.12	1
protein disulphide isomerase	-1.01	0.25	1
putative rhamnogalacturonase	(-1.1) - (-1.5)	0.16 - 0.43	2
peritrophin 1	-1.37	0.27	1
ornithine decarboxylase	-2.09	0.16	1
Total			62

Putative Functions were assigned using the best or most informative hit from BlastX searches of GenBank nr. Experimental details are available at NCBI GEO, accession #s GPL 551, GSM14812-14815

a) Values indicate differences in expression in fed vs. starved midguts using a log base 2 scale.

Real Time PCR

The recent acquisition of sequences of several mevalonate pathway genes (see above, transcript cataloging) permitted us to use quantitative real-time PCR to examine feeding-induced changes in expression of genes in this pathway.[82]   This allowed us to more accurately study how the pheromone-biosynthetic pathway responds to feeding.   Transcript levels of seven mevalonate pathway genes represented by our ESTs were examined in the midguts of beetles of both sexes that had fed on host phloem for 4, 8, 16 and 32 hours.

Although HMG-R is usually considered the key enzyme for regulation of the mevalonate pathway in most organisms,[28,83] all genes examined were coordinately regulated by feeding in the *I. pini* midgut.[82]   Interestingly, most of the genes are similarly regulated in both sexes, even though only males produce ipsdienol. However, there are clear differences between the sexes that determine the sex-specificity of pheromone biosynthesis. First, genes in the latter steps of the pathway, *GPPS* and *FPPS*, appear un-regulated or down-regulated in females. Second, unfed males have significantly higher transcript levels than females, ranging from 5 to 41-fold higher (Fig. 3.3).   *GPPS* shows the largest difference in basal expression between sexes (41-fold), consistent with its role in directing products of the mevalonate pathway to the monoterpenoid ipsdienol. Thus, although these genes respond similarly to feeding in males and females, females start out with lower basal levels and do not appear to shunt carbon into the ipsdienol pathway upon feeding. These differences suggest that males are developmentally "primed" for pheromone production and also respond differently than females to feeding-induced cues.[82] This experiment illustrates the advantage of knowing which genes are induced as well as which are already expressed basally/developmentally.   Combining the two perspectives provides a greater understanding of gene regulation for pheromone biosynthesis.

SUMMARY

Part of the attraction of entomology is the surprises that insects provide. Here, we have seen how each advance in knowledge has caused an evolution in how we think about different bark beetle processes.   Early results showing that bark beetle tissues can synthesize monoterpenes were unexpected.   It was even more surprising to learn that monoterpene synthesis occurs in male *I. pini* midguts, because that tissue was assumed to be fully occupied with the phloem meal.[33]   The male *I. pini* midgut has become an unusual model tissue that performs both digestive and pheromone-biosynthetic functions.   Our recent application of functional genomics tools to this model has afforded a glimpse into these processes. It is now evident that multiple developmental, environmental, and endocrine signals appear to

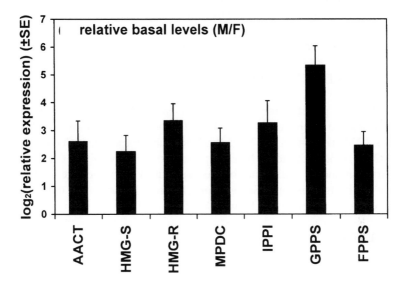

Fig. 3.3: Relative basal expression levels of mevalonate pathway genes in adult male and female *Ips pini*. For all genes, mRNA levels were significantly higher in males (t-test, $P<0.001$). AACT, acetoacetyl-CoA thilase; HMG-S, HMG-CoA synthase; HMG-R, HMG-CoA reductase; MPDC, mevalonate-5-diphosphate decarboxylase; IPPI, isopentenyldiphosphate isomerase; GPPS, geranyldiphosphate synthase; FPPS, farnesyldiphosphate synthase. Figure reproduced from Keeling *et al.*,[82] with permission.

act on numerous genes in the pheromone-biosynthetic pathway. Perhaps we should not be surprised that regulatory schemes can be active post-transcriptionally as well.[29] A third possible function of bark beetle midguts is detoxification of phloem and resin. Surprisingly, little is known about the biochemical tools bark beetles may use to tolerate the hostile environment of the host tree. Two novel genes that are highly transcribed in *I. pini* midguts have characteristics that suggest they may contribute to this process, and we are working to characterize them. Certainly, this aspect of insect/plant interaction deserves more attention, especially considering the close relationships that bark beetles form with their host trees. As more information becomes available, it may well develop that bark beetle midguts are superb metabolic multi-taskers -- and we can reliably await even more surprises.

ACKNOWLEDGEMENTS

We are grateful for support from the USDA (2001-35302-11035), the National Science Foundation (IBN-0316370), the Nevada Biomedical Infrastructure Network (P20 RR16464), the Nevada Agricultural Experiment Station, (publication #303042975), and a Canadian NSERC postdoctoral fellowship to C.I.K. C.T. thanks J. T. Arnason and B. Philogene for the opportunity to participate in the 2004 joint meeting of the International Society for Chemical Ecology and the Phytochemical Society of North America, in Ottawa, Canada.

REFERENCES

1. TURUNEN, S., CRAILSHEIM, K. Lipid and sugar absorption. *in*: Biology of the Insect Midgut (M.J. Lehane, P.F. Billingsley, eds), Chapman and Hall, London, U.K. 1996, pp. 293-320.
2. TILLMAN, J.A., SEYBOLD, S.J., JURENKA, R.A., BLOMQUIST, G.J., Insect pheromones--an overview of biosynthesis and endocrine regulation, *Insect Biochem. Molec. Biol.*, 1999, **29**, 481-514.
3. SEYBOLD, S.J., OHTSUKA, T., WOOD, D.L., KUBO, I., The enantiomeric composition of ipsdienol: a chemotaxonomic character for North American populations of *Ips* spp. in the *pini* subgeneric group (Coleoptera: Scolytidae), *J. Chem. Ecol.*, 1995, **21**, 995-1016.
4. MILLER, D.R., GIBSON, K.E., RAFFA, K.F., SEYBOLD, S.J., TEALE, S.A., WOOD, D.L., Geographic variation in response of pine engraver, *Ips pini*, and associated species to pheromone, lanierone, *J. Chem. Ecol.*, 1997, **23**, 2013-2031.
5. SEYBOLD, S.J., BOHLMANN, J., RAFFA, K.F., Biosynthesis of coniferophagous bark beetle pheromones and conifer isoprenoids: Evolutionary perspective and synthesis, *Can. Entomol.*, 2000, **132**, 697-753.
6. SEYBOLD, S.J., QUILICI, D.R., TILLMAN, J.A., VANDERWEL, D., WOOD, D.L., BLOMQUIST, G.J., *De novo* biosynthesis of the aggregation pheromone components ipsenol and ipsdienol by the pine bark beetles *Ips paraconfusus* Lanier and *Ips pini* (Say) (Coleoptera: Scolytidae), *Proc. Natl. Acad. Sci. USA*, 1995, **92**, 8393-8397.
7. VANDERWEL, D., OEHLSCHLAGER, A.C. Biosynthesis of pheromones and endocrine regulation of pheromone production in Coleoptera. *in*: Pheromone Biochemistry (G.D. Prestwich,G.J. Blomquist, eds), Academic Press, Orlando. 1987, pp. 175-215.
8. BLOMQUIST, G.J., JURENKA, R.A., SCHAL, C., TITTIGER, C. Biochemistry and molecular biology of pheromone production. *in*: Comprehensive Insect

Physiology, Biochemistry, Pharmacology, and Molecular Biology (L.I. Gilbert, K. Iatrou, S.S. Gill, eds), Elsevier. 2005, 705-751.

9. HUGHES, P.R., RENWICK, A.A., Neural and hormonal control of pheromone biosynthesis in the bark beetle, *Ips paraconfusus*, *Physiol. Entomol.*, 1977, **2**, 117-123.

10. RENWICK, J.A.A., HUGHES, P.R., Oxidation of unsaturated cyclic hydrocarbons by *Dendroctonus frontalis*, *Insect Biochem.*, 1975, **17**, 459-463.

11. RENWICK, J.A.A., HUGHES, P.R., TY, T.D., Oxidation products of pinene in the bark beetle, *Dendroctonus frontalis.*, *J. Insect. Physiol.*, 1973, **19**, 1735-1740.

12. HUGHES, P.R., Myrcene: a precursor of pheromones in *Ips* beetles, *J. Insect. Physiol.*, 1974, **20**, 1274-1275.

13. BYERS, J., Pheromone biosynthesis in the pine bark beetle, *Ips paraconfusus*, during feeding or exposure to vapors of host plant precursors., *Insect Biochem.*, 1981, **11**, 563-569.

14. BYERS, J., BIRGERSSON, G, Pheromone production in a bark beetle independent of myrcene precursor in host pine species, *Naturwissenschaften*, 1990, **77**, 385-387.

15. IVARSSON, P., SCHLYTER, F., BIRGERSSON, G, Demonstration of *de novo* pheromone biosynthesis in *Ips duplicatus* (Coleoptera: Scolytidae): Inhibition of ipsdienol and *E*-myrcenol production by compactin., *Insect Biochem. Molec. Biol.*, 1993, **23**, 655-662.

16. BARKAWI, L.S., FRANCKE, W., BLOMQUIST, G.J., SEYBOLD, S.J., Frontalin: *de novo* biosynthesis of an aggregation pheromone component by *Dendroctonus* spp. bark beetles (Coleoptera: Scolytidae), *Insect Biochem. Molec. Biol.*, 2003, **33**, 773-788.

17. DICKENS, J.C., OLIVER, J.E., HOLLISTER, B., DAVIS, J.C., KLUN, J.A., Breaking a paradigm: Male-produced aggregation pheromone for the Colorado potato beetle, *J. Exp. Biol.*, 2002, **205**, 1925-1933.

18. WIYGUL, G., SIKOROWSKI, P.P., The effect of glucose and ATP on sex pheromone production in fat bodies from male boll weevils *Anthonomus grandis* Boheman (Coleoptera: Curculionidae), *Comp. Biochem. Physiol. B.*, 1985, **81**, 1073-1075.

19. LANNE, B.S., IVARSSON, P., JOHNSSON, P., BERGSTROM, G., WASSGREN, A.B., Biosynthesis of 2-methyl-3-buten-2-ol, a pheromone component of *Ips typographus* (Coleoptera: Scolytidae), *Insect Biochem.*, 1989, **19**, 163-167.

20. SEYBOLD, S.J., TITTIGER, C., Biochemistry and molecular biology of *de novo* isoprenoid pheromone production in the Scolytidae, *Annu. Rev. Entomol.*, 2003, **48**, 425-453.

21. OLDHAM, N.J., VEITH, M., BOLAND, W., Iridoid monoterpene biosynthesis in insects: evidence for a *de novo* pathway occurring in the defensive glands of *Phaedon armoraciae* (Chrysomelidae) leaf beetle larvae, *Naturwissenschaften*, 1996, **83**, 470-473.

22. WEIBEL, D.B., OLDHAM, N.J., FELD, B., GLOMBITZA, G., DETTNER, K., BOLAND, W., Iridoid biosynthesis in staphylinid rove beetles (Coleoptera: Staphylinidae, Philonthinae), *Insect Biochem. Mol. Biol.*, 2001, **31**, 583-591.

23. TILLMAN, J.A., HOLBROOK, G.L., DALLARA, P.L., SCHAL, C., WOOD, D.L., BLOMQUIST, G.J., SEYBOLD, S.J., Endocrine regulation of *de novo* aggregation pheromone biosynthesis in the pine engraver, *Ips pini* (Say) (Coleoptera: Scolytidae), *Insect Biochem. Mol. Biol.*, 1998, **28**, 705-715.

24. CHEN, N.M., BORDEN, J.H., PIERCE JR., H.D., Effect of juvenile hormone analog fenoxycarb, on pheromone production by *Ips paraconfusus* (Coleoptera: Scolytidae), *J. Chem. Ecol.*, 1988, **14**, 1087-1098.

25. TITTIGER, C., BLOMQUIST, G.J., IVARSSON, P., BORGESON, C.E., SEYBOLD, S.J., Juvenile hormone regulation of HMG-R gene expression in the bark beetle *Ips paraconfusus* (Coleoptera: Scolytidae): Implications for male aggregation pheromone biosynthesis, *Cell. Molec. Life Sci.*, 1999, **55**, 121-127.

26. QUILICI, D.R., *De novo* biosynthesis of aggregation pheromone components by the pine bark beetles, *Ips paraconfusus* (Lanier) and *Ips pini* (Say) (Coleoptera: Scolytidae), and identification of an interruptant and a synergist produced by *Ips pini.*, Ph.D., 1997, University of Nevada, Reno.

27. BRIDGES, J.R., Effects of juvenile hormone on pheromone synthesis in *Dendroctonus frontalis*, *Environ. Entomol.*, 1982, **11**, 417-420.

28. GOLDSTEIN, J.L., BROWN, M.S., Regulation of the mevalonate pathway, *Nature*, 1990, **343**, 425-430.

29. TILLMAN, J.A., LU, F., STAEHLE, L., DONALDSON, Z., DWINELL, S.C., TITTIGER, C., HALL, G.M., STORER, A.J., BLOMQUIST, G.J., SEYBOLD, S.J., Juvenile hormone regulates *de novo* isoprenoid aggregation pheromone biosynthesis in pine bark beetles, *Ips* spp. (Coleoptera: Scolytidae), through transcriptional control of HMG-CoA reductase, *J. Chem. Ecol.*, 2004, **30**, 2335-2358.

30. TITTIGER, C., BARKAWI, L.S., BENGOA, C.S., BLOMQUIST, G.J., SEYBOLD, S.J., Structure and juvenile hormone-mediated regulation of the HMG-CoA reductase gene from the Jeffrey pine beetle, *Dendroctonus jeffreyi*, *Molec. Cell. Endocrinol.*, 2003, **199**, 11-21.

31. BORDEN, J.H., NAIR, K.K., SLATER, C.E., Synthetic juvenile hormone: induction of sex pheromone production in *Ips confusus*, *Science*, 1969, **166**, 1626-1627.

32. IVARSSON, P., TITTIGER, C., BORGESON, C.E., SEYBOLD, S.J., BLOMQUIST, G.J., Pheromone precursor synthesis is localized in the metathorax of *Ips paraconfusus* Lanier (Coleoptera: Scolytidae), *Naturwissenschaften*, 1998, **85**, 507 - 511.

33. HALL, G.M., TITTIGER, C., ANDREWS, G.L., MASTICK, G.S., KUENZLI, M., LUO, X., SEYBOLD, S.J., BLOMQUIST, G.J., Midgut tissue of male pine engraver, *Ips pini*, synthesizes monoterpenoid pheromone component ipsdienol *de novo*, *Naturwissenschaften*, 2002, **89**, 79-83.

34. HALL, G.M., TITTIGER, C., BLOMQUIST, G.J., ANDREWS, G.L., MASTICK, G.S., BARKAWI, L.S., BENGOA, C., SEYBOLD, S.J., Male jeffrey pine beetle, *Dendroctonus jeffreyi*, synthesizes the pheromone component frontalin in anterior midgut tissue, *Insect Biochem. Molec. Biol.*, 2002, **32**, 1525-1532.

35. PERCY-CUNNINGHAM, J.E., MCDONALD, J.A. Biology and ultrastructure of sex pheromone producing glands. *in*: Pheromone Biochemistry (G.J. Blomquist, G.D. Prestwich, eds), Academic Press, Orlando, FL. 1987, pp. 22-75.

36. WIYGUL, G., DICKENS, J.C., SMITH, J.W., Effect of juvenile hormone and beta-bisabolol on pheromone production in fat bodies of male boll weevils, *Anthonomus grandis* Boheman (Coleoptera; Curculionidae). *Comp. Biochem. Physiol. B*, 1990, **95B**, 4898-4491.

37. NARDI, J., YOUNG, A., UJHELYI, E., TITTIGER, C., LEHANE, M., BLOMQUIST, G.J., Specialization of midgut cells for synthesis of male isoprenoid pheromone components in two scolytid beetles, *Dendroctonus jeffreyi* and *Ips pini*, *Tissue and Cell*, 2002, **34**, 221-231.

38. CHIN, D.J., LUSKEY, K.L., ANDERSON, R.G., FAUST, J.R., GOLDSTEIN, J.L., BROWN, M.S., Appearance of crystalloid endoplasmic reticulum in compactin-resistant Chinese hamster cells with a 500-fold increase in 3-hydroxy-3-methylglutaryl-coenzyme A reductase, *Proc. Natl. Acad. Sci. USA*, 1982, **79**, 1185-1189.

39. PATHAK, R.K., LUSKEY, K.L., ANDERSON, R.G., Biogenesis of the crystalloid endoplasmic reticulum in UT-1 cells: evidence that newly formed endoplasmic reticulum emerges from the nuclear envelope, *J. Cell Biol.*, 1986, **102**, 2158-2168.

40. WRIGHT, R., RINE, J., Transmission electron microscopy and immunocytochemical studies of yeast: analysis of HMG-CoA reductase overproduction by electron microscopy, *Methods Cell Biol*, 1989, **31**, 473-512.

41. FRIESEN, J.A., RODWELL, V.W., Protein engineering of the HMG-CoA reductase of *Pseudomonas mevalonii*. Construction of mutant enzymes whose activity is regulated by phosphorylation and dephosphorylation, *Biochemistry*, 1997, **36**, 2173-2177.

42. KELLEY, S.T., FARRELL, B.D., MITTON, J.B., Effects of specialization on genetic differentiation in sister species of bark beetles, *Heredity*, 2000, **84 (Pt 2)**, 218-227.

43. SMITH, R.H., Toxicity of pine resin vapors to three species of *Dendroctonus* bark beetles, *J. Econ. Entomol.*, 1963, **56**, 827-831.

44. STURGEON, K.B., MITTON, J.B. Evolution of bark beetle communities, *in* Bark Beetle Communities: A System for the Study of Evolutionary Biology (J.B.Mitton, K.B. Sturgeon, eds.) University of Texas Press, Austin TX, 1982, pp. 350-384.

45. LANGENHEIM, J.H., Plant Resins: Chemistry, Evolution, Ecology, Ethnobiology, Timber Press, Portland, OR, 2003, 586 p.

46. PAINE, T.D., RAFFA, K.F., HARRINGTON, T.C., Interactions among scolytid bark beetles, their associated fungi, and live host conifers, *Annu. Rev. Entomol.*, 1997, **42**, 179-206.

47. ANDERSON, A.B., Monoterpenes, fatty and resin acids of *Pinus ponderosa* and *Pinus jeffreyi*, *Phytochemistry*, 1969, **8**, 873-875.

48. ANDERSON, A.B., RIFFER, R., WONG, A., Monoterpenes, fatty and resin acids of *Pinus edulis* and *Pinus albicaulis*, *Phytochemistry*, 1969, **8**, 1999-2001.

49. PAINE, T.D., BLANCHE, C.A., NEBEKER, T.E., STEPHEN, F.M., Composition of loblolly pine resin defenses: Comparison of monoterpenes from induced lesion and sapwood resin, *Can. J. Forest Res.*, 1987, **17**, 1202-1206.

50. STEELE, C.L., LEWINSOHN, E., CROTEAU, R., Induced oleoresin biosynthesis in grand fir as a defense against bark beetles, *Proc. Natl. Acad. Sci. USA*, 1995, **92**, 4164-4168.

51. STEELE, C.L., KATOH, S., BOHLMANN, J., CROTEAU, R., Regulation of oleoresinosis in grand fir (*Abies grandis*). Differential transcriptional control of monoterpene, sesquiterpene, and diterpene synthase genes in response to wounding, *Plant Physiol.*, 1998, **116**, 1497-1504.

52. PHILLIPS, M.A., CROTEAU, R.B., Resin-based defenses in conifers, *Trends Plant. Sci.*, 1999, **4**, 184-190.

53. WARREN, J.M., ALLEN, H.L., BOOKER, F.L., Mineral nutrition, resin flow and phloem phytochemistry in loblolly pine, *Tree Physiol.*, 1999, **19**, 655-663.

54. TRAPP, S., CROTEAU, R., Defensive resin biosynthesis in conifers, *Annu. Rev. Plant Physiol. Molec. Biol.*, 2001, **52**, 689-724.

55. DE MORAES, C.M., MESCHER, M.C., Biochemical crypsis in the avoidance of natural enemies by an insect herbivore, *Proc. Natl. Acad. Sci. USA*, 2004, **101**, 8993-8997.

56. SABOURAULT, C., GUZOV, V.M., KOENER, J.F., CLAUDIANOS, C., PLAPP, F.W., JR., FEYEREISEN, R., Overproduction of a P450 that metabolizes diazinon is linked to a loss-of-function in the chromosome 2 ali-esterase (MdalphaE7) gene in resistant house flies, *Insect Molec. Biol.*, 2001, **10**, 609-618.

57. LI, X., SCHULER, M.A., BERENBAUM, M.R., Jasmonate and salicylate induce expression of herbivore cytochrome P450 genes, *Nature*, 2002, **419**, 712-715.

58. SMITH, J.R., OSBORNE, T.F., BROWN, M.S., GOLDSTEIN, J.L., GIL, G., Multiple sterol regulatory elements in promoter for hamster 3-hydroxy-3-methylglutaryl-coenzyme A synthase, *J. Biol. Chem.*, 1988, **263**, 18480-18487.

59. DELORME, L., LIEUTIER, F., Monoterpene composition of the preformed and induced resins of scots pine, and their effect on bark beetles and associated fungi, *Eur. J. Forest Path.*, 1990, **20**, 304-316.

60. RAFFA, K.F., SMALLEY, E.B., Interaction of pre-attack and induced monoterpene concentrations in host conifer defense against bark beetle-fungal complexes, *Oecologia*, 1995, **102**, 285-295.

61. WERNER, R.A., Toxicity and repellency of 4-allylanisole and monoterpenes from white spruce and tamarack to the spruce beetle and eastern larch beetle (Coleoptera: Scolytidae), *Environ. Entomol.*, 1995, **24**, 372-379.

62. HUGHES, P.R., Pheromones of *Dendroctonus*: Origin of alpha-pinene oxidation products present in emergent adults, *J. Insect. Physiol.*, 1975, **21**, 687-691.

63. MARTIN, D., BOHLMANN, J., GERSHENZON, J., FRANCKE, W., SEYBOLD, S.J., A novel sex-specific and inducible monoterpene synthase activity associated with a pine bark beetle, the pine engraver, *Ips pini*, *Naturwissenschaften*, 2003, **90**, 173-179.

64. STURGEON, K.B., ROBERTSON, J.L., Microsomal polysubstrate monooxygenase activity in western and mountain pine beetles (Coleoptera: Scolytidae), *Ann. Entomol. Soc. Am.*, 1985, **78**, 1-4.

65. TITTIGER, C. Molecular biology of bark beetle pheromone production and endocrine regulation. *in*: Insect Pheromone Biochemistry and Molecular Biology (ed. G.J. Blomquist, Vogt, R., eds), Elsevier, San Diego. 2003, pp. 201-230.
66. EIGENHEER, A.L., KEELING, C.I., YOUNG, S., TITTIGER, C., Comparison of gene representation in midguts from two phytophagous insects, *Bombyx mori* and *Ips pini*, using expressed sequence tags, *Gene*, 2003, **316C**, 127-136.
67. FERNANDES, J., BRENDEL, V., GAI, X., LAL, S., CHANDLER, V.L., ELUMALAI, R.P., GALBRAITH, D.W., PIERSON, E.A., WALBOT, V., Comparison of RNA expression profiles based on maize expressed sequence tag frequency analysis and micro-array hybridization, *Plant Physiol.*, 2002, **128**, 896-910.
68. HWANG, D.M., DEMPSEY, A.A., LEE, C.Y., LIEW, C.C., Identification of differentially expressed genes in cardiac hypertrophy by analysis of expressed sequence tags, *Genomics*, 2000, **66**, 1-14.
69. OKANO, K., SHIMADA, T., MITA, K., MAEDA, S., Comparative expressed-sequence-tag analysis of differential gene expression profiles in BmNPV-infected BmN cells, *Virology*, 2001, **282**, 348-356.
70. ADAMS, M.D., KERLAVAGE, A.R., FLEISCHMANN, R.D., FULDNER, R.A., BULT, C.J., LEE, N.H., KIRKNESS, E.F., WEINSTOCK, K.G., GOCAYNE, J.D., WHITE, O., *ET AL.*, Initial assessment of human gene diversity and expression patterns based upon 83 million nucleotides of cDNA sequence, *Nature*, 1995, **377**, 3-174.
71. GIRARD, C., JOUANIN, L., Molecular cloning of cDNAs encoding a range of digestive enzymes from a phytophagous beetle, *Phaedon cochleariae*, *Insect Biochem. Molec. Biol.*, 1999, **29**, 1129-1142.
72. UZAKOVA, D.U., KOLESNIK, A.A., ZHEREBIN, Y.L., EVSTIGNEEVA, R.P., SARYCHEVA, I.K., Lipids of mulberry leaves and of mulberry silkworm excreta, *Chem. Nat. Comp.*, 1988, **23**, 419 - 422.
73. PEDRA, J.H., BRANDT, A., WESTERMAN, R., LOBO, N., LI, H.M., ROMERO-SEVERSON, J., MURDOCK, L.L., PITTENDRIGH, B.R., Transcriptome analysis of the cowpea weevil bruchid: identification of putative proteinases and alpha-amylases associated with food breakdown, *Insect Molec. Biol.*, 2003, **12**, 405-412.
74. MELO, F.R., RIGDEN, D.J., FRANCO, O.L., MELLO, L.V., ARY, M.B., GROSSI DE SA, M.F., BLOCH, C., JR., Inhibition of trypsin by cowpea thionin: Characterization, molecular modeling, and docking, *Proteins*, 2002, **48**, 311-319.
75. ZHU-SALZMAN, K., KOIWA, H., SALZMAN, R.A., SHADE, R.E., AHN, J.E., Cowpea bruchid *Callosobruchus maculatus* uses a three-component strategy to overcome a plant defensive cysteine protease inhibitor, *Insect Molec. Biol.*, 2003, **12**, 135-145.
76. KACHARE, D.P., CHAVAN, J.K., KADAM, S.S., Nutritional quality of some improved cultivars of cowpea, *Plant Food Hum. Nut.*, 1988, **38**, 155-162.
77. SUJATHAMMA, P., DANDIN, S.B., Leaf quality evaluation of mulberry (*Morus* spp.) genotypes through chemical analysis, *Indian J. Seric.*, 2000, **39**, 117-121.

78. EDGAR, R., DOMRACHEV, M., LASH, A.E., Gene Expression Omnibus: NCBI gene expression and hybridization array data repository, *Nucl. Acids Res.*, 2002, **30**, 207-210.
79. COLANTUONI, C., HENRY, G., ZEGER, S., PEVSNER, J., SNOMAD Standardization and normalization of microarray data: Web-accessible gene expression data analysis, *Bioinformatics*, 2002, **18**, 1540-1541.
80. TUSHER, V.G., TIBSHIRANI, R., CHU, G., Significance analysis of microarrays applied to the ionizing radiation response, *Proc. Natl. Acad. Sci. USA*, 2001, **98**, 5116-5121.
81. SANDERS, H.R., EVANS, A.M., ROSS, L.S., GILL, S.S., Blood meal induces global changes in midgut gene expression in the disease vector, *Aedes aegypti*, *Insect Biochem. Mol. Biol.*, 2003, **33**, 1105-1122.
82. KEELING, C.I., BLOMQUIST, G.J., TITTIGER, C., Coordinated gene expression for pheromone biosynthesis in the pine engraver beetle, *Ips pini* (Coleoptera: Scolytidae). *Naturwissenschaften*, 2004, **91**, 324-328.
83. HAMPTON, R., DIMSTER-DENK, D., RINE, J., The biology of HMG-CoA reductase: The pros of contra-regulation, *Trends Biochem. Sci.*, 1996, **21**, 140-145.

Chapter Four

INTERACTIONS AMONG CONIFER TERPENOIDS AND BARK BEETLES ACROSS MULTIPLE LEVELS OF SCALE: AN ATTEMPT TO UNDERSTAND LINKS BETWEEN POPULATION PATTERNS AND PHYSIOLOGICAL PROCESSES

Kenneth F. Raffa,[1*] Brian H. Aukema,[1,2] Nadir Erbilgin,[3] Kier D. Klepzig,[4] Kimberly F. Wallin[5]

* Author for correspondence, email: raffa@entomology.wisc.edu

1 Dept. Entomology, University of Wisconsin, Madison, WI, USA.
2 Natural Resources Canada, Canadian Forest Service, Victoria, British Columbia, Canada
3 Dept. Environmental Science, Policy & Management, University of California, Berkeley, CA, USA
4 Southern Research Station, USDA Forest Service, Pineville, LA, USA.
5 Dept. For. Sci., Oregon State University, Corvallis, OR, USA

INTRODUCTION

A major challenge confronting ecologists involves scaling up and down across various levels of biological organization.[1,2] The ability to conduct such scaling is important, because there is often a gap between the level at which information is most needed or best described versus the level at which it is most reliably generated or best explained. Many patterns are most appropriately addressed at the landscape level, such as how to manage eruptive insect herbivores or understand their roles in ecosystem processes like fire and succession. However, the mechanisms that guide our understanding are often best suited for experimentation at the individual or suborganismal levels. In addition, there are many examples where system properties change dramatically with the scale at which they are examined. Failure to recognize this has resulted in some costly lessons, such as with fire eradication, predator exclusion, and calendar application of pesticides.

There are two general approaches to this problem. Landscape approaches describe patterns at a large scale, and try to infer mechanisms based on emergent "signatures". Mechanistic approaches first characterize specific processes, and try to link them across various levels. Both approaches have their advantages and limitations, and their relative applicability varies with the system and objectives. However, our ability to merge these two approaches remains limited. This chapter attempts to integrate interactions of phytochemicals and their derivatives with herbivores from the molecular through landscape levels. Such an understanding could be applied to epidemiology, ecosystem function, and natural resource management. However, we wish to emphasize that our attempt does not resolve this challenge, and the interface between process - and pattern - oriented approaches provides a rich and needed area for future research.

Our approach is to focus on one group of compounds in one system. We believe this can help identify key gaps in both our knowledge of underlying mechanisms and our ability to construct relevant linkages. This will hopefully facilitate studies of other plant-herbivore-community relationships. Our efforts are at synthesis, not comprehensive review, as thousands of primary papers and many outstanding reviews have been written on this model. Any synthesis suffers from the need to resort to "apples-to-oranges" comparisons, so we have tried to provide examples from a few common systems. This necessarily emphasizes our own work.

BARK BEETLES AND ASSOCIATED MICROORGANISMS IN HOST CONIFERS

Bark beetles provide an ideal system for scaling across layers of biological organization, both because they are intensively studied and because a single phytochemical group, monoterpenes, has been shown to exert major roles at multiple levels. Some examples of studies conducted from the molecular through landscape scales, and brief descriptors, are illustrated in Table 4.1. We ask the reader to refer to this table and its citations throughout each scale. Few systems have been studied at such a diversity of levels. This opportunity owes largely to the extensive economic losses and dramatic landscape-level changes that bark beetles exert during outbreaks.[3-5]

Bark beetles reproduce in the subcortical region of trees. Adults disperse from brood trees, land on a potential host, and if they deem it suitable, chew through the bark. Otherwise, they resume flight and land again. The sex responsible for host selection varies with genus. As beetles bore through the bark, they produce pheromones that attract mates.[6] They excavate a nuptial chamber, copulate, and dig a long gallery along which the female oviposits. The male assists in clearing wood shavings and frass out through the entrance gallery. Larvae develop in the subcortical tissues, excavating tunnels from the main ovipositional gallery as they feed on phloem and fungi. Emerging adults exit the tree and repeat the process. Development requires one month to two years, depending on the system and temperature.

As adults enter the tree, they introduce a variety of microorganisms, mostly fungi. Beetles possess elaborate mechanisms for transporting some fungi, and there can be intense competition among fungal species.[7] Fungal associates play a variety of roles that appear to vary among systems.[8] Some fungi assist with larval nutritional physiology or serve as food,[9,10] some appear to assist in overcoming tree defenses,[11] some metabolize plant monoterpenes into oxygenated pheromones,[12] and some compete directly or indirectly with larvae for their resource.[13,14] Not only do different fungal species exert mutualistic and antagonistic effects on their vectors, it seems likely that single species have multiple and opposing effects, and hence may be conditional mutualists.[15,16] The composition of fungi can affect bark beetle population dynamics, and hence any phytochemical influence on this community does likewise.

Three features of bark beetle relationships with conifers are particularly germane to their interactions with host phytochemicals:[4,6,17] 1) Bark beetles spend almost their entire life history within the plant. Eggs, larvae, and pupae have no opportunity to leave the host if it becomes unsuitable due to induced phytochemical changes or other causes; 2) They must kill their host (or colonize a dead host) to reproduce. An exception occurs with "strip" or "top" killing, in which certain species sometimes kill portions of a tree, although reproduction in such instances can be reduced; 3) They usually exhaust their resource within a single generation. Hence each generation must undergo the process of locating dead trees or killing live trees in which to breed.

Table 4.1: Representative studies of bark beetle - conifer - fungal associations conducted at various levels of scale. This list is not intended to be comprehensive, but rather to illustrate the range of previously conducted work. Almost all of the examples below include terpene – mediated effects.

Level of Biological Organization	References
Molecular Biology & Biochemistry	
Terpene synthesis in conifers	21,22,28,29
Pheromone synthesis by bark beetles	23,90
Genetics of bark beetles	91-94
Genetics of bark beetle - associated fungi	95-97
Genetics of host resistance	98,99
Histology	
Conifer Responses to Attack	24-26
Tissue – specific pheromone synthesis by bark beetles	100-102
Physiology	
Physiology of bark beetle - associated fungi	9,10,103
Chemosensory physiology of bark beetles	104-107
Constitutive chemical barriers to bark beetle colonization	18,19,108,109
Localized Induced chemical responses to bark beetle colonization	20,30,31,110-112
Systemic Induced chemical responses to bark beetle colonization	113,114
Effects of host compounds on beetle survival	31,33-35
Effects of host compounds on beetle - associated fungi	37,38,115
Behavior	
Effects of host compounds on beetle host selection behavior	38,48,116
Bark beetle attraction to pheromones	6,39,40,117
Host compound mediation of beetle responses to pheromones	6,42,43
Ecology	
Associations of fungi with bark beetles	7,8,95,118-120
Interactions between fungi & bark beetles	9,11-16,121
Colonization dynamics at the whole tree level	36,122
Spatial components of tree killing by bark beetles	44,45,53,123-125
Inter-Guild Interactions with folivores & root insects / pathogens	27,46,74,126
Tritrophic interactions:	
Attraction of natural enemies to beetle pheromones	6,77,78,84
Population dynamics of bark beetles	3,4,49-51,85,86,127
Landscape, Ecosystem	
Ecosystem impacts & landscape ecology of bark beetles	5,32,52,54,76,128
Anticipated responses of bark beetles to global atmospheric change	56,68,69
Evolution	
Coevolution of bark beetles and fungi	95
Coevolution of bark beetles and conifers	32,129-132

LOCALIZED REACTIONS: CONSTITUTIVE AND INDUCED DEFENSES

Conifer defenses include histological responses at two temporal scales. First, preformed ducts within the cambium respond to wounding by exuding resin.[18,19] Traumatic resin ducts form quickly thereafter, and assist in transporting phytochemicals to the beetle's entry site.[17,20-23] Necrotic lesions form in advance of and contain the beetle-fungal complex.[24-26] These lesions continue to expand as long as this complex progresses.[17] All trees form necrotic lesions in response to controlled inoculations, as non-recognition does not appear highly operative in a system accompanied by such extensive mechanical damage. However, relatively resistant trees form lesions more rapidly. They typically have longer lesions during the first few days, but ultimately shorter ones once the invaders are confined.[27] Primary resin can serve as a partial physical barrier that prevents or delays entry. Monoterpene concentrations are sometimes high enough to kill beetles.

This rapid, usually localized, response also includes biochemical changes, in which monoterpene, diterpene, and sesquiterpene concentrations rise. These terpenes are derived from isoprenoids synthesized via the mevalonate or 1-deoxy-D-xylulose-5-phosphate pathways[21,23,28,29] in the cytosol, endoplasmic reticulum, and plastids.[21] A diverse array of terpenoid synthethases yield the parent compounds, and a number of genes have been characterized.[21,23] Induction can be elicited by applying methyl jasmonates.[22]

The area around the entry site may show a several hundred-fold increase in monoterpenes within two weeks.[26,30,31] The phloem becomes saturated, and liquid resin exudes from the entry site. The rate and extent of this reaction varies markedly among individuals. Induction may include compositional changes, in which relative proportions of constitutive compounds change. Compositional changes are more prominent in *Abies* and *Picea* than *Pinus*.[26] There is generally a higher increase in those compounds having the most biological activity.[32] *De novo* appearance of previously absent monoterpenes appears relatively uncommon during conifer induction. Specificity varies among tree species.[26] For example, red pine, *Pinus resinosa* (Aiton) shows marked responses to *Ophiostoma ips* and *Leptographium terebrantis*, but only responds to *Ophiostoma nigrocarpum* as to aseptic wounds. In contrast, jack pine, *Pinus banksiana* Lamb., responses to *O. nigrocarpum* are intermittent between these fungi and mechanical wounds.

High concentrations of monoterpenes are toxic to bark beetle adults,[33-35] eggs,[36] and presumably larvae, although the latter have not been tested. The dynamics of this interaction are shown in Fig. 4.1. Based on known rates of localized induction in healthy red pine, and toxicities of corresponding compounds to the pine engraver, *Ips pini* (Say), we can estimate the percentage of adults that would die within two days at the monoterpene concentration present at each time following challenge inoculation. The constitutive concentration is sufficient to kill only 60% of the beetles, indicating some tolerance to these compounds. Within only 3 days, however, the concentration is high enough to kill 90% of them (Fig. 4.1). Hence there is little likelihood that the adults, let alone their eggs, larvae, pupae, and teneral adults, would survive and complete development in this environment. The data in Fig. 4.1 also illustrate how constitutive and induced monoterpene - based defenses function in an integrated fashion.

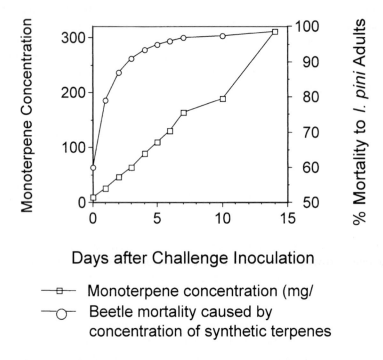

Days after Challenge Inoculation

—□— Monoterpene concentration (mg/
—○— Beetle mortality caused by
        concentration of synthetic terpenes

Fig. 4.1: Effect of monoterpene induction in red pine on survival of pine engravers, *Ips pini*. The squares indicate monoterpene concentration of phloem tissue vs. days after challenge inoculation. The circles indicate the mortality that occurs to adult *I. pini* following exposure to synthetic monoterpenes (in a 2-day assay) at the monoterpene concentration present at a given time after inoculation. For example, at one day post-inoculation the monoterpene concentration is approximately 25 mg/gm, which kills approximately 80% of the adults in a controlled assay. The data indicate that *I. pini* would not be able to survive and reproduce in red pines unless this induced response is prevented.[31] (with permission from Springer-Verlag).

Conifer monoterpenes also inhibit the germination and mycelial growth of fungi vectored by bark beetles.[37,38] In general, the effects are stronger on germination (25% - 60% reduction) than mycelial growth (7% - 55%). There tends to be more variation among different monoterpenes in their effects on fungi than on beetles.

Monoterpenes do not function alone, but act in concert with other chemical groups, particularly diterpene acids and stilbene phenolics (Table 4.2). Diterpene acids appear to have the highest anti-fungal activity, but relatively little activity against beetles. These compounds are highly inducible, and some are only detectable following induction. Stilbene phenolics have intermediate activity against both beetles and fungi, and are weakly inducible.[38] Other preformed structures such as lignified stone cell masses, periderm layers, and calcium oxalate crystals can contribute to defense.

In summary, the histological and biochemical defenses of conifers pose a formidable barrier against bark beetles and their fungal associates. In particular, inducible reactions raise monoterpene concentrations above the physiological tolerance of adult beetles and their endophytic brood. Unless these responses are interrupted, bark beetles have little chance of reproducing in live trees.

WHOLE TREES: INDIVIDUAL TREE DEFENSES AND GROUP COLONIZATION

Bark beetles can kill and colonize vigorous, well-defended trees despite the above defenses, owing to their cooperative behavior.[17] Using oxygenated terpenes as aggregation pheromones, they engage in joint attacks that collectively exhaust host resistance.[39,40] This can be visualized as a dose-response relationship, in which the tree's negative effect on brood production varies from 100% to nearly 0% with increasing beetle densities. Three lines of evidence support this interpretation. First, trees' resin flow in response to a mechanical wound, and their ability to form necrotic lesions in response to a challenge inoculation, diminish markedly within only 2 - 3 days during natural attacks.[17] Second, when natural attacks are artificially interrupted, there is a clear density-based demarcation between killed vs. surviving trees and corresponding surviving vs. killed brood.[17] Third, in some cases increasing the density of challenge inoculations with beetle-vectored fungi decreases induced monoterpene accumulation, reduces trees' ability to confine fungi within lesions, and in one system, kills trees.[11,17]

Bark beetles use oxygenated terpenes as aggregation pheromones. These may be metabolized from host compounds, produced *de novo* following stimulation by host compounds, converted from host compounds by beetle-associated microorganisms, or various combinations thereof.[23] The relevant enzymes, bio-

Table 4.2: Multiple components of red pine defense against bark beetle – fungal associates

Biological Effect	Monoterpenes	Diterpene Acids	Stilbene Phenolics
Adult beetle repellency	++	-	++
Adult beetle toxicity	++	-	?
Larval beetle toxicity	?	-	?
Fungal spore germination inhibition	+	+++	++
Fungal mycelia growth inhibition	+	+++	++
Inducibility	+++	+++	+

chemical pathways, and underlying genes have become increasingly well described.[41] Moreover, there may be some overlap in the biochemical mechanisms by which beetles synthesize oxygenated terpenes and detoxify host terpenes.[23] The attraction of beetles to their pheromones is often synergized by host terpenes.[42,43]

Although the ability to conduct coordinated mass attacks might appear to render every tree susceptible, eliciting aggregation in nature can be difficult. Laboratory studies at the scale of individual beetles yield a rather deterministic picture, but observations at the whole-tree scale demonstrate that the first beetle to enter a tree may fail to elicit attraction (but far less likely when switching from adjacent trees undergoing mass attack).[17,44,45] This may arise in part from a concentration-dependent effect of monoterpenes on beetle attraction to pheromones (Fig. 4.2). Low amounts of alpha-pinene, the predominant monoterpene in *P. resinosa*, synergize attraction of *I. pini* to their pheromone, racemic ipsdienol plus lanierone. Conversely, concentrations similar to those in induced phloem inhibit attraction to pheromone. This bimodal relationship may be widespread, as similar results have been obtained with Norway spruce, *Picea abies* (L.) Karsten, and *Ips typographus* L. (Erbilgin, Krokene, Christiansen, Raffa unpublished data).

Beetle ability to overcome tree defenses via mass attack, and trees' ability to interfere with beetle communication, are incorporated into a tree defense model based on terpene content in Fig. 4.3. These relationships yield no stable outcome,

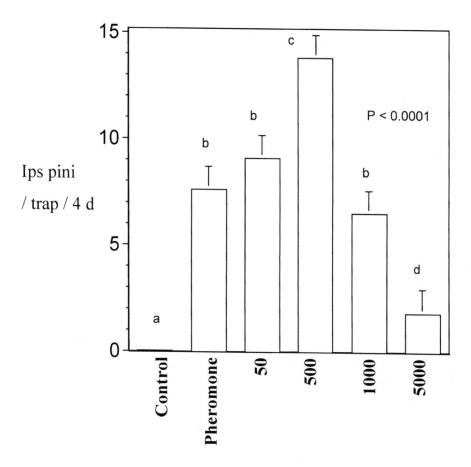

Monoterpene : Pheromone Ratio

Fig. 4.2: Variable effect of host monoterpene concentration on *I. pini* attraction to its pheromones. Lower concentrations, equivalent to those in constitutive host tissue, synergize the attractiveness of pine engraver pheromones. High concentrations, equivalent to those occurring during the first few days of induction, inhibit attraction to pine engraver pheromones.[73] (with permission from Blackwell).

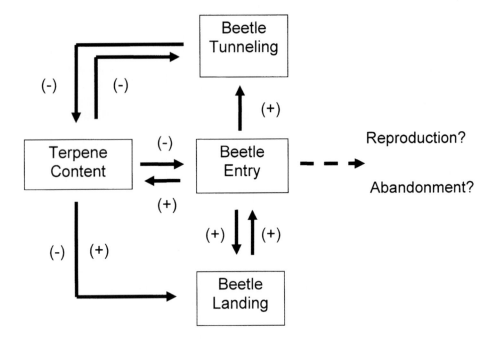

Fig. 4.3: Summary of conifer - bark beetle - fungal interactions at tissue- and whole- tree levels. Beetle entry induces pheromone production (Beetle Entry → Beetle Landing), which in combination with host monoterpenes (Terpene Content → Beetle Landing), attracts other beetles. Their tunneling diminishes the amount of resin (Terpene Content) in the host, and so reproduction (Final output on right) can proceed (Beetle Entry → Beetle Tunneling and Oviposition); (Terpene Content → (-) Beetle Tunneling and Oviposition). However, beetle entry also elicits an induced accumulation of monoterpenes (Beetle Entry → Terpene Content), which if high enough can inhibit the attraction of flying beetles to pheromones (Terpene Content → (-) Beetle Landing), inhibit beetles that have landed from entering (Terpene Content → (-) Beetle Entry), and inhibit tunneling via repellency or toxicity (Terpene Content → (-) Beetle Tunneling). This results in the abandonment or failure of colonization attempts (Final output on right).

leading to either successful beetle reproduction or failure of colonization attempts. The rates of these various opposing processes determine the outcome. Almost all of these relationships involve terpenes (Table 4.1).

Beetle choices during host selection can be categorized as two different strategies, along a continuum. One strategy is to only enter poorly defended trees. Resistance is often compromised by physiological stresses such as severe drought, crowding, disease, and old age.[18,27,46,47] The advantage of this strategy is that beetles incur little risk. The disadvantages are that such trees are relatively rare, are accessible to competing species, and provide a relatively poor nutritional substrate. The alternate strategy is to enter and attempt to initiate mass colonization of trees spanning a broader physiological range. The advantages are that such trees are plentiful, there are fewer competitors, and the nutritional quality is often high. The disadvantage is that beetles may be killed or repelled in their attempt. The ability of bark beetles to make such decisions is based largely on their behavioral responses to host monoterpene content. For example, entry by *I. pini* into denatured phloem-based media amended with alpha-pinene decreases with increasing concentration (Fig. 4.4). Different monoterpenes yield different entry vs. concentration relationships, with some compounds eliciting higher than control entry at low concentrations.[48] The same monoterpene may elicit different relationships for entry and continued excavation. All monoterpenes, however, inhibit entry and tunneling at high concentrations.

POPULATION- AND LANDSCAPE- LEVEL DYNAMICS: BIMODAL EQUILIBRIA, ALLEE EFFECTS, AND EXTENDED PHENOTYPES

Populations of some bark beetle species undergo dramatic changes in abundance through time.[49-51] They can remain at low levels within an area for several decades, during which reproductive gains are largely offset by losses during dispersal, establishment, and development. Mortality occurs to individual trees, but ecosystem-level effects consist largely of canopy thinning, gap formation, and increased nutrient cycling. Populations can rise suddenly, however, and during these eruptions there can be nearly 100% mortality to the host population, at least among diameter classes that can support brood development. These eruptions occur on scales of millions of hectares and over several years. They cause major alterations of forest age structure and species composition, and can redirect or amplify successional processes.[52] Southern pine beetle outbreaks can convert pine to oak hickory forests, mountain pine beetle outbreaks can provide ignition fuels for fires that favor reproduction of host lodgepole pines whose serotinous cones remain viable after death, and spruce beetle can convert extensive areas from spruce to birch.[5,52-54] Largescale eruptions by bark beetles cause significant economic losses, stimulate

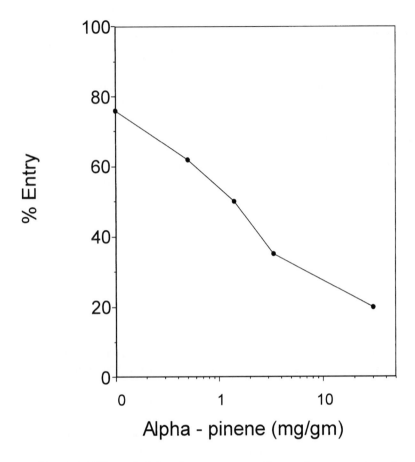

Fig. 4.4: Effect of alpha pinene concentration on host acceptance by
Ips pini.[48] (with permission from the Entomol. Soc. America).

political and socioeconomic challenges at the forest-human habitation and
wilderness- managed forest interfaces, and can cause particular hardship to rural
communities.

The underlying processes behind these dynamics are not well understood.
One descriptive model that has been proposed for bark beetles and other eruptive
herbivores is known as "dual equilibria theory" (Fig. 4.5).[55-60] According to this
view, population growth rates follow the standard discretized nonlinear curve

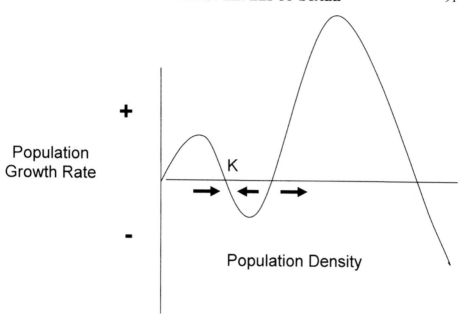

Fig. 4.5: Hypothetical dual-equilibrium replacement curve of eruptive species. Populations are typically within an area dominated by negative feedback, $K/2 - K_1$. Any increase to densities above K leads to negative growth. However, if a sudden improvement in environmental quality, or immigration, raises the population above K_1, positive feedback dominates and eruptive population growth occurs.[55,57,59,60]

governed by negative feedback across most densities. That is, growth is positive below a carrying capacity K, and the growth rate reaches a maximum at K/2. Theory presents the carrying capacity as a stable equilibrium, because any increases above this value are followed by decreases. Bimodal equilibria theory proposes that if the population somehow rises far above K, past a value K_1, then the growth rate again becomes positive. Populations above K_1 are dominated by positive feedback, and grow continuously to explosive levels.

Validation of this concept is problematic, however. The first part, how populations can rise from K to K_1, is fairly well understood for bark beetles: Severe drought, windstorms, wide scale defoliation, etc., can suddenly make the habitat more suitable for reproduction (Table 4.1). It is less clear, however, why some species do not return to equilibrium once the perturbation is removed, but rather continue to expand indefinitely until the resource is exhausted. Empirically, it is difficult to distinguish whether the growth curve is truly as depicted in Fig. 4.5, versus whether the environment has changed to dramatically increase K. Both could yield similar data, a weakness addressed in part by Dynamic Systems Theory.[61,62] Mechanistically, it is not clear how numerical responses could behave so differently unless there are qualitative differences between low- and high- density populations. An understanding of whether population changes represent mere responses to environmental quality, or fundamental changes in the system's feedback structure, could strongly improve management decisions.

Knowledge of the tree-insect interaction (Fig. 4.3) offers some guidance in integrating pattern and process to resolve this difficulty. That is, certain tradeoffs governing host acceptance behavior by individuals generate a potentially important interface between beetle characteristics and numbers. Specifically, we considered whether beetle decisions to enter or avoid well-defended trees vary with population density. Our rationale was that when populations are low, there is low probability that beetles that enter healthy trees will be joined by enough conspecifics to overcome host defenses. In contrast, when populations are high, there is little penalty for entering well-defended trees, due to the high likelihood of successful cooperative attack. Moreover, entering such trees incurs significant benefits. These trees are plentiful, relatively free of interspecific competitors, and nutritious. Further, any beetles that would exclusively accept weakened trees would have a high risk of dying before finding an acceptable host, as such trees are largely depleted during the rising phase of an outbreak (Fig. 4.5). All three components of this interface, tree defense, beetle communication, and host acceptance behavior, are mediated primarily by terpenes.

We tested the theory that responses by individual herbivores to host compounds both reflect and contribute to population increase using the spruce beetle, *Dendroctonus rufipennis* (Kirby), as a model. This insect shows the above pattern of remaining within an area at stable densities for lengthy periods, and then suddenly undergoing outbreaks.[63] For example, in Alaska the spruce beetle killed only 4000 ha / yr from 1955 to 1974, compared to 290,000 ha / yr during 1992 to 1999.[64,65] This outbreak arose when several years of warm weather reduced development times. However, it persisted, including some of the most damage-laden years, after those conditions ended.[64-66] Several million hectares of near total tree mortality resulted, with a nearly pure spruce forest being converted to deciduous species (Fig. 4.6).

Fig. 4.6: Former spruce forest converted to deciduous (light colored) stand, primarily birch, by *Dendroctonus rufipennis* on Kenai Pen., Alaska. These stands were naturally occurring spruce monocultures, in which near total mortality occurred within approximately three years. Spruce trees in these forests typically range from several meters above water levels to tree line. Note dark trees are young spruces that were too small for spruce beetle during the outbreak. These trees will ultimately replace birch and reconvert to a spruce forest. Outbreak by spruce beetle will eventually follow when favorable conditions occur. Photo by Kirsten Haberkern.

We considered three questions: 1) Do populations colonize trees of different physiological conditions during endemic versus eruptive phases? 2) Are behavioral responses to phytochemicals heritable? 3) Do beetles from eruptive vs. endemic populations differ in responses to phytochemicals? We conducted this research in 29 sites in three regions: Alaska, Yukon Territory in Canada, and Utah. The 15 eruptive and 14 endemic sites were evenly distributed among the 3 regions, all sites were nearly pure spruce monocultures separated by more than the effective dispersal distance of *D. rufipennis*, and contained trees that did not differ in monoterpene concentration or stem diameter.

We addressed the first question by conducting a two - part experiment. First, we labeled three types of trees: randomly selected live trees, trees which we felled (thereby removing resistance), and where present, trees that contained failed attacks from the previous year. After one year, we recorded beetle entry and colonization. Second, we collected adult progeny from these trees prior to emergence. We brought the beetles to the lab, and assayed their entry into media amended with varying amounts of alpha-pinene, the predominant monoterpene in spruce.

Beetles entered and successfully colonized all of the felled trees at both endemic and eruptive sites. However, patterns of live-tree colonization differed markedly between population phases.[66] No live trees in endemic plots were entered. In contrast, beetles entered 65% of the trees in eruptive plots, all of which were successfully colonized.   Beetles entered and successfully colonized 95% of previously entered trees in eruptive plots, and there were no such trees in endemic plots.   These results are consistent with the hypothesis that beetles in eruptive populations have a broader range of host acceptance than those in endemic populations.   An alternate possibility is that after endemic beetles colonized the downed trees there were no beetles left for subsequent attacks. It seems unlikely, however, that we "trapped-out" populations from multiple sites.

In the laboratory phase of this experiment, progeny beetles from trees that were colonized while alive differed from those collected from trees in the same eruptive sites that were dead before colonization.[66] Both groups exhibited relatively low entrance rates when there were no monoterpenes in the test medium, and at high concentrations. However at the concentration typical of live trees, entry rates by beetles collected from live trees (*i.e.*, the progeny of adults that selected well-defended hosts) had an entry rate twice that of beetles from felled trees.

We considered whether host acceptance behavior is heritable by conducting two sets of experiments: mother-daughter correlations and breeding line selection. In the former, we collected pre-emergent adults in the field, established them on logs in a common environment and density to reduce environmental effects, and bred them for one generation. We then assayed the adult female (the host selecting sex) progeny of these beetles at a concentration of alpha-pinene mimicking host trees. We then established independent male - female breeding lines, and again bioassayed the adult female progeny.   There was a strong correlation between mother and daughter gallery construction (Fig. 4.7a), with heritabilities of 0.64 and 0.36 among those derived from endemic and eruptive sites, respectively.[66]

Because spruce beetle generations are at least one year, we used *I. pini* (@30-day generation time) as a surrogate for directional selection studies. We assayed 200 males (the host-selecting sex) at a discriminating concentration of alpha-pinene, and then established independent breeding lines in logs based on whether they did or did not enter the medium. These males were paired with random females. Both groups of males readily bored into logs and reproduced, indicating those that did not enter the medium were not incapable of tunneling, but rather refused to enter at the monoterpene concentration provided. We then assayed the progeny males (F_0) at the same concentration. Those that did or did not enter were again established on logs, in full-sib mating lines with their sisters. We repeated the process for F_1 to F_3 progeny. The results (Fig. 4.7b) show a strong heritable component ($h^2 = 0.78$). Within 3 generations, the percentage of entering beetles varied by 3X between selected lines.[67] Controls remained stable.

We compared responses to host terpenes between spruce beetles from endemic vs. eruptive populations by collecting females from the above sites across the three geographic regions, and bringing them to the lab for controlled entry bioassays.[66] There was a strong effect of alpha-pinene concentration on whether beetles entered the medium, no effect due to geographic region, and no effect due to year in the Alaskan sites that were assayed twice. When beetles were tested singly, eruptive beetles were not less discriminating, contrary to our prediction. However, beetles from endemic and eruptive sites showed an important distinction. Those from endemic stands made host entry decisions independent of whether other beetles were present in the arena (Fig. 4.8). In contrast, beetles from eruptive sites were less repelled by high alpha-pinene concentrations when more beetles were present. There was a strong population phase X local density interaction. Thus, when spruce beetle populations are high, and large numbers of beetles are likely to simultaneously land on potential hosts, they are more likely to enter a tree they would otherwise reject. Because of the high background populations available for subsequent pheromone-based attraction, these attempts are likely to succeed. The physiological bases for these differences are not known.

Based on these results, we can construct linkages between within- and whole-tree processes and landscape- level processes. Figure 4.9 shows the same feedback structure as Figure 4.3, but includes density- dependent feedback between individual host selection and population size. The key feature is the terpene- mediated linkage at the level of host selection, which suggests a basis for qualitative differences between populations below versus above K_1 in Fig. 4.5. This behavioral difference could contribute to the dual equilibria structure, and result in eruptive population dynamics, proposed by this model. This model proposes a strong Allee effect in host procurement (individuals benefit from conspecifics) at high densities. Together,

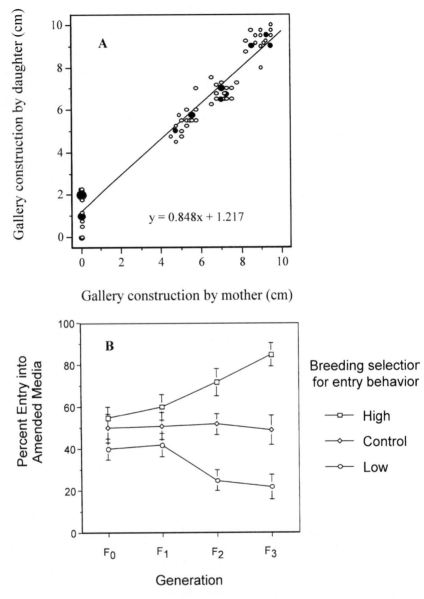

Fig. 4.7: Heritability of host selection behavior in bark beetles: a) Mother-daughter correlations of gallery construction in endemic *D. rufipennis*. Closed circles indicate multiple data points[66] (with permission from the Ecol. Soc. America); b) Selection for degree of aversion to alpha-pinene in *I. pini* [67] (with permission from the Entomol. Soc. America).

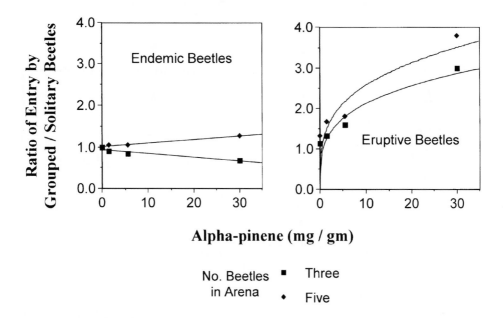

Fig. 4.8: Effect of alpha-pinene concentration, number of other beetles in assay chamber, and population phase on spruce beetle entry into amended media.[66] (with permission from the Ecol. Soc. America).

these interactions provide evidence of an extended phenotype,[2] in which bark beetles function as keystone species and heritable production of terpenes by conifers is at the foundation of landscape-level effects. This also has implications to management, because it implies that for some species, habitat suitability must be kept below a threshold, K_1. This feedback not only supports the view that anticipated global warming will favor bark beetle outbreaks,[56,68,69] but further suggests outbreaks can sometimes become self-sustaining even if warm years are not consecutive (Fig. 4.9).

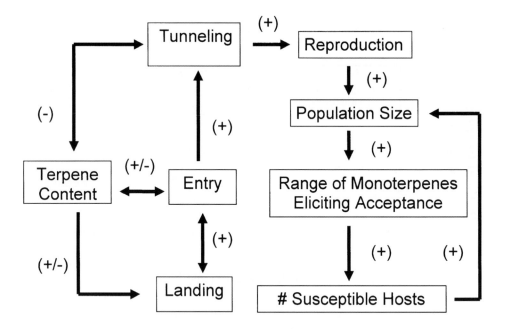

Fig. 4.9: Feedback among tree and beetle population processes involving terpenes. Beetles enter trees, produce pheromones, and deplete host resins, and trees respond with induces accumulation of monoterpenes that inhibit beetle entry, communication, and tunneling, as in Figure #3. However, beetle host acceptance behavior is plastic, and responds to interactions of regional (population phase) and local (plant surface) population density (Population Size → Range of Monoterpenes Eliciting Acceptance). This initiates a positive feedback loop, in which rising beetle populations expand their own host range (Range of Monoterpenes Eliciting Acceptance → # Susceptible Hosts), and hence populations grow (# Susceptible Hosts → Population Size). The linkage between host acceptance at the individual and population scales can contribute to eruptive behavior.

CONSTRAINTS ON POPULATION ERUPTIONS

Many conifer-bark beetle-fungal systems possess elements of the terpene-based dynamics shown in Figures 4.3 and 4.9. However, most species do not undergo extensive population eruptions. We do not have a clear understanding of the circumstances under which various competing processes will dominate, and hence whether the landscape-level outcomes will be canopy thinning, gap formation, or forest conversion. We consider this question by evaluating results from two other systems.

Jack pine forests in the Great Lakes region of North America recently underwent extensive defoliation by the jack pine budworm, *Choristoneura pinus pinus* Freeman. Defoliation compromises resistance against subcortical beetles, reducing constitutive resin, slowing fungal confinement via autonecrosis, and reducing monoterpene induction following challenge inoculations (Fig. 4.10a). Beetles responded accordingly, with high colonization rates and death of defoliated trees (Fig. 4.10b). Populations of *Ips grandicollis* (Eichhoff) rose dramatically in response to this increased resource. However, these beetles never spread to the healthy-tree resource (Fig. 4.11). Because the jack pine budworm is monophagous, it did not affect red pine, a common host of *I. grandicollis*.[70-73] Once the pool of defoliated trees was exhausted, the population crashed. This behavior differs substantially from the behavior of spruce beetle described above, and from mountain pine beetle and southern pine beetle, which often expand onto less favored tree species during outbreaks.[63,70]

The second system involves interactions between below- and above- ground processes, specifically a diverse community of insects colonizing roots of red pines in the Great Lakes region. Six species of weevils and bark beetles partition this resource based on the particular section of root tissue, and host physiology.[72,74] Each species is attracted to combinations of alpha-pinene plus ethanol, but displays a unique combination of preferred chirality, component ratios, and gender-based responses. These beetles vector *Leptographium* fungi, which colonize the roots and also grow through root grafts. Colonization by root beetles and associated fungi does not kill mature red pines. However, it compromises tree defenses against lethal attacks on the main stem by *I. pini* and its *Ophiostoma* associates. Root-colonized trees have lower constitutive resin flow at the base of the crown than healthy trees, and exhibit less rapid accumulation of monoterpenes in response to challenge inoculation than trees with healthy roots.[47] Extracts of induced stem tissue from root-infested trees are preferred over extracts of induced stem tissue from healthy trees by *Ips pini* in amended- diet assays.[38] Likewise, *I. pini* caged onto red pines that have healthy roots typically die without entering, whereas those caged onto root-colonized tress enter more readily.

RAFFA, et al.

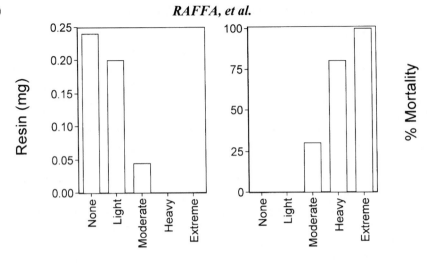

Defoliation Level

Fig. 4.10: Effects of folivory on jack pine resistance against subcortical insects a) Compromised host defenses against subcortical insects following defoliation by jack pine budworm. b) Defoliated trees also accumulate lower monoterpene concentrations during active induced responses.[27] Light: <25%; Moderate: 26-50%; Heavy: 51-75%; Extreme: >76% (with permission from the Ecol. Soc. America).

Fig. 4.11: Mortality of jack pine (light trees) following defoliation by jack pine budworm and subsequent stem attack by wood boring beetles. Note living (dark trees) red pines in plantations (Photo courtesy Wisconsin DNR).

These interactions among root herbivory, fungal infection, host physiology, and beetle behavior generate a positive feedback that yields a specific spatial pattern (Fig. 4.12). A tree's likelihood of being colonized and killed by *I. pini* is closely related to its proximity to the perimeter of previously killed trees, being nearly 100% along the margin and only slightly above 0% 7 m away (Fig. 4.13a). This pattern corresponds closely to the below-ground distribution of *Leptographium*, which proceeds approximately 7 m in advance of above-ground symptoms (Fig. 4.13b). There is an approximate three-year lag between root infection and colonization by *I. pini*. In one declining stand observed over 16 years, 90% of trees killed by *I. pini* had prior infestation by root insects and or fungi. The number of killed trees varied by 23X among years. During drought years both the percentage of trees killed without prior root infestation, *i.e.*, further from the margin, and overall *I. pini* populations, rise. However, the population does not expand onto the healthy tree resource. Rather it returns to almost total reliance on the root-colonized resource, and the concentrically expanding pattern of mortality resumes (Fig. 4.12).

What are the key differences between systems that do vs. do not generate the positive feedback proposed in Fig. 4.5? Each of these systems includes all of the terpene-based processes shown in Fig. 4.3, and potentially all of those in Fig. 4.9. Unfortunately, we can only describe these as "case studies" rather than "model systems" in regard to this question, because each was pursued to test a different hypothesis and hence employed different methodologies. Still, we can consider some possibilities. One is phylogeny: Overall, there are more eruptive *Dendroctonus* than *Ips* species.[4,6,39] However, this is only partly explanatory because most *Dendroctonus* are not eruptive,[63,70] some *Ips* are eruptive,[11,65] and some eruptive *Dendroctonus* species never undergo outbreaks in portions of their range, including regions with favorable weather.[70,75] Moreover, Figure 4.7b indicates that *I. pini*, which does not expand beyond the stressed tree resource in the field (Fig. 4.13a), appears to have the genetic capacity to do so.

A second possibility is habitat favorability. We and others have argued that a large contiguous area with low host species diversity is a requirement for bark beetle outbreaks.[3,32,76] However, all three of the above systems are monocultures, with the spruce and jack pine occurring naturally and the red pine planted. Moreover, the red pine system is even-aged and genetically homogenous, which should further favor outbreaks.[76] Thus, habitat homogeneity appears to be a necessary but insufficient condition for bark beetle outbreaks. Similar observations have been made in southern ecosystems.[4,76] A third possibility is that differences in weather separate eruptive from noneruptive patterns. However, favorable weather appears, like habitat homogeneity, to be a necessary but insufficient condition for beetle outbreaks. For example, unusually low precipitation generated higher *I. pini* populations as its resource expanded, but it did not proceed to outbreak behavior. In

Fig. 4.12: Declining even-aged red pine plantation due to interactions between below- and above- ground herbivory.   This photograph shows a side view of a circular zone of tree mortality that reached a road and so can be seen in two dimensions.   The photograph shows only the eastern half of this mortality.   The area on the right contains dead trees that were killed several years ago, and have lost all of their foliage. Further into the stand are tress killed even sooner which have since blown down and been replaced by herbaceous vegetation. Adjacent to the defoliated trees are trees which were killed during the current year and have red foliage.   Further from the epicenter, live trees show reduced growth, have thinner crowns, and appear faded. These are infected with root fungi and herbivores.   Trees on the left appear healthy and their roots are not yet infected. Every year, rings of trees showing each set of symptoms radially from the epicenter.

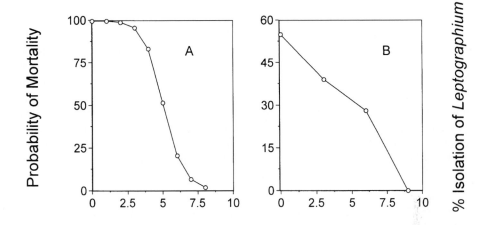

Fig. 4.13: Effects of belowground herbivory on susceptibility to bark beetles: a) Mortality of red pines in 17 plantations during 1997 – 2000.[126] (with permission from Elsevier); b) Spatial pattern of infection of red pine roots with *Leptographium*.[72] (with permission from Elsevier). Trees with *Leptographium* infection show altered monoterpene profiles.

contrast, unusually high temperatures released *D. rufipennis* populations in south-central Alaska, but these outbreaks persisted even after conditions became more normal. A further illustration is provided by *D. rufipennis* in central Alaska, where summer temperatures are always warm enough to support univoltine development, yet outbreaks do not occur.[64,65]

A common feature of all three systems is that natural enemies are strongly attracted to their prey's pheromones.[77-79] Orientation by predators to these oxygenated terpenes is often synergized by host monoterpenes, and this attraction can be even stronger than by the bark beetle to its own pheromones.[80] However there appear to be substantial differences in the degrees of predation and competition among these systems (Table 4.3). Gara and coworkers[81] quantified densities of *D. rufipennis*, predators, and competitors in infested spruce trees in Alaska. In south central Alaska, where populations intermittently reach outbreak levels,[65,81] they observed 17 times as many spruce beetles as predators, and nearly half as many competitors as spruce beetles. This is likely a common condition, as, anecdotally, entomologists there find no need to use insecticide strips to prevent destruction of pheromone-trap contents by predators, a necessity in other systems. In contrast, *I. pini* typically experiences much higher predation and competition (Table 4.3).[82]

There are over twice as many predators, and twice as many competitors, as pine engravers, based on colonization data in red pine. This also seems a common condition, as it is consistent with population trend and impact studies.[73,83] Similarly, *I. grandicollis* colonization appears to result in higher relative numbers of predators and competitors than *D. rufipennis*.[27,71,73] The predator data for *I. grandicollis* in Table 4.3 are not entirely comparable to the others, as our only available information is from attraction to ipsenol, but as before, population analyses support this impression. Competition can be an even more important factor. Ninety-five percent of the trees colonized by *I. grandicollis* following defoliation by jack pine budworm were also colonized by *Monochamus* species (Coleoptera: Cerambycidae).[27] *Monochamus* galleries typically overrun and destroy *Ips* galleries, and the larvae are

Table 4.3: Comparisons of Natural Enemy Pressures among Conifer – Bark Beetle systems

Bark Beetle	Beetles/Predator	Beetles/Competitor	Ref.
D. rufipennis			
Eruptive (South-Central AK)	17.1	0.5	[81]
Endemic (Interior AK)	2.8	0.24	[81]
Endemic (great Lakes)	0.5	0.0005	[75]
I. pini	0.4	0.5	[82]
I. grandicollis	0.9	0.6	[27,71,133]

facultative predators. Interestingly, predation and especially competition appear much higher in interior Alaska, where outbreaks are rare despite the temperatures being warmer than the outbreak-prone south.[64] Likewise, predation and especially competition are extremely high in spruce forests of the Great Lakes region, again, where outbreaks are rare.[70,75]

We propose that there is feedback between the host selection behavior of bark beetles, the spatial and temporal pattern of predisposing stress agents, and the impacts of natural enemies.[79] That is, when beetles track a highly predictable predisposing agent, such as root colonizers (Figs. 4.12, 4.13), they are likewise highly predictable to natural enemies (Table 4.3). Similarly, when beetles rely on trees that are severely stressed (Fig. 4.11), it is difficult to escape competitors that can also acquire this resource. These conditions make it less likely that populations will move from K to K_1 (Fig. 4.5). However, numbers alone may not explain these dynamics, as *I. grandicollis* populations were extremely high following the budworm outbreak (Fig. 4.11). Rather, the X-axis in Fig. 4.5 should incorporate a spatial component to distinguish population increases above K_1 that result from regional versus more localized events. Secondly, there is likely an optimal window of host physiological stress for primary bark beetles, with healthy trees being too well defended, and severely stressed and dead trees being available to competitors. The breadth of this physiological window helps define the distance between K and K_1 in Fig. 4.5. This might explain, for example, why large-scale windthrows often do not result in sustained outbreaks by bark beetles.

We have incorporated this hypothesis into Fig. 4.14, which is an expanded tritrophic version of Fig. 4.9. The top of this diagram now includes feedback between tunneling, which elicits oxygenated terpene and monoterpene emission and resulting arrival by predators and competitors, and impacts of these predators and competitors on beetle reproduction. Both of these relationships are well supported in the literature.[4,6,39,51,83,84] However, this model also proposes that the spatial and temporal patterns of host availability, in terms of both compromised tree defenses and beetle perception of these alterations, affect natural enemy numbers. Natural enemy populations in turn affect beetle population size, which in turn affects host acceptance behavior (Fig. 4.14 center).

This hypothesis is unvalidated at present, because these case studies employed different methods, are not replicated across genera, variability in natural enemies numbers is complex, and we cannot adequately separate cause and effect. Also, ratios of predators to prey are highly plastic within systems,[73,85,86] and predation and competition are not independent owing to dilution effects.[83,87] We also lack information on its applicability to other systems. Predaceous checkered beetles cause greater proportionate mortality to mountain pine beetle, *Dendroctonus ponderosae* Hopkins during endemic than eruptive conditions,[88] which is consistent with our model, but not validating without information on the pre-attack chemistry of killed trees. Likewise, predation of southern pine beetle, *Dendroctonus frontalis* Zimmermann, can be extremely high.[85,86,89] This eruptive insect relies greatly on lightning-struck trees during its endemic stage, and this resource tends to be poorly defended, heavily colonized by competing species, temporally and spatially clustered, and chemically apparent to natural enemies due to terpene emission.[4,76] Fig. 4.14 allows for key natural enemy species to vary from system to system. For example, the fungus *Ophiostoma minus* inhibits *D. frontalis* development by competing with its mutualistic fungi, an effect intensified by phoretic mites.[14] The effects of the constitutive and inducible host compounds in Table 4.2 on these mites are unknown. Despite these constraints, the combination of validated individual components of Fig. 4.14 and *post hoc* comparisons in Table 4.3 identify a particular need for chemically-informed, spatially explicit studies on interactions among natural enemies, population dynamics, and predisposing agents.

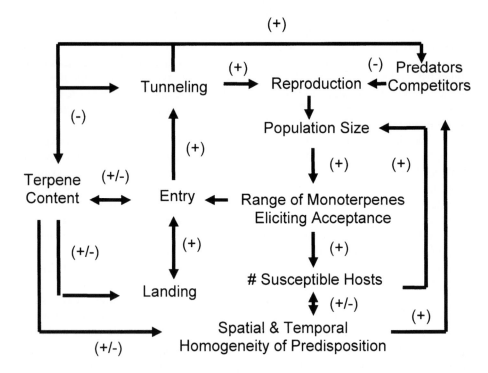

Fig. 4.14: Terpene - mediated links in population process across three trophic levels. The same feedbacks among terpenes, bark beetle behavior, tree defense, and population dynamics in Figs. 4.3 and 4.9 are present. This figure also includes attraction of predators and competitors to tunneling beetles emitting oxygenated terpene (and other) pheromones (Tunneling → Predators, Competitors) and host monoterpenes (Terpene Content → Predators, Competitors), and the impacts of these natural enemies on bark beetle reproduction (Predators, Competitors → (-) Reproduction). It also proposes that the spatial and temporal homogeneity of resource availability to the herbivore (lower)., and the herbivore's behavior responses to tree physiology (Spatial & Temporal Homogeneity of Predisposition → #Susceptible Hosts), structure the magnitude of impacts by natural enemies (Spatial & Temporal Homogeneity of Predisposition → Predators, Competitors).

Table 4.4: Thresholds in Conifer – Bark Beetle – Fungal Interactions

Discrete Threshold	Continuous Variables Affecting Whether Threshold is Surpassed
Host Entry	Concentration & composition of host terpenes & phenolics Beetle age, Prior trials, Lipid content Beetle density, genotype, population phase
Aggregation	Resin flow, Monoterpenes Beetle density, Terpenoid pheromones
Establishment	Constitutive & induced terpenes & phenolics Beetle density
Eruptive Phase	Behavioral responses to monoterpenes Predators and competitors exploiting terpenes to locate prey Spatial & temporal distributions of agents that compromise terpene – based defenses

HOW TO LINK THE SCALES?

To link these various scales, it is necessary to recognize both that each level of conifer - bark beetle - fungal interaction is characterized by a discrete threshold, and that the outcome at each level depends on feedback among multiple variables (Table 4.4). For example, a beetle can either enter or not enter a tree. However, that discrete outcome is determined by monoterpene and phenolic concentrations and composition, beetle age, the number of rejections already made by a beetle, beetle lipid content, beetle density on the plant surface, beetle genotype, beetle population phase, and presumably other factors. Similar relationships characterize thresholds at the levels of aggregation, host establishment, and population eruption (Table 4.4).

The presence of multiple discrete thresholds governed by complex interactions among continuous variables poses a significant challenge to attempts to link various levels of scale. Landscape approaches that emphasize the detection of "signatures" to reveal mechanisms have proven quite powerful in some systems. However, the very nature of a threshold is that its "signature" is erased as soon as it is surpassed. Likewise, mechanist approaches are powerful at characterizing one level of scale, but linking across levels is especially challenging when the system is dominated by thresholds, which introduce nonlinear dynamics. Thus, a second major lesson of this well-studied bark beetle - conifer - microbial model (Table 4.1) is that integrated approaches incorporating both landscape- and mechanistic-methodologies are needed.

SUMMARY

Transferring information from specific components of a plant - herbivore interaction to population and landscape- level impacts poses a major challenge to ecologists. Bark beetle- conifer - microbial interactions comprise a valuable model for addressing this issue, because host plant compounds are known to affect multiple components of these relationships. In particular, terpenoids play important roles in host acceptance, beetle aggregation, host defense, establishment of microbial symbionts, exposure to and avoidance of predators, and other functions. Some bark beetle species undergo dramatic population eruptions in which they convert from relatively stable to outbreak dynamics. These eruptions both play major roles in ecosystem processes, and pose significant economic and natural resource management challenges. A wealth of information has been developed for each individual component of bark beetle-fungal-conifer interactions. However, we have limited ability to scale across multiple layers of biological organization, which is essential for an integrated understanding of the system and for judicious management decisions. We propose that focusing on one group of compounds that plays an important role at each stage of colonization, and whose effects are density - dependent, can provide a useful approach to achieving integration. We also identify biological thresholds, whose outcomes are qualitative but whose determinant inputs are quantitative, as a major challenge to both mechanistic and landscape approaches, and which need to be addressed in an integrated fashion. Based on these analyses, it appears that linkages among plant defense physiology, individual host acceptance decisions, cooperative behavior, and beetle density can constrain or generate eruptions in a fashion consistent with bimodal equilibria theory, including Allee effects. Moreover, chemically mediated interactions with predators and competitors can constrain these eruptions, but their ability to do so may be linked to the spatial and temporal distribution of agents compromising tree defenses, which in turn both reflects and contributes to host selection behavior. A narrow set of host, climatic, and natural enemy conditions, and distribution patterns of each, is needed to release populations to eruptive levels.

Our specific conclusions are: 1) Individual compounds can affect interactions across multiple levels of scale, from molecular through landscape; 2) At each level of scale, the same compounds can be sources of both positive and negative feedback. Their interactions across scales can be amplified or buffered, depending on these feedback processes; 3) Host selection behavior can be an important link between physiological and population processes, particularly where responses to phytochemicals are plastic; 4) Tritrophic interactions mediated by chemical cues can be either important or ineffective constraints on eruptive behavior, depending on how prey are spatially and temporally distributed, which in turn reflects their host selection behavior; 5) Each level of scale is characterized by thresholds, whose qualitative outcome is determined by quantitative factors.

Based on these conclusions, we identify two areas in particular need of future research: 1) Chemically informed, spatially explicit studies on interactions among natural enemies, population dynamics, and predisposing agents that affect host tree chemistry and physiology, can improve both our understanding of linkages across multiple trophic levels, and how single chemical groups function at multiple levels of scale; 2) Integrated studies incorporating landscape and mechanistic approaches are needed to bridge our understanding of pattern and process.

ACKNOWLEDGMENTS

We appreciate support from NSF DEB-9629776, DEB-0314215 & DEB-0080609, Natural Resources Canada: Can. FS MPB Initiative, USDA-NRI WIS04746, McIntire-Stennis, and Univ. WI CALS. Many of these experiments would have been impossible without extensive help from the USDA FS and WI DNR. Statistical (Erik Nordheim, Murray Clayton), genetic (Jack Rutledge), and mycological (Barbara Illman, Eugene Smalley) assistance by our colleagues at UW-Madison is greatly appreciated. Much of this work was performed by undergraduates, who provided enormous energy and talent. Special thanks from KFR to two Wash. St. Univ. mentors: Alan Berryman who inspired the quest for theoretical interpretation, and Rodney Croteau for his introduction to the significance and analysis of terpenes.

REFERENCES

1. HATCHER, P.E., MOORE, J., TAYLOR, J.E., TINNEY, G.W., PAUL, N.D., Phytohormones and plant-herbivore-pathogen interactions: Integrating the molecular with the ecological, *Ecology*, 2004, **85**, 59-69 .

2. WHITHAM, T.G., YOUNG, W.P., MARTINSEN, G.D., GEHRING, C.A., SCHWEITZER, J.A., SHUSTER, S.M., WIMP, G.M., FISCHER, D.G., BAILEY, J.K., LINDROTH, R.L., WOOLBRIGHT, S., KUSKE, C.R., Community and ecosystem genetics: A consequence of the extended phenotype, *Ecology*, 2003, **84**, 559-573.

3. SAFRANYIK, L., SHRIMPTON, D.M., WHITNEY, H.S., An interpretation of the interaction between lodgepole pine, the mountain pine beetle and its associated blue stain fungi in Western Canada, *in*: Management of Lodgepole Pine Ecosystems Symposium Proceedings (D.M. Baumgartner, ed,), Washington State University Cooperative Extension Service, Pullman, Washington. 1975, pp. 406-428.

4. COULSON, R.N., Population dynamics of bark beetles., *Annu. Rev. Entomol.*, 1979, **24**, 417-447.

5. ROMME, W.H., KNIGHT, D.H., YAVITT, J.B., Mountain pine beetle outbreaks in the Rocky Mountains - Regulators of primary productivity, *Am. Nat.*, 1986, **127**,

484-494.

6. WOOD, D.L., The role of pheromones, kairomones, and allomones in the host selection and colonization behavior of bark beetles, *Annu. Rev. Entomol.*, 1982, **27**, 411-446.

7. KLEPZIG, K.D. , SIX, D.L., Bark beetle fungal symbiosis: Context dependency in complex interactions, *Symbiosis*, 2004 , **37**, 189-206.

8. PAINE, T.D., RAFFA, K.F., HARRINGTON, T.C., Interactions among scolytid bark beetles, their associated fungi, and live host conifers, *Annu. Rev. Entomol.*, 1997, **42**, 179-206.

9. SIX, D.L., PAINE, T.D., Effects of mycangial fungi and host tree species on progeny survival and emergence of *Dendroctonus ponderosae* (Coleoptera: Scolytidae), *Environ. Entomol.*, 1998, **27**, 1393-1401.

10. AYRES, M.P., WILKENS, R.T., RUEL, J.J., LOMBARDERO, M.J., VALLERY, E., Nitrogen budgets of phloem-feeding bark beetles with and without symbiotic fungi, *Ecology*, 2000, **81**, 2198-2210.

11. CHRISTIANSEN, E., *Ips/Ceratocystis* infection of Norway spruce: What is a deadly dosage, *Z. ang. Ent.*, 1985, **99**, 6-11.

12. BRAND, J.M., BRACKE, J.W., BRITTON, L.N., MARKOVETZ, A.J., BARRAS, S.J., Bark beetle pheromones: production of verbenone by a mycangial fungus of *Dendroctonus frontalis*, *J. Chem. Ecol.*, 1976 , **2**, 195-199.

13. BARRAS, S.J. , Antagonism between *Dendroctonus frontalis* and the fungus *Ceratocystis minor*, *Ann. Entomol. Soc. Amer.*, 1970, **63**, 1187-1190.

14. LOMBARDERO, M.J., AYRES, M.P., HOFSTETTER, R.W., MOSER, J.C., LEPZIG, K.D., Strong indirect interactions of *Tarsonemus* mites (Acarina: Tarsonemidae) and *Dendroctonus frontalis* (Coleoptera: Scolytidae), *Oikos*, 2003, **102**, 243-252.

15. KOPPER, B.J., KLEPZIG, K.D., RAFFA, K.F., Components of antagonism and mutualism in *Ips pini*-fungal interactions: relationship to a life history of colonizing highly stressed and dead trees, *Environ. Entomol.*, 2004, **33**, 28-34.

16. ECKHARDT, L.G., GOYER, M.A., KLEPZIG, K.D., JONES, J.P. , Interactions of *Hylastes* species (Coleoptera: Scolytidae) with *Leptographium* species associated with loblolly pine decline, *J. Econ. Entomol.*, 2004, **97**, 468-474.

17. RAFFA, K.F., BERRYMAN, A.A., The role of host plant resistance in the colonization behavior and ecology of bark beetles (Coleoptera: Scolytidae), *Ecol. Mongr.*, 1983, **53**, 27-49.

18. LORIO, JR.P.L., HODGES, J.D., Tree water status affects induced southern pine beetle attack and brood production, *USDA FS South. For. Exper. Stn. Res. Pap. SO-135*, 1977.

19. NEBEKER, T.E., HODGES, J.D., BLANCHE, C.A., HONEA, C.R., TISDALE, R.A., Variation in the constitutive defensive system of loblolly pine in relation to bark beetle attack, *For. Sci.*, 1992 , **38**, 457-466.

20. RAFFA, K.F., BERRYMAN, A.A., Physiological differences between lodgepole pines resistant and susceptible to the mountain pine beetle (Coleoptera, Scolytidae) and associated microorganisms, *Environ. Entomol.*, 1982, **11**, 486-492.

21. BOHLMANN, J., GERSHENZON, J., AUBOURG, S., Biochemical, molecular

genetic, and evolutionary aspects of defense-related terpenoids in conifers., *Rec. Adv. Phytochem.*, 2000, **34**, 109-149.

22. MARTIN, D., THOLL, D., GERSHENZON, J., BOHLMANN, J., Methyl jasmonate induces traumatic resin ducts, terpenoid resin biosynthesis, and terpenoid accumulation in developing xylem of Norway spruce stems, *Plant Physiol.*, 2002, **129**, 1003-1018.

23. SEYBOLD, S.J., BOHLMANN, J., RAFFA, K.F., Biosynthesis of coniferophagous bark beetle pheromones and conifer isoprenoids: evolutionary perspective and synthesis, *Can. Entomol.*, 2000, **132**, 697-753.

24. PAINE, T.D., STEPHEN, F.M., Induced defenses of loblolly pine, *Pinus taeda*: a potential impact on *Dendroctonus frontalis* within-tree mortality, *Entomol. Exper. Applic.*, 1988, **46**, 39-46.

25. FRANCESCHI, V.R., KROKENE P, KREKLING T., CHRISTIANSEN E., Phloem parenchyma cells are involved in local and distant defense response to fungal inoculation or bark-beetle attack in Norway Spruce (Pinaceae), *Am. J. Bot.*, 2000, **87**, 314-326.

26. LIEUTIER, F., BERRYMAN, A.A., Preliminary histological investigations of the defense reactions of three pine to *Ceratocystis clavigera* and two chemical elicitors, *Can. J. For. Res.*, 1988, **18**, 1243-1247.

27. WALLIN, K.F., RAFFA, K.F., Effects of folivory on subcortical plant defenses: can defense theories predict interguild processes?, *Ecology*, 2001, **82**, 1387-1400.

28. BOHLMANN, J., MEYER-GAUEN, G., CROTEAU, R., Plant terpenoid synthases: Molecular biology and phylogenetic analysis, *Proc. Natl. Acad. Sci. USA*, 1998, **95**, 4126-4133.

29. MARTIN, D., BOHLMANN, J., GERSHENZON, J., FRANKE, W., SEYBOLD, S., A novel sex-specific and inducible monoterpene synthase activity associated with a pine bark beetle, the pine engraver, *Ips pini, Naturwissenschaften*, 2003, **90**, 173-179.

30. COOK, S.P., HAIN, F.P., Defensive mechanisms of loblolly and shortleaf pine against attack by southern pine beetle, *Dendroctonus frontalis* Zimmerman, and its fungal associate, *Ceratocystis minor* (Hedgecock) Hunt, *J. Chem. Ecol.*, 1986, **12**, 1397-1406.

31. RAFFA, K.F., SMALLEY, E.B., Interaction of pre-attack and induced monoterpene concentrations in host conifer defense against bark beetle-fungal complexes., *Oecologia*, 1995, **102**, 285-295.

32. RAFFA, K.F., BERRYMAN, A.A., Interacting selective pressures in conifer-bark beetle systems: A basis for reciprocal adaptations?, *The Am. Nat.*, 1987, **129**, 234-262.

33. COYNE, J.F., LOTT, L.H., Toxicity of substances in pine oleoresin to southern pine beetles, *J. Georg. Entomol. Soc.*, 1961, **11**, 301-305.

34. SMITH, R.H., Toxicity of pine resin vapors to three species of *Dendroctonus* bark beetles, *J. Econ. Entomol.*, 1963, **56**, 827-831.

35. COOK, S.P., HAIN, F.P., Toxicity of host monoterpenes to *Dendroctonus frontalis* and *Ips calligraphus* (Coleoptera: Scolytidae), *J. Entomol. Sci.*, 1988, **23**, 287-292.

36. RAFFA, K.F., BERRYMAN, A.A., Physiological aspects of lodgepole pine wound

responses to a fungal symbiont of the mountain pine beetle *Dendroctonus ponderosae* (Coleoptera: Scolytidae), *Can. Entomol.*, 1983, **115**, 723-734.

37. COBB, F.W., KRSTIC, JR.M., ZAVARIN, E., BARBER, JR.H.W., Inhibitory effects of volatile oleoresin components on *Fomes annosus* and four *Ceratocystis* species, *Phytopathlogy*, 1968, **58**, 1327-1335.

38. KLEPZIG, K.D., SMALLEY, E.B., RAFFA, K.F., Combined chemical defenses against an insect-fungal complex, *J. Chem. Ecol.*, 1996, **22**, 1367-1388.

39. BORDEN, J.H., Aggregation pheromones, *Comprehensive Insect Physiol., Biochem., & Pharmacol.*, 1985, 257-285.

40. BYERS, J.A., Chemical ecology of bark beetles, *Experientia*, 1989, **45**, 271-283.

41. SEYBOLD, S.J., TITTIGER, C., Biochemistry and molecular biology of *de novo* isoprenoid pheromone production in the Scolytidae, *Annu. Rev. Entomol.*, 2003, **48**, 425-453.

42. MILLER, D.R., BORDEN, J.H., The use of monoterpenes as kairomones by *Ips latidens* (Leconte) (Coleoptera: Scolytidae), *Can. Entomol.*, 1990, **122**, 301-307.

43. MILLER D.R., BORDEN J.H., beta-Phellandrene: kairomone for pine engraver, *Ips pini* (Say) (Coleoptera: Scolytidae)., *J. Chem. Ecol.*, 1990, **16**, 2519-2531.

44. GEISZLER, D.R., GALLUCCI, V.F., GARA, R.I., Modeling the dynamics of mountain pine beetle aggregation in a lodgepole pine stand, *Oecologia*, 1980, **46**, 244-253.

45. GEISZLER, D.R., GARA, R.I., Mountain pine beetle attack dynamics in lodgepole pine, Forest, Wildlife and Range Experiment Station, University of Idaho & USDA Forest Service, Forest Insect and Disease Research/Theory and Practice of Mountain Pine Beetle Management in Lodgepole Pine Forests Symposium, pp. 182-187.

46. WRIGHT, L.C., BERRYMAN, A.A., GURUSIDDAIAH, S., Host resistance to the fir engraver beetle, *Scolytus ventralis* (Coleoptera: Scolytidae): 4. Effect of defoliation on wound monoterpene and inner bark carbohydrate concentrations, *Can. Entomol.*, 1979, **111**, 1255-1262.

47. KLEPZIG, K.D., KRUGER, E.L., SMALLEY, E.B., RAFFA, K.F., Effects of biotic and abiotic stress on induced accumulation of terpenes and phenolics in red pines inoculated with bark beetle-vectored fungus., *J. Chem. Ecol.*, 1995, **21**, 601-626.

48. WALLIN, K.F., RAFFA, K.F., Influences of host chemicals and internal physiology on the multiple steps of postlanding host acceptance behavior of *Ips pini* (Coleoptera: Scolytidae), *Environ. Entomol.*, 2000, **29**, 442-453.

49. BERRYMAN, A.A., Dynamics of bark beetle populations: towards a general productivity model, *Environ. Entomol.*, 1974, **3**, 579-585.

50. BERRYMAN, A.A., Theoretical explanation of mountain pine beetle dynamics in lodgepole pine forests, *Environ. Entomol.*, 1976, **5**, 1225-1233.

51. STEPHEN, F.M., TAHA, H.A., Area-wide estimation of southern pine beetle populations, *Environ. Entomol.*, 1979, **8**, 850-855.

52. AMMAN, G.D., The role of the mountain pine beetle in lodgepole pine ecosystems: Impact on succession, *in*: The Role of Arthropods in Forest Ecosystems (W.J. Mattson, ed,), Springer-Verlag. 1977, pp. 3-18.

53. SCHOWALTER, T.D., COULSON, R.N., CROSSLEY, D.A., Role of southern pine beetle *Dendroctonus frontalis* Zimmermann (Coleoptera, Scolytidae) and fire

in maintenance of structure and function of the southeastern coniferous forest, *Environ. Entomol.*, 1981, **10**, 821-825.

54. SCHOWALTER, T.D., POPE, D.N., COULSON, R.N., FARGO, W.S., Patterns of southern pine beetle (*Dendroctonus frontalis* Zimm.): Infestation Enlargement, *For. Sci.*, 1981, **27**, 837-849.

55. BERRYMAN, A.A., Towards a theory of Insect epidemiology, *Res. Pop. Ecology*, 1978, **19**, 181-196.

56. CARROLL, A.L., TAYLOR, S.W., RÉGNIÈRE, J., SAFRANYIK, L., Effects of climate change on range expansion by the mountain pine beetle in British Columbia, *in*: Challenges and Solutions: Proceedings of the Mountain Pine Beetle Symposium. Kelowna, British Columbia, Canada October 30-31, 2003. Information Report BC-X-399 (T.L. Shore, J.E. Brooks, and J.E. Stone, eds,), Canadian Forest Service, Pacific Forestry Centre, Victoria, British Columbia, Canada. 2004, p. 221-230.

57. CAMPBELL, R.W., SLOAN, R.J., Release of Gypsy Moth populations from innocuous levels, *Environ. Entomol.*, 1977, **6**, 323-329.

58. MAWBY, W.D., HAIN, F.P., DOGGETT, C.A., Endemic and epidemic populations of southern pine beetle: Implications of the two-phase model for forest managers, *For. Sci.*, 1989, **35**, 1075-1087.

59. LARSSON, S., EKBOM, B., BJORKMAN, C., Influence of plant quality on pine sawfly population dynamics, *Oikos*, 2000, **89**, 440-450.

60. SOUTHWOOD, T.R.E., Habitat, The template for ecological strategies?, *J. Anim. Ecol.*, 1977, **46**, 337-365.

61. BELOVSKY, G.E., BOTKIN, D.B., CROWL, T.A., CUMMINS, K.W., FRANKLIN, J.F., HUNTER, M.L., JOERN, A., LINDENMAYER, D.B., MACMAHON, J.A., MARGULES, C.R., SCOTT, J.M., Ten suggestions to strengthen the science of ecology, *Bioscience*, 2004, **54**, 345-351.

62. BERRYMAN, A.A., ARCE, M.L., HAWKINS, B.A., Population regulation, emergent properties, and a requiem for density dependence, *Oikos*, 2002, **99**, 600-606.

63. FURNISS, R.L., CAROLIN, V.M., Western Forest Insects. USDA FS Misc. Pub. No. 1339. US Government Printing Office, Washington, D.C., 1977, p.

64. WERNER, R.A., HOLSTEN, E.H., Factors influencing generation times of spruce beetles in Alaska, *Can. J. For. Res.*, 1985, **15**, 438-443.

65. WERNER, R.A., RAFFA, K.F., ILLMAN, B.L., Insect and pathogen dynamics, *in*: Alaska's Changing Boreal Forest (F.S. Chapin III, M. Oswood, K. Van Cleve, L.A. Viereck, and D. Verbyla, eds,), Oxford University Press, Oxford. 2005, p. in press .

66. WALLIN, K.F. , RAFFA, K.F., Feedback between individual host selection behavior and population dynamics in an eruptive herbivore, *Ecol. Mongr.*, 2004, **74**, 101-116.

67. WALLIN, K.F. , RUTLEDGE, J., RAFFA, K.F., Heritability of host acceptance and gallery construction behaviors of the bark beetle *Ips pini* (Coleoptera: Scolytidae), *Environ. Entomol.*, 2002, **31**, 1276-1281.

68. LOGAN, J.A., POWELL, J.A., Ghost forests, global warming, and the mountain pine beetle (Coleoptera: Scolytidae), *Am. Entomol.*, 2001, **47**, 160-173.

69. LOGAN, J.A., REGNIERE, J., POWELL, J.A., Assessing the impacts of global

warming on forest pest dynamics, *Frontiers Ecol. Environ.*, 2003, **1**, 130-137.

70. DROOZ, A.T., Insects of Eastern Forests, *USDA FS Misc. Pub. No. 1426*, 1985, 608 pp.

71. ERBILGIN, N., RAFFA, K.F., Effects of host tree species on attractiveness of tunneling pine engravers, *Ips pini*, to conspecifics and insect predators, *J. Chem. Ecol.*, 2000, **26**, 823-840.

72. ERBILGIN, N., RAFFA, K.F., Association of declining red pine stands with reduced populations of bark beetle predators, seasonal increases in root colonizing insects, and incidence of root pathogens, *For. Ecol. Manage.*, 2002, **164**, 221-236.

73. ERBILGIN, N., NORDHEIM, E.V., AUKEMA, B.H., RAFFA, K.F., Population dynamics of *Ips pini* and *Ips grandicollis* in red pine plantations in Wisconsin: Within- and between-year associations with predators, competitors, and habitat quality, *Environ. Entomol.*, 2002, **31**, 1043-1051.

74. KLEPZIG, K.D., RAFFA, K.F., SMALLEY, E.B., Association of insect-fungal complexes with Red Pine Decline in Wisconsin., *For. Sci.*, 1991, **41**, 1119-1139.

75. HABERKERN, K.E., ILLMAN, B.L., RAFFA, K.F., Bark beetles and fungal associates colonizing white spruce in the Great Lakes region, *Can. J. For. Res.*, 2002, **32**, 1137-1150.

76. COULSON, R.N., MCFADDEN, B.A., PULLEY, P.E., LOVELADY, C.N., FITZGERALD, J.W., JACK, S.B., Heterogeneity of forest landscapes and the distribution and abundance of the southern pine beetle, *For. Ecol. Manage*, 1999, **114** , 471-485.

77. MIZELL, R.F. , FRAZIER, J.L., NEBEKER, T.E., Response of the clerid predator *Thanasimus dubius* (F.) to bark beetle pheromones and tree volatiles in a wind tunnel, *J. Chem. Ecol.*, 1984, **10**, 177-187.

78. GREGOIRE, J.C., BAISIER, M., DRUMONT, A., DAHLSTEN, D.L., MEYER, H., FRANCKE, W., Volatile compounds in the larval frass of *Dendroctonus valens* and *Dendroctonus micans* (Coleoptera, Scolytidae) in relation to oviposition by the predator, *Rhizophagus grandis* (Coleoptera, Rhizophagidae), *J. Chem. Ecol.*, 1991, **17**, 2003-2019.

79. RAFFA, K.F., Mixed messages across multiple trophic levels: the ecology of bark beetle chemical communication systems, *Chemoecology*, 2001, **11**, 49-65.

80. AUKEMA, B.H. , DAHLSTEN, D.L., RAFFA, K.F., Improved population monitoring of bark beetles and predators by incorporating disparate behavioral responses to semiochemicals, *Environ. Entomol.*, 2000, **29**, 618-629.

81. GARA, R.I., WERNER, R.A., WHITMORE, M.C., HOLSTEN, E.H., Arthropod associates of the spruce beetle *Dendroctonus rufipennis* (Kirby) (Col., Scolytidae) in spruce stands of south-central and interior Alaska, *J. Appl. Entomol.*, 1995, **119**, 585-590.

82. AUKEMA, B.H. , RICHARDS, G.R., KRAUTH, S.J., RAFFA, K.F., Species assemblage arriving at and emerging from trees colonized by *Ips pini* in the Great Lakes region: Partitioning by time since colonization, season, and host species, *Ann. Entomol. Soc. Am*, 2004, **97**, 117-129.

83. AUKEMA, B.H., RAFFA, K.F., Relative effects of exophytic predation, endophytic predation, and intraspecific competition on a subcortical herbivore: Consequences to

the reproduction of *Ips pini* and *Thanasimus dubius*, *Oecologia*, 2002, **133**, 483-491.

84. AUKEMA, B.H., DAHLSTEN, D.L., RAFFA, K.F., Exploiting behavioral disparities among predators and prey to selectively remove pests: maximizing the ratio of bark beetles to predators removed during semiochemically based trap-out, *Environ. Entomol.*, 2000, **29**, 651-660.

85. TURCHIN, P., LORIO, P.L.JR., TAYLOR, A.D., BILLINGS, R.F., Why do populations of southern pine beetles (Coleoptera: Scolytidae) fluctuate?, *Environ. Entomol.*, 1991, **20**, 401-409.

86. TURCHIN, P., TAYLOR, A.D., REEVE, J.D., Dynamical role of predators in population cycles of a forest insect: an experimental test., *Science*, 1999, **285**, 1068-1070.

87. AUKEMA, B.H. , RAFFA, K.F., Does aggregation benefit bark beetles by diluting predation?   Links between a group-colonization strategy and the absence of emergent multiple predator effects., *Ecol. Entomol.*, 2004, **29**, 129-138.

88. AMMAN, G.D., Mountain pine beetle (Coleoptera: Scolytidae) mortality in three types of infestations., *Environ. Entomol.*, 1984, **13**, 184-191.

89. THATCHER, R.C., SEARCY, J.L., COSTER, J.E., HERTEL, G.D., The Southern Pine Beetle. USDA FS Science and Education Administration Tech. Bull. 1631. USDA, Washington, DC, 1981, 267 pp..

90. SEYBOLD, S.J., QUILICI, D.R., TILLMAN, J.A., VANDERWEL, D., WOOD, D.L., BLOMQUIST, G.J., *De novo* biosynthesis of the aggregation pheromone components ipsenol and ipsdienol by the pine bark beetles *Ips paraconfusus* Lanier and *Ips pini* (Say) (Coleoptera: Scolytidae), *Proc. Natl. Acad. Sci. USA* , 1995 , **92**, 8393-8397.

91. STAUFFER, C. , A molecular method for differentiating sibling species within the genus *Ips*, *Proc: Integrating cultural tactics into the management of bark beetle and reforestation pests. USDA FS Gen. Tech. Rept. NE-236*, 1997, 87-91.

92. COGNATO, A.I., SEYBOLD, S.J., SPERLING, F.A.H., Incomplete barriers to mitochondrial gene flow between pheromone races of the North American pine engraver, *Ips pini* (Say) (Coleoptera: Scolytidae), *Proc. Roy. Soc.f London – Ser. B: Biol. Sci.*, 1999, **266**, 1843-1850.

93. COGNATO, A.I., HARLIN, A.D., FISHER, M.L., Genetic structure among pinyon pine beetle populations (Scolytinae: *Ips confusus*), *Environ. Entomol.*, 2003, **32**, 1262-1270.

94. ZUNIGA, G., CISNEROS, R., HAYES, J.L., MACIAS-SAMANO, J., Karyology, geographic distribution, and origin of the genus *Dendroctonus* Erichson (Coleoptera : Scolytidae), *Ann. Entomol. Soc. Am*, 2002, **95**, 267-275.

95. SIX, D.L., PAINE, T.D., Allozyme diversity and gene flow in *Ophiostoma clavigerum* (Ophiostomatales: Ophiostomataceae), the mycangial fungus of the jeffrey pine beetle, *Dendroctonus jeffreyi* (Coleoptera: Scolytidae), *Can. J. For. Res.*, 1999, **29**, 324-331.

96. GORTON, C., WEBBER, J.F., Reevaluation of the status of the bluestain fungus and bark beetle associate *Ophiostoma minus*, *Mycologia*, 2000, **92**, 1071-1079.

97. SIX, D.L., HARRINGTON, T.C., STEIMEL, J., MCNEW, D., PAINE, T.D., Genetic relationships among *Leptographium terebrantis* and the mycangial fungi of

116   *RAFFA, et al.*

three western Dendroctonus bark beetles, *Mycologia*, 2003, **95**, 781-792.

98. VARGAS, C.C. , LOPEZ, A., SANCHEZ, H., RODRIGUEZ, B., Allozyme analysis of host selection by bark beetles in central Mexico, *Can. J. For. Res.*, 2002, **32**, 24-30.

99. ROBERDS, J.H., STROM, B.L., HAIN, F.P., GWAZE, D.P., MCKEAND, S.E., LOTT, L.H., Estimates of genetic parameters for oleoresin and growth traits in juvenile loblolly pine, *Can. J. For. Res.*, 2003, **33**, 2469-2476.

100. HALL, G.M., TITTIGER, C., ANDREWS, G.L., MASTICK, G.S., KUENZLI, M., LUO, X., SEYBOLD, S.J., BLOMQUIST, G.J., Midgut tissue of male pine engraver, *Ips pini*, synthesizes monoterpenoid pheromone component ipsdienol *de novo*, *Naturwissenschaften*, 2002, **89**, 79-83.

101. NARDI, J.B. , YOUNG, A.G., UJHELYI, E., TITTIGER, C., LEHANE, M.J., BLOMQUIST, G.J., Specialization of midgut cells for synthesis of male isoprenoid pheromone components in two scolytid beetles, *Dendroctonus jeffreyi* and *Ips pini*, *Tissue & Cell*, 2002, **34**, 221-231.

102. SILVA-OLIVARES, A., DIAZ, E., SHIBAYAMA, M., TSUTSUMI, V., CISNEROS, R., ZUNIGA, G., Ultrastructural study of the midgut and hindgut in eight species of the genus *Dendroctonus* Erichson (Coleoptera: Scolytidae), *Ann. Entomol. Soc. Am*, 2003, **96**, 883-900.

103. KLEPZIG, K.D., FLORES-OTERO, J., HOFSTETTER, R.W., AYRES, M.P., Effects of available water on growth and competition of southern pine beetle associated fungi, *Mycol. Res.*, 2004, **108**, 183-188.

104. PAYNE, T.L., Bark Beetle olfaction. III. Antennal olfactory responsiveness of *Dendroctonus frontalis* Zimmerman and *D. brevicomis* LeConte (Coleoptera: Scolytidae) to aggregation pheromones and host tree terpene hydrocarbons, *J. Chem. Ecol.*, 1975, **1**, 233-242.

105. MUSTAPARTA, H., ANGST, M.E., LANIER, G.N., Receptor discrimination of enantiomers of the aggregation pheromone ipsdienol, in two species of *Ips.*, *J. Chem. Ecol.*, 1980, **6**, 689-701.

106. MUSTAPARTA, H., TØMMERAS, B.Å., LANIER, G.N., Pheromone receptor cell specificity in interpopulational hybrids of *Ips pini* (Coleoptera: Scolytidae)., *J. Chem. Ecol.*, 1985, **11**, 999-1007.

107. ASCOLICHRISTENSEN, A., SALOM, S.M., PAYNE, T.L., Olfactory receptor cell responses of *Ips grandicollis* (Eichhoff) (Coleoptera: Scolytidae) to intraspecific and interspecific behavioral chemicals, *J. Chem. Ecol.*, 1993, **19**, 699-712.

108. HODGES, J.D., ELAM, W.W., WATSON, W.F., NEBEKER, T.E., Oleoresin characteristics and susceptibility of four southern pines to southern pine beetle (Coleoptera: Scolytidae) attacks, *Can. Entomol.*, 1979, **111**, 889-896.

109. PAINE, T.D., BLANCHE, C.A., NEBEKER, T.E., STEPHEN, F.M., Composition of loblolly pine resin defenses: comparison of monoterpenes from induced lesion and sapwood resin, *Can. J. For. Res.*, 1987, **17**, 1202-1206.

110. LIEUTIER, F., GARCIA, J., ROMARY, P., YART, A., JACTEL, H., SAUVARD, D., Inter-tree variability in the induced defense reaction of Scots pine to single inoculations by *Ophiostoma brunneo-ciliatum*, a bark-beetle-associated fungus, *For. Ecol. Manage.*, 1993, **59**, 257-270.

111. LIEUTIER, F., BRIGNOLAS, F., SAUVARD, D., YART, A., GALET, C., BRUNET, M., VAN DE SYPE, H., Intra- and inter-provenance variability in phloem phenols of *Picea abies* and relationship to a bark beetle-associated fungus, *Tree Physiol.*, 2003, **23**, 247-256.

112. TISDALE, R.A., NEBEKER, T.E., HODGES, J.D., The role of oleoresin flow in the induced response of loblolly pine to a southern pine beetle associated fungus, *Can. J. Bot.*, 2003, **81**, 368-374.

113. KROKENE, P., CHRISTIANSEN, E., SOLHEIM, H., FRANCESCHI, V.R., BERRYMAN, A.A., Induced Resistance to Pathogenic Fungi in Norway Spruce, *Plant Physiol.*, 1999, **121**, 565-569.

114. HUDGINS, J.W., CHRISTIANSEN, E., FRANCESCHI, V.R., Methyl jasmonate induces changes mimicking anatomical defenses in diverse members of the Pinaceae, *Tree Physiol.*, 2003, **23**, 361-371.

115. PAYNE, C., WOODWARD, S., PETTY, J.A., Modification of the growth habit of the softwood disfiguring fungus *Ophiostoma piceae* by monoterpene vapors, *For. Prod. J.*, 2001, **51**, 89-92.

116. ELKINTON, J.S., WOOD, D.L., Feeding and boring behavior of the bark beetle *Ips paraconfusus* (Coleoptera: Scolytidae) on the bark of a host and non-host tree species, *Can. Entomol.*, 1980, **112**, 797-809.

117. BORDEN, J.H., RYKER, L.C., CHONG, L.J., PIERCE, H.D., JOHNSTON, B.D., OEHLSCHLAGER, A.C., Response of the mountain pine beetle, *Dendroctonus ponderosae* Hopkins (Coleoptera: Scolytidae), to five semiochemicals in British Columbia lodgepole pine forests, *Can. J. For. Res.* 1986, **17**, 118-128.

118. WINGFIELD, M.J., SEIFERT, K.A., WEBBER, J.F., *Ceratocystis* and *Ophiostoma*: Taxonomy, Ecology, and Pathogenicity. APS Press, St. Paul, MN, 1993, 293 pp.

119. KROKENE, P., SOLHEIM, H., Fungal associates of five bark beetle species colonizing Norway spruce, *Can. J. For. Res.*, 1996, **26**, 2115-2122.

120. JACOBS, K., WINGFIELD, M.J., *Leptographium* species: Tree Pathogens, Insect Associates, and Agents of Blue Stain. APS Press, St. Paul, MN, 2001, 207 pp.

121. YEARIAN, W.C., GOUGER, R.J., WILKINSON, R.C., Effects of the bluestain fungus, *Ceratocystis ips*, on development of *Ips* bark beetles in pine bolts., *Ann. Entomol. Soc. Am*, 1972, **65**, 481-487.

122. BENTZ, B.J., POWELL, J.A., LOGAN, J.A., Localized spatial and temporal attack dynamics of the mountain pine beetle in lodgepole pine, *USDA FS Intermountain Res. Stn. Res. Pap*, 1996.

123. MITCHELL, R.G., PREISLER, H.K., Analysis of spatial patterns of lodgepole pine attacked by outbreak populations of the mountain pine beetle, *For. Sci.*, 1991, **37**, 1390-1408.

124. LOGAN, J.A., WHITE, P., BENTZ, B.J., POWELL, J.A., Model analysis of spatial patterns in mountain pine beetle outbreaks, *Theor. Popul. Biol.*, 1998, **53**, 236-255.

125. DOAK, P., The impact of tree and stand characteristics on spruce beetle (Coleoptera : Scolytidae) induced mortality of white spruce in the Copper River basin, Alaska, *Can. J. For. Res.*, 2004, **34**, 810-816.

126. ERBILGIN, N., RAFFA, K.F., Spatial analysis of forest gaps resulting from bark

beetle colonization of red pines experiencing belowground herbivory and infection, *For. Ecol. Manage*, 2003, **177**, 145-153.

127. BERRYMAN, A.A., Biological control, thresholds, and pest outbreaks, *Environ. Entomol.*, 1982, **3**, 544-549.

128. WILSON, J.S., ISAAC, E.S., GARA, R.I., Impacts of mountain pine beetle (*Dendroctonus ponderosae*) (Col., Scolytidae) infestation on future landscape susceptibility to the western spruce budworm (*Choristoneura occidentalis*) (Lep., Tortricidae) in north central Washington, *J. Appl. Entomol.*, 1998, **122**, 239-245.

129. KELLEY, S.T., FARRELL, B.D., Is specialization a dead end? The phylogeny of host use in *Dendroctonus* bark beetles (Scolytidae), *Evolution*, 1998, **52**, 1731-1743.

130. STURGEON, K.B., MITTON, J.B., Allozyme and morphological differentiation of mountian pine beetles *Dendroctonus ponderosae* Hopkins associated with host tree, *Evolution*, 1986, **40**, 290-302.

131. KELLEY, S.T., MITTON, J.B., PAINE, T.D., Strong differentiation in mitochondrial dna of *Dendroctonus brevicomis* (Coleoptera: Scolytidae) on different subspecies of ponderosa pine, *Ann. Entomol. Soc. Am*, 1999, **92**, 193-197.

132. SALINAS-MORENO, Y., MENDOZA, M.G., BARRIOS, M.A., CISNEROS, R., MACIAS-SAMANO, J., ZUNIGA, G., Areography of the genus *Dendroctonus* (Coleoptera: Curculionidae: Scolytinae) in Mexico, *J. Biogeo.*, 2004, **31**, 1163-1177.

133. ERBILGIN, N., RAFFA, K.F., Modulation of predator attraction to pheromones of two prey species by stereochemistry of plant volatiles, *Oecologia*, 2001, **127**, 444-453.

Chapter Five

MOLECULAR BIOLOGY AND BIOCHEMISTRY OF INDUCED INSECT DEFENSE IN *POPULUS*

C. Peter Constabel* and Ian T. Major

Centre for Forest Biology and Department of Biology
University of Victoria,
Victoria BC, Canada

*Author for correspondence, email: cpc@uvic.ca

INTRODUCTION

All plants are faced with a wide variety of potential pests and pathogens. Being long-lived and large organisms, trees in particular must be adapted to defend themselves against these threats in order to persist in the environment. Plant defense is thought to be a driving force in the evolution of diverse secondary metabolites, and many phytochemicals with strong anti-herbivore or anti-microbial activities have been identified.[1] The characterization of many anti-herbivore proteins and their corresponding genes by using the tools of biochemistry and molecular biology has contributed to our understanding of plant defense, in particular as many of these genes are upregulated by herbivore stress. Within the last decade, the rise of plant genomics has accelerated the rate of gene discovery dramatically; it now permits the rapid identification of specific defense-related genes, as well as the analysis of genome-wide patterns of gene expression in response to herbivory.

The objective of this review is to provide an overview of our knowledge of the induced anti-herbivore defenses in *Populus*, a model tree for plant molecular biology and genomics. Unlike other model plants, *Populus* are often ecological keystone species in forest ecosystems, and therefore have complex interactions with symbionts, pathogens, and pests. Furthermore, the genus is found throughout the northern hemisphere and has widely distributed species with unique adaptations. These characteristics will make *Populus* a unique system for studies of plant-environment interactions, in particular herbivore defense, to which powerful genomics tools can be applied. In this chapter, recent progress in the area of induced defense of *Populus* will be reviewed with an emphasis on molecular biology approaches, and placed in the context of earlier chemical ecology studies. Most of the studies involving plant defense to date have focused on leaf defenses against folivore pests, but this will likely change in the future as *Populus* as a model system is explored in greater detail.

THE BIOLOGY OF *POPULUS*, THE MODEL TREE

The genus *Populus* includes hardwood trees commonly referred to as poplars, cottonwoods, and aspen, which are found throughout North America, Europe, and northern Asia. They are particularly dominant in the boreal forest and in the parkland zones of Canada and the American midwest.[2] Widespread North American species include *P. balsamifera* (balsam poplar), with a range from Western Alaska to northern Quebec and Newfoundland, and *P. tremuloides* (trembling or quaking aspen), which is found from Alaska to Mexico. Both are prominent species of the boreal forest of Canada. *P. trichocarpa* (black cottonwood) is found along the Pacific coast from California to Alaska. The range of *P. tremula* (European aspen) extends across Europe into northern Asia, and *P. nigra* (black poplar) is also widely distributed and planted.[2]

Ecologically, poplars tend to be early successional species, which is reflected in their rapid growth rates, and is one reason for their widespread use in plantation forestry. Poplars tend to reproduce asexually and form clones by vegetative propagation, an ecological adaptation for competing in favorable environments. This feature also makes them useful trees for forestry and plant biology research, as a particular clone or hybrid with desirable characteristics can be replicated easily. Many species of *Populus* occupy the riparian zones along creeks, rivers, and wet areas, exemplified by the majestic *P. deltoides* (Eastern and plains cottonwoods). Drier upland environments are a habitat for many of the aspens, and some species such as the Asian *P. euphratica* can tolerate extreme drought.[2]

The genus *Populus* has been divided into six taxonomic sections based on ecological and morphological characteristics, of which sections Aigeiros (cottonwoods; *i.e.*, *P. deltoides*), Tacamahaca (balsam poplars; *i.e.*, *P. trichocarpa*), and Populus (aspens and white poplars; *i.e.*, *P. tremuloides*) are most prominent. Eckenwalder (1996)[3] conservatively defines 29 *Populus* species; the exact number is obscured by the propensity of poplars to form interspecific hybrids within taxonomic sections where ranges overlap. The relative ease with which poplars can be crossed has led to breeding programs for rapid-cycling poplars suitable for plantation forestry and shelterbelts. There is a long history of poplar breeding in Europe and North America, and interspecific hybrids between species in the same or in different sections have been generated. Prominent examples of hybrids are *P. deltoides* x *P. nigra* (*P. x canadensis*), *P. trichocarpa* x *P. deltoides* (*P. x generosa*), and *P. nigra* x *P. macimowiczii*.[4] There are also several different aspen hybrids *(P. tremula* x *P. tremuloides, P. tremula* x *P. alba)*. Many of these are used in research programs as well as in plantations. In the north-western United States, *P. trichocarpa* x *deltoides* crosses are well-adapted and produce impressive gains in intensive culture.

More than 300 pest insects have been recorded on poplar in North America, and in Europe the number of pest species is greater than 500.[5,6] Nevertheless, only a few species cause significant damage. The polyphagous forest tent caterpillar (*Malacosoma disstria*) causes dramatic defoliations of large tracts of aspen forest, with regular outbreaks approximately every 10 years in the lake states of the US and the boreal forest and parkland in Canada. While seldom lethal to healthy trees, repeated defoliation leads to reduced growth and wood formation, and under poor conditions leads to increased tree mortality.[5] The large aspen tortrix (*Choristoneura conflictana*) is also a major defoliator of aspen, causing outbreaks in the northern regions of the boreal forest. The cottonwood leaf beetle (*Chrysomela scripta*) causes significant damage to cottonwood and hybrid poplar plantations throughout North America, with the exception of the west coast.[5] These defoliators clearly have the potential for reducing productivity or growth of affected trees; simulated cottonwood leaf beetle defoliation of hybrid poplar resulted in reduction of growth and biomass of up to 33%.[7] In addition to defoliators, there are also many boring insects (*i.e.*, *Saperda calcarata*), which cause significant damage to poplar via secondary pathogen infections or wind breakage. Overall, the number of herbivores attacking *Populus* is high relative to other tree species, which may represent a trade-off with poplar's high growth rates. In general, poplar-insect interactions studied at the molecular level should prove to be a promising area of research.

The ecological importance of *Populus* in the northern hemisphere and the extremely rapid growth of poplar hybrids in plantation forestry has generated wide interest in its use for wood products, biomass, phytoremediation, and carbon sequestration.[8, 9] Furthermore, *Populus* has become the model system for scientists interested in wood formation and tree biology in general.[10] Key characteristics of poplars which make them excellent experimental plants are their rapid growth, ease of vegetative or clonal propagation, and tractability to genetic engineering.[11] Poplar was the first woody plant to be genetically transformed, and a number of poplar and aspen genotypes are readily transformable by *Agrobacterium*.[12,13] In addition, detailed molecular genetic maps for *Populus* have been generated.[14,15] These permit specific regions and loci of the genome to be linked with physiological traits such as *Melampsora* rust resistance.[15, 16]

Extensive molecular and genomic tools, such as expressed sequence tag (EST) libraries and DNA microarrays, have been developed for poplar. Since the publication of the first *Populus* EST project,[17] over 200,000 ESTs from several different *Populus* species have been deposited in public databases (http://www.ncbi.nlm.nih.gov/). Most significantly, the recent elucidation of the complete *P. trichocarpa* genome sequence will have tremendous impact on poplar and other plant biologists. This is the first completed tree genome sequence and one of only three sequenced plant genomes, and will produce a plethora of new information for *Populus*.[9,11] Whereas *Arabidopsis* will continue to be the model plant of choice for many questions of fundamental cell biology and plant physiology, the rich environmental interactions of *Populus* will make this an attractive system for ecological studies. The large number of insect herbivores that utilize *Populus* species and the diversity of phenolic phytochemicals in *Populus* has already stimulated much research on poplar insect defense by chemical ecologists.[18-20] In addition, early molecular biology research on tree defense has been carried out in hybrid poplar,[8,21] further stimulating work on these trees.[22-24] Poplar is thus poised to become a key experimental system for studies of plant-insect interactions, where phytochemical, ecological, and genomic approaches can be integrated.

INDUCED DEFENSE AND ITS EFFECTIVENESS FOR POPLAR PEST RESISTANCE

Plant defense mechanisms can be preformed, or primarily expressed following herbivory, that is, induced. Ecologists first coined the term "induced resistance" for the observation that previous herbivory can result in heightened resistance to subsequent attacks by the same or different herbivores. This phenomenon has been described in dozens of plant species and taxa including conifers, grasses, herbs, and deciduous trees.[25] Interestingly, most examples come from perennial plants, consistent with a predicted evolutionary strategy of investing more heavily in defense compared to annuals. Most biochemical studies on induced defense have been carried out with herbaceous plants, and the defense mechanisms that underlie this resistance typically include anti-nutritive proteins, such as protease inhibitors and oxidative enzymes that reduce the assimilation of essential amino acids or otherwise destroy nutrients.[26-30] In addition, enzymes that lead to the synthesis of toxic or anti-feedant secondary metabolites can be induced, including alkaloids, phenolics, or terpenoids.[31] Recently, the release of volatiles during induced defense has been the focus of intense investigation.[32] These can act as direct defenses to discourage oviposition by the herbivore,[33] and as an indirect defense by attracting predators and parasitoid insects.[34-36]

The overall importance of the induced defense response in protecting the plant has been elegantly shown using jasmonate signaling mutants or transgenic plants that cannot induce a defense response; these plants show a hypersusceptible phenotype.[37-39] Such increased susceptibility extends to natural conditions, and in field experiments with transgenic *Nicotiana attenuata*, Kessler *et al.* (2004) were able to show that the induced defense is effective in limiting damage by both adapted and generalist herbivores.[40]

In poplar, the direct effects of induced defenses on insect pests has been extensively investigated by Raffa and coworkers. Using both choice and no-choice insect bioassays with a series of poplar hybrids, Havill and Raffa (1999) found that saplings damaged by previous herbivory were poorer hosts for gypsy moth larvae in subsequent attacks, sustaining up to 71% less damage. These changes were induced not only by live herbivores, but also by mechanical wounding and simulated herbivory.[41] Damage-induced protection of hybrid poplar to subsequent pest attacks was also demonstrated in experiments with forest tent caterpillar and whitemarked tussock moth (*Orygia leucostigma*).[42,43] Significantly, the induced resistance is observed not only in the wounded leaf but also in unwounded leaves of the plant. This indicates a systemic, or plant-wide, response, consistent with the systemic induction of defense-related genes (see below). A systemic defense system is well documented in other model plants such as tomato.[26, 44]

Interestingly, the same poplar hybrids can show substantial differences in the degree of induced resistance when different clones are compared. Some clones had dramatic induction of resistance, whereas others showed no effect.[41,42] Such genotype- and hybrid-dependent differences in pest susceptibility have also been documented when trees were not previously exposed to damage or other inducing treatments. In common garden experiments with *P. tremuloides*, large clonal differences in forest tent caterpillar and gypsy moth palatability are also observed.[45] Strong correlations of pest resistance with high levels of phenolic glycosides and, to some extent with condensed tannins, suggest that phytochemical defenses could be key factors for host suitability, at least in preformed defense. By contrast, the mechanisms which underlie induced resistance phenomena are just now beginning to be dissected, and are a major focus for research in poplar defense. These mechanisms appear to have both phytochemical and protein components.

PHYTOCHEMICAL DEFENSE IN *POPULUS*

Phytochemicals, also called secondary metabolites, have been extensively studied in *Populus* from both chemical and ecological perspectives. Like other Salicaceae, the genus *Populus* is characterized by a diversity of phenolics; in particular, the salicin-based phenolic glycosides are found in leaves and bark of all species examined.[18] These phytochemicals consist of glucosides of salicyl alcohol, which are generally further esterified and benzoylated.[46] Tremulacin and salicortin are particularly common in *P. tremuloides*, but related glycosides are present at varying levels in other species.[18] They are potent herbivore toxins, with negative effects on forest tent caterpillar, tiger swallowtail (*Papilio glauca*), gypsy moth, and large aspen tortrix larvae feeding on *P. tremuloides*.[20] In general, high levels of phenolic glycosides correlate closely with increased larval development times and reduced pupal weights. Since salicortin and tremulacin are unstable and may degrade during herbivore feeding,[46,47] the biologically active chemicals are not definitively known. However, during leaf maceration these glycosides decompose to release 6-hydroxy-2-cyclohexenone, which subsequently gives rise to phenol or catechol.[47,48] Ruuhola *et al.* (2001) found that salicortin is entirely degraded to salicin and catechol after passage through the alkaline gut of a lepidopteran herbivore (*Operophtera brumata*).[49] Work in the authors' laboratory also demonstrated the release of catechol from salicortin at high pH. Catechol is an excellent substrate for polyphenol oxidase, an induced defense enzyme in hybrid poplar and aspen.[22,50] The enzyme-mediated activation of phenolic glycoside-derived products could thus enhance their toxic effects (see below). To date there is little evidence that the synthesis of these phenolics is induced by insect damage. Although Lindroth and Kinney (1998) found a small increase in tremulacin and salicortin following leaf damage,[51] later experiments failed to corroborate this.[52, 53] This inconsistency may be due to environmentally-induced variability in phenolic levels.

A second class of defensive phytochemicals often found in poplar and aspen at substantial levels are the proanthocyanidins, or condensed tannins (CTs) (Fig. 5.1). Unlike the phenolic glycosides that are found exclusively in the Salicaceae, CTs are widespread in the plant kingdom.[54] These flavonoid polymers consist of mostly 4,8-linked flavan-3,4-diols and flavan-3-ols, ranging in size from 1440 to over 4500 Da depending on the species.[55,56]   Flavonoids are derived from the general phenylpropanoid pathway by a series of enzymes beginning with the enzyme

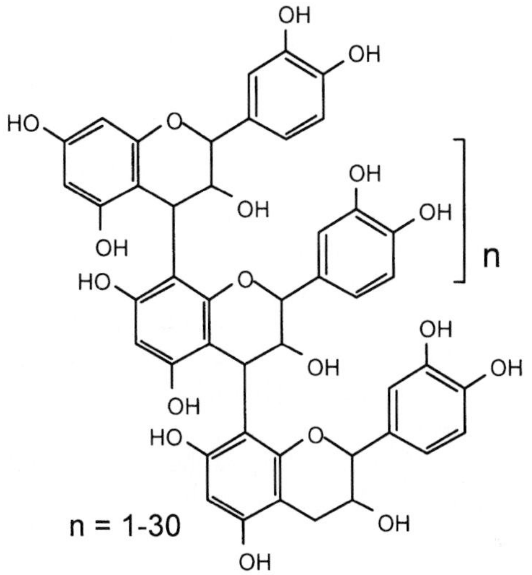

Fig. 5.1: Structure of condensed tannin. *P. tremuloides* condensed tannins are composed of an average of seven flavonoid monomers (Ayres *et al*).[56]

chalcone synthase (Fig. 5.2), and essentially all major flavonoid enzymes have been cloned from *Arabidopsis* or other plants.[57] The final condensation steps of CT biosynthesis are not yet elucidated, however.[55] The availability of heterologous sequence information for genes involved in flavonoid synthesis has facilitated a molecular approach in the analysis of CTs in poplar defense (see below).

CTs can be effective anti-nutrients against insects, especially at high concentrations as can be found in *P. tremuloides* (up to 18 % DW).[20] Tannin effectiveness and mechanism of action has been debated,[58] but it clearly depends both on variations in chemical structure and plant source, as well as on the biochemical conditions in the gut of the particular insect species.[56,59,60]   In *P. tremuloides*, CT levels were shown to be negatively correlated with gypsy moth and forest tent caterpillar larvae performance, although the effects were smaller than with the phenolic glycosides (C. P. Constabel and J. Spence, unpublished data).[45, 61]

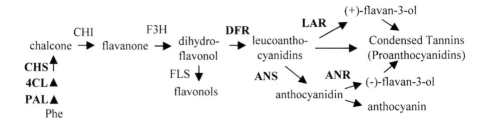

Fig, 5.2. Biosynthesis of flavonoids and proanthocyanidins (condensed tannins). Enzymes in bold have been cloned from *P. tremuloides* and show induction by herbivory (Peters and Constabel, 2002[62]; R. Mellway and C. P. Constabel, unpublished data). Abbreviations are as follows: Phe, phenylalanine; PAL, phenylalanine ammonia lyase; 4CL, 4-coumarate CoA Ligase; CHS, chalcone synthase; CHI, chalcone isomerase; F3H, flavanone 3-hydroxylase; FLS, flavonol synthase; DFR, dihydroflavonol reductase; ANS, anthocyanin synthase; ANR, anthocyanidin reductase; LAR, leucoanthocyanidin reductase.

CT synthesis is induced by herbivory of *P. tremuloides*, supporting the idea that they are important chemicals for defense.[52,62] The cloning and expression analysis of *P. tremuloides* dihydroflavonol reductase (DFR) showed that wounding and herbivory induces DFR transcripts as well as DFR activity as part of a systemic response.[62] This demonstrated that flavonoid and CT synthesis are part of the induced defense response in aspen leaves. Transcripts encoding the key phenylpropanoid and flavonoid enzymes phenylalanine ammonia lyase (PAL), 4-coumarate CoA ligase (4CL), and CHS were also induced by wounding.[62] Interestingly, specific isoforms of PAL and 4CL are associated with wound-induction and CT accumulation, whereas other isoforms appear to be responsible primarily for lignin synthesis.[53] We recently cloned two additional genes putatively encoding enzymes downstream from DFR in the CT pathway, leucoanthocyanidin reductase (LAR) and anthocyanidin reductase (ANR), by searching *Populus* EST databases with new sequence data from heterologous species.[63,64] LAR reduces flavan-3,4-diols to 2,3-*trans*-flavan-3-ols ((+)-catechin), whereas the latter is in a newly discovered pathway from anthocyanidins to 2,3-*cis*-flavan-3-ols ((-)-

epicatechin). Flavan-3-ols are required as starter units in the CT polymer, although *Arabidopsis* appears to use only the ANR-derived (-)-epicatechin.[65] Both ANR and LAR are expressed and induced by wounding in aspen leaves, suggesting both types of monomer are synthesized in *Populus* (R. Mellway and C. P. Constabel, unpublished). This is consistent with C-13 NMR analysis of the CT polymer isolated from aspen leaves, which demonstrated it contains both types of subunits.[56] Overall, the herbivore-induced pattern of expression of genes involved in CT synthesis points to a role of these chemicals in defense, and is the first step in the identification of regulatory genes modulating this pathway.

In contrast to the leaf phytochemical constituents, the study of poplar volatiles in the context of herbivore defense is only now beginning. Leaf damage by forest tent caterpillar herbivory leads to the release of E-ß-ocimene, (-)-germacrene D, and other terpenoid volatiles.[24] These emissions are substantially higher following insect damage than after mechanical wounding. Work in cotton and maize first reported that the mixture of volatiles released from wounded leaves is distinct from leaves subjected to live herbivores, and that different herbivores induce the release of different blends.[66, 67] Plant volatile bouquets are used by parasitic wasps to locate their hosts, and the release of volatiles in response to herbivory can be considered an indirect defense mechanisms. [35] In poplar, the induced emissions are preceded by increased expression of the germacrene D synthase gene, *PtdTPS1*. Therefore, volatile induction appears to be a component of the transcriptional response of poplar leaves to herbivory. Like other induced defenses, the induction of this gene has been observed in unwounded leaves of a wounded plant, indicating a systemic response.[24] It will be interesting to determine if the release of poplar volatiles from poplar influences the behavior of tent caterpillar parasitoids and predators, that is, if these induced volatiles can act as an indirect defense.

A characteristic feature of phytochemicals in natural populations is the variation in concentration and profile that is found among different individuals. Extensive work by Lindroth and others has documented such genotype-dependant variation in phenolics in *P. tremuloides* genotypes.[20,45,61,68] As mentioned above, high levels of phenolics correlate with increased pest resistance in this system. Comparisons of different *Populus* species and hybrids have also demonstrated a significant variation in levels and types of phenolic phytochemicals.[18,19] Further variability is often due to environmental conditions, as aspen plants grown with high soil nutrient availability contain reduced levels of condensed tannins[52,69,70] Such phytochemical variability is likely to have profound impacts on the effectiveness of defenses and evolution of resistance. Elucidating the genetic mechanisms for phytochemical variation in natural populations will be an important advance in understanding how defense mechanisms work in nature.

MOLECULAR ANALYSIS OF INDUCED DEFENSE

The induced defense response of plants is characterized by major changes in transcriptional patterns. By isolating herbivore or wound-inducible genes using molecular differential screening techniques, a number of defense-related genes have been isolated. Large-scale genomics and microarray studies indicate that hundreds of genes respond to herbivory, and genome-wide surveys of these responses will soon be possible in poplar and other plants (see below). How many induced genes contribute to defense is not clear, however. In this section, genes for which a defensive role has emerged or is likely will be highlighted.

Kunitz Trypsin Inhibitors

The pioneering work of Milton Gordon and coworkers first demonstrated wound-induced gene expression in hybrid poplar and led to the isolation of several wound-responsive genes by differential cDNA screening.[8,21] Genes with strong similarity to Kunitz trypsin inhibitors (TIs), chitinases, and vegetative storage protein (VSP) genes were cloned and shown to respond rapidly and systemically to damage. TIs are common defense proteins in many plants that interfere with protein digestion in insects and may have other toxic effects. Kunitz TIs are one of eight proteinase inhibitor families.[27,71] Their discovery in wounded poplar leaves was significant because it demonstrated the adaptiveness of the induced defense response. Subsequently, poplar TI-overexpressing tobacco leaves were found to reduce growth of tobacco budworm.[72]

Recent work from the author's laboratory has identified three additional hybrid poplar genes (*PtdTI3, PtdTI4, PtdTI5*), which belong to the Kunitz TI family, but show less than 50 % identity among themselves or with the original poplar TI gene.[21, 23] This sequence divergence suggested redundancy or multiplicity in defense; a preliminary scan of the completed *P. trichocarpa* genome indicates that there are over 30 distinct Kunitz TI genes (I. Major and C. P. Constabel, unpublished data). This number is similar to that estimated from potato tubers, where 21 distinct Kunitz TIs were identified.[73] We speculate that multiple protease inhibitors, perhaps with different protease specificities, have evolved to match the numerous gut proteases in insect pests.[74] The protease inhibitory activity of three TIs has been confirmed *in vitro* using recombinant proteins, and provides preliminary evidence for distinct specificities for these inhibitors (I. Major, C. Melnyk, and C. P. Constabel, unpublished data).[75] Given the observation that pests can express inhibitor-resistant proteases, it is likely that the multiplicity of poplar TIs reflects functional specialization within the TI gene family.[76-78]

The idea that Kunitz TIs have diversified in response to selective pressure by insect pests is further supported by preliminary evidence that the TI genes are evolving rapidly. In *P. tremuloides*, Haruta *et al.*, (2001) found that native aspen genotypes contain more restriction fragment polymorphisms for two TI genes (*PtTI1/2* and *PtTI3*) than other genes.[75] Later comparative studies of Kunitz TI orthologs among *P. trichocarpa*, *P. tremula*, and related hybrids demonstrated a greater than expected sequence divergence and high ratio of non-synonymous to synonymous nucleotide changes, suggesting positive selection of these genes.[79] In the context of plant defense, such sequence divergence could indicate an evolutionary arms race with insect herbivores.

Polyphenol Oxidase

In hybrid poplar and trembling aspen, leaf damage by wounding or herbivory induces polyphenol oxidase activity (PPO; EC 1.10.3.2).[22,50,80] This enzyme can oxidize a variety of *ortho*-diphenolic compounds to their respective reactive quinones using molecular oxygen.[81] Leaf PPOs have been cloned from hybrid poplar and trembling aspen, and their wound- and herbivore-induced expression suggests a defensive function.[22,50] Quinones produced by PPO during insect feeding can lead to protein and amino acid alkylation and cross-linking, with concomitant loss of essential amino acid assimilation by leaf-eating herbivores.[82,83] In addition, reverse disproportionation of quinones to semiquinones, ultimately resulting in H_2O_2, could contribute to oxidative stress for the insects. Poplar PPO is stable and not only resists proteolysis in the gut of forest tent caterpillar, but is activated by passage through the gut, consistent with a role in defense against these insects.[84] Which substrates poplar or aspen PPO oxidizes during defense reactions is not yet known; as mentioned earlier, catechol is an excellent PPO substrate and can be produced by decomposition of the phenolic glycoside salicortin and tremulacin.[47,48,50] Other possible substrates include caffeic acid derivatives or catechins, common phytochemicals that have been reported from *Populus*.[18,19]

Three poplar PPO gene families have been described to date, and members of at least two of these families show induction by wounding and expression in different plant organs.[84] To test if the inducible poplar PPO is important in defense against lepidopteran insects, the hybrid poplar *PtdPPO1* cDNA was overexpressed in hybrid aspen. Feeding on PPO-overexpressing leaves reduced forest tent caterpillar performance relative to control leaves; however the effect was visible only with older egg masses that gave rise to less vigorous larvae.[84] Nevertheless, this direct proof of the anti-herbivore effects of PPO is the first such demonstration for a poplar defense gene in transgenic *Populus*, and is a first step in the functional analysis of defense genes in poplar defense.

Chitinases

Genes with similarity to chitinases were among the first wound-induced defense genes from *Populus* to be characterized. Two distinct chitinase genes named *win6* and *win8*, sharing approximately 50% amino acid sequence identity, are strongly upregulated in hybrid poplar leaves following wounding.[21, 85] Leaf chitinase activity increases following wounding, and transgenic tobacco expressing *win6* also show enhanced chitinase activity.[86] The *win6* and *win8* chitinases are unusual in having close resemblance to class 1B basic chitinases from other species, yet having an acidic pI. Other class 1B chitinases are also wound-induced, unlike the typical acidic chitinases.[87] EST sequencing of a wound-induced cDNA library identified two other chitinase-like genes, at least one of which is wound-induced.[23] Chitinases have been shown to act as pathogen defenses, and their wound-induction may prevent wound site ingress of opportunistic pathogens. In addition, it is possible that they also act directly against insect herbivores. The abundance of chitinase transcripts in the EST library and their strong induction on macroarrays supports this idea.[23] A possible site of action of anti-herbivore chitinase activity might be the peritrophic membrane, which functions in protecting the underlying cells from digestive enzymes.[88] Microbial and insect-derived chitinases have been shown to be detrimental to herbivorous insects.[89, 90]

Other Induced Genes and Proteins

Several other known defense-related genes have been found to be upregulated by wounding. Lipoxygenase and hydroperoxide lyase are rapidly induced, most plausibly for biosynthesis of jasmonate and other octadecanoid signals.[23] Genes involved in phenylpropanoid and lignin biosynthesis, such as phenylalanine ammonia lyase, cinnamoyl Co-A reductase, are also upregulated (I. Major and C. P. Constabel, unpublished), possibly contributing to leaf toughening via lignin-like depositions.[91] Some of these genes may also contribute to the accumulation of other phenolic phytochemicals such as the condensed tannins. The strong induction of cytochrome P450 genes, which often encode hydroxylases, may also be related to secondary metabolism.

Other wound- or herbivore-induced transcripts encode proteins previously identified as vegetative storage proteins. For example, Davis *et al.* (1991) identified a strongly inducible cDNA (*win4*) with high similarity to a bark storage protein.[92] This so-called vegetative storage protein (VSP) is also expressed at low levels in young, growing shoots, and is often found in association with vascular tissue.[93] Another highly inducible gene in poplar leaves encodes an acid phosphatase, which in *Arabidopsis* and soybean acts as a VSP.[94] How these function in defense is unclear, but it may be related to nitrogen remobilization as part of a plant-wide defense strategy similar to senescence.

IMPACT OF GENOMICS ON STUDIES OF *POPULUS* DEFENSE

Since the first report of a poplar expressed sequence tag (EST) project,[17] *Populus* genomics has grown rapidly, and there are now several high-throughput sequencing projects in Europe and North America.[9] EST collections have provided a crucial resource for the cloning of genes and gene families, and are the basis for the construction of current poplar microarrays. Poplar arrays permit the simultaneous analysis of gene expression of thousands of genes, and have been used to document the complexity of gene expression underlying developmental processes such as wood formation and resource allocation,[95, 96] as well as environmental adaptation and stress responses.[23,91,97,98] Poplar fully entered the genomics era with the complete elucidation of the nucleotide sequence of the *P. trichocarpa* genome (released in September 2004). Many new resources will soon be available, including arrays for monitoring gene expression of essentially the entire poplar genome. These will allow for faster discovery of genes important for developmental and adaptive processes.

EST Libraries as a Resource for Poplar Genomics

With well over 200,000 ESTs from *Populus* species in the public domain, poplar EST databases are useful in many experiments. First, they provide immediate access to whole sets of genes of interest for specific tissues and organs. Second, ESTs from several *Populus* species are available, which permits comparisons of orthologous genes that can provide insight into molecular evolution of genes. As described earlier, comparative analyses of ESTs encoding Kunitz TIs from several different *Populus* species and hybrids suggest these genes are subject to positive selection and evolving rapidly, perhaps due to herbivore pressure. Third, the EST collections can be used to generate tissue-specific expression profiles, or "digital northerns." Since the ESTs are derived from separate cDNA libraries constructed from a range of tissues and experimental treatments, the abundance of a given transcript in the library provides an indication of its level of expression in that tissue.[99] This approach has been effectively used to study lipid metabolism, and can be used to obtain preliminary gene expression data on any gene.[100] The relative representation of genes within an EST library can also be used to obtain a snapshot of gene expression, for example during the herbivore defense response of hybrid poplar.[23] In this study, the inducible chitinases and trypsin inhibitors were found to be among the ten most abundant predicted gene products in the EST library, together with many photosynthesis-related and primary metabolism genes.[23] This result suggests that wound stress has had a significant impact on the leaf transcriptome as a whole.

Transcript Profiling of Poplar Induced Herbivore Defense

Global changes in gene expression in response to wounding and insect herbivory have been studied using cDNA macro- or micro-arrays. In addition to providing insight into signal transduction and global patterns of gene expression, these approaches are identifying many new wound- and herbivore-responsive genes. Macroarrays constructed from a small EST set of 569 genes were used in the authors' laboratory to profile gene expression in *P. trichocarpa x P. deltoides* (TD) hybrid poplar. Mechanical wounding and treatment with the defense hormone methyl jasmonate significantly upregulated 107 and 163 genes after 24 h, respectively, including the defense genes discussed previously. An experiment using a much larger array (approximately 10,000 genes) found 947 significantly upregulated transcripts 14 days after wounding TD hybrid poplar leaves by abrasion.[91] While it is difficult to directly compare array experiments carried out under different conditions and using partial genome arrays, the proportions of induced genes are similar to reports from *Arabidopsis*. Cheong *et al.* (2002) identified 657 wound-responsive transcripts on arrays with 8200 genes (8%),[101] and Schenk *et al.* (2000) found that approximately 9% of genes on a 2375 gene array were upregulated by methyl jasmonate.[102]

Such array studies produce interesting lists of genes, but it is not always easy to establish their biological significance or roles. As more studies of plant defenses employ microarrays, broad comparisons between experiments and plant species should help to narrow the search for major defense genes. For example, there are common features of the induced defense response in poplar and *Arabidopsis,* including expression of jasmonate biosynthesis genes, trypsin inhibitors (though from different families), chitinases, and cytochrome P450 enzymes. Interestingly, Smith *et al.* (2004) identified a number of genes encoding cysteine proteases, which were also highly represented in a wounded leaf EST library.[23, 91] Cysteine proteases were recently found to have direct toxicity to insects.[103,104] Wound- or herbivore-induced phenylpropanoid gene expression in *Populus* (see below) appears to be a common feature of the defense response and is seen in *Arabidopsis, Nicotiana attenuata*, and other species.[101,105] Other wound-induced poplar genes have not been described from other plants in a defense context; these include genes encoding apyrase, lipases, invertase, and predicted proteins without any known functions (I. Major and C. P. Constabel, unpublished data).[23] How these might function in plant defense is being investigated further, but their strong response to herbivore stress suggests a direct involvement. Many of the 14-day wound-induced genes identified in poplar by Smith *et al.* (2004) are metabolism-related, but those transcripts that encode gene products related to photosynthesis and protein synthesis typically decreased in response to wounding.[91] The repression of photosynthesis-related genes is presumably indicative of resource reallocation for defense and has also been described in *N. attenuata*.[105,106]

DNA array experiments are clearly powerful tools that are beginning to have an impact in studies of poplar defense. Whole genome oligonucleotide-based arrays will soon be available, which will increase the volume of gene expression data further. We can expect an unprecedented expansion of our knowledge of poplar defense at the level of transcription. Controlled cross-species experiments will provide comparisons of global gene expression, and determine how variable induced defense is among species. Genomics tools will also be crucial in the drive to understand the molecular basis for variation in pest resistance among poplar genotypes.

DEFENSE SIGNALING IN *POPULUS*

How cellular signaling pathways activate the defense response is a major question for poplar and other model plants.[29,30,107] As has been established for other plants, a central component in defense signaling in poplar appears to jasmonic acid and related compounds, collectively known as jasmonates. These are fatty acid derived plant hormones that act as both developmental and stress signals.[108] Treatment of poplar and aspen with methyl jasmonate has a strong effect on inducing defense gene expression (PPO and TIs) and resistance to gypsy moth larvae.[22,41,50,75] Conversely, wounding of poplar leaves induces key enzymes of jasmonate biosynthesis such as lipoxygenase and allene oxide cyclase, suggesting involvement of jasmonate biosynthesis as an amplification of a primary defense signal.[23] Other components of herbivore defense signaling pathways of poplar remain unknown, but this will likely change rapidly as the poplar genome database is mined for genes with candidate signaling functions. In *Arabidopsis* the wound-induced transcriptome includes a large number of transcription factor- and signaling protein-encoding genes that may modulate the induced response;[101] the wealth of information available from this model plant will facilitate rapid progress in poplar.

Poplar induced defense has features that could make it a valuable system for studies of whole-plant signaling. Parsons *et al.* (1989) first demonstrated that hybrid poplar herbivore defense is systemic, and that wounding of lower leaves induces gene expression in the upper, unwounded leaves.[21] Subsequent work demonstrated that the strongest systemic induction occurred in orthostichous leaves that have direct vascular connections.[109] The wound signal appears to be transmitted from source to sink leaves, and is dependent on photoassimilate transport in phloem, since it can be disrupted by shading of the source leaf.[109] Arnold and Schultz (2002) reported that wounding and jasmonic acid increase the sink strength of young poplar leaves, and that the concomitant increase of imported carbohydrates in these leaves is used for the synthesis of condensed tannins.[110] It thus appears that source-sink relations and wound signaling are related, and that signaling correlates with sink import of sucrose. However, in *P. deltoides*, leaf resistance to folivore beetles could be induced both above and below the damaged leaves.[111] Likewise in *N. attenuata*, wound signal transmission was reported from both sink and source leaves.[112] In our laboratory, wounding of poplar leaves was found to cause increased TI gene expression in roots, thus suggesting that signal movement does occur from shoots to roots (I. Major and C. P. Constabel, unpublished data). The systemic wound signal in poplar has not been identified, but in tomato, recent evidence suggests that it is a jasmonate-related compound that acts in concert with a short 18 amino acid peptide called systemin.[44,113] While jasmonates are universally found in the plant kingdom, systemin-like peptides have only been found in the Solanaceae.

Many defense responses can be triggered by mechanical damage as well as by live insect herbivores. Other responses, for example volatile production, require insect-specific cues.[114] The analysis of volatiles from maize infested with different leaf-eating pests suggests that there are pest species-specific effects, even when these feed on the same host plant.[67] The effects of live caterpillars can be mimicked with insect regurgitant from feeding caterpillars, which has led to the purification of insect elicitors from insect saliva and regurgitant. A fatty acid conjugate known as volicitin was first isolated from beet army worm regurgitant, and identified as N-(17-hydroxylinolenoyl)-L-glutamine.[115] Similar fatty acid conjugates that stimulate plant defense reactions have been purified from saliva in other leaf-eating insects.[29,116] Regurgitant can activate many wound-induced defenses in the absence of significant mechanical damage, and can variously modulate the wound response.[39,116,117]

The ability of herbivorous insect regurtitant to elicit plant defense responses have also been observed in poplar. Havill and Raffa (1999) first showed for poplar that gypsy moth and tent caterpillar larvae regurgitant stimulates induced resistance to subsequent pest attack, much like mechanical wounding or MeJa treatment.[41] At the level of gene expression, macroarray experiments have demonstrated that forest tent caterpillar regurgitant to leaves is a strong inducer of all the major wound-induced defense genes, even in the absence of significant leaf damage (I. Major and C. P. Constabel, unpublished data).

SUMMARY

Populus species are susceptible to many insect pests. The induced pest defense of various *Populus* species or hybrids has been studied at the phytochemical, molecular, and genomic levels. This response involves rapid changes in gene expression, which has provided insight into many defense processes. Induced defenses include flavonoid gene expression and condensed tannin accumulation, release of terpenoid and other volatiles, and synthesis of trypsin inhibitors and enzymes such as chitinase and polyphenol oxidase. Induced defense mechanisms thus include both protein-based and phytochemical defense mechanisms. Transgenic studies have demonstrated that induced defense proteins, such as trypsin inhibitors and polyphenol oxidase, can have a negative effect on leaf-eating insects. In the future, the availability of genomics tools will greatly facilitate the identification of the complete herbivore-induced suite of genes, but a major challenge will be to determine the function of individual genes in defense. Comparing herbivore-induced gene suites of poplar to *Arabidopsis* and other plants will help to define general defense responses and conserved signaling processes, and comparisons of responses to different herbivore guilds will be also be instructive. In addition, genomics approaches should lead to rapid progress in our understanding of defense signaling, such as identifying transcription factors that regulate sets of induced genes and signal transduction cascades.

ACKNOWLEDGMENTS

The authors gratefully acknowledge research funding from the Natural Sciences and Engineering Research Council of Canada (NSERC), Alberta Agricultural Research Institute, and the University of Victoria. We also thank Robin Mellway for critical reading of the manuscript.

REFERENCES

1. HARBORNE, J.B., Introduction to Ecological Biochemistry. 4th ed., Academic Press, London. 1993, 318 p.
2. DICKMAN, D.I., An overview of the genus *Populus, in:* Poplar Culture in North America (J.G. Isebrands, D.I. Dickmann, J.E. Eckenwalder, J. Richardson, eds.), NRC Research Press, Ottawa. 2001, pp. 1-42
3. ECKENWALDER, J.E., Systematics and evolution of *Populus, in:* Biology of *Populus* and Its Implications For Management and Conservation (H.D. Bradshaw, R.F. Stettler, P.E. Heilman, T.M. Hinckley, eds.), NRC Research Press, Ottawa. 1996, pp. 7-32
4. ECKENWALDER, J.E., Descriptions of clonal characteristics, *in:* Poplar Culture in North America (J.G. Isebrands, D.I. Dickmann, J.E. Eckenwalder, J. Richardson, eds.), NRC Research Press, Ottawa. 2001, pp. 331-382
5. MATTSON, W.J., HART, E.A., VOLNEY, W.J.A., Insect pests of *Populus*: coping with the inevitable, *in:* Poplar Culture in North America (J.G. Isebrands D.I. Dickmann, J.E. Eckenwalder, J. Richardson, eds.), NRC Research Press, Ottawa. 2001, pp. 219-248
6. YVES, W.G.H., WONG, H.R., Tree and Shrub Insects of the Prairie Provinces.Canadian Forest Service Information Report NOR-X-292., Edmonton. 1988.
7. REICHENBACKER, R.R., SCHULTZ, R.C., HART, E.R., Artificial defoliation effect on *Populus* growth, biomass production, and total nonstructural carbohydrate concentration, *Environ. Entomol.*, 1996, **25**, 632-642.
8. BRADSHAW, H.D., PARSONS, T.J., GORDON, M.P., Wound-responsive gene expression in poplars, *For. Ecol. Manag.*, 1991, **43**, 211-224.
9. TUSKAN, G.A., DIFAZIO, S.P., TEICHMANN, T., Poplar genomics is getting popular: The impact of the poplar genome project on tree research, *Plant Biol.*, 2004, **6**, 2-4.

10. WULLSCHLEGER, S.D., JANSSON, S., TAYLOR, G., Genomics and forest biology: *Populus* emerges as the perennial favorite, *Plant Cell*, 2002, **14**, 2651-2655.

11. BRUNNER, A.M., BUSOV, V.B., STRAUSS, S.H., Poplar genome sequence: Functional genomics in an ecologically dominant plant species, *Trends Plant Sci.*, 2004, **9**, 49-56.

12. PENA, L., SEGUIN, A., Recent advances in the genetic transformation of trees, *Trends Biotechnol.*, 2001, **19**, 500-506.

13. HAN, K.H., MEILAN, R., MA, C., STRAUSS, S.H., An *Agrobacterium tumefaciens* transformation protocol effective on a variety of cottonwood hybrids (genus Populus), *Plant Cell Rep.*, 2000, **19**, 315-320.

14. TAYLOR, G., *Populus: Arabidopsis* for forestry. Do we need a model tree?, *Ann. Bot.*, 2002, **90**, 681-689.

15. YIN, T.M., DIFAZIO, S.P., GUNTER, L.E., JAWDY, S.S., BOERJAN, W., TUSKAN, G.A., Genetic and physical mapping of *Melampsora* rust resistance genes in *Populus* and characterization of linkage disequilibrium and flanking genomic sequence, *New Phytol.*, 2004, **164**, 95-105.

16. VILLAR, M., LEFEVRE, F., BRADSHAW, H.D., DUCROS, E.T., Molecular genetics of rust resistance in poplars (*Melampsora larici-populina* Kleb Populus sp) by bulked segregant analysis in a 2x2 factorial mating design, *Genetics*, 1996, **143**, 531-536.

17. STERKY, F., REGAN, S., KARLSSON, J., HERTZBERG, M., ROHDE, A., HOLMBERG, A., AMINI, B., BHALERAO, R., LARSSON, M., VILLARROEL, R., VAN MONTAGU, M., SANDBERG, G., OLSSON, O., TEERI, T.T., BOERJAN, W., GUSTAFSSON, P., UHLEN, M., SUNDBERG, B., LUNDEBERG, J., Gene discovery in the wood-forming tissues of poplar: Analysis of 5,692 expressed sequence tags, *Proc. Natl. Acad. Sci. USA*, 1998, **95**, 13330-13335.

18. PALO, R.T., Distribution of birch (*Betula* spp), willow (*Salix* spp), and poplar (*Populus* spp) secondary metabolites and their potential role as chemical defense against herbivores, *J. Chem. Ecol.*, 1984, **10**, 499-520.

19. JULKUNEN-TIITTO, R., A chemotaxonomic survey of phenolics in leaves of northern Salicaceae species, *Phytochemistry*, 1986, **25**, 663-667.

20. LINDROTH, R.L., HWANG, S.-Y., Diversity, redundancy, and multiplicity in chemical defense systems of aspen, *in:* Phytochemical Diversity and Redundancy in Ecological Interactions (J.A. Saunders J.T. Romeo, P. Barbosa, eds.), Plenum, New York. 1996, pp. 25-56

21. PARSONS, T.J., BRADSHAW, H.D., GORDON, M.P., Systemic accumulation of specific messenger RNAs in response to wounding in poplar trees, *Proc. Natl. Acad. Sci. USA*, 1989, **86**, 7895-7899.

22. CONSTABEL, C.P., YIP, L., PATTON, J.J., CHRISTOPHER, M.E., Polyphenol oxidase from hybrid poplar. Cloning and expression in response to wounding and herbivory, *Plant Physiol.*, 2000, **124**, 285-295.

23. CHRISTOPHER, M.E., MIRANDA, M., MAJOR, I.T., CONSTABEL, C.P., Gene expression profiling of systemically wound-induced defenses in hybrid poplar, *Planta*, 2004, **219**, 936-947.

24. ARIMURA, G., HUBER, D.P.W., BOHLMANN, J., Forest tent caterpillars (*Malacosoma disstria*) induce local and systemic diurnal emissions of terpenoid volatiles in hybrid poplar (*Populus trichocarpa* x *deltoides*): cDNA cloning, functional characterization, and patterns of gene expression of (-)-germacrene D synthase, PtdTPS1, *Plant J.*, 2004, **37**, 603-616.

25. KARBAN, R., BALDWIN, I.T., eds., Induced Responses to Herbivory, University of Chicago Press, Chicago. 1997, 319 p.

26. BERGEY, D.R., HOI, G.A., RYAN, C.A., Polypeptide signaling for plant defensive genes exhibits analogies to defense signaling in animals, *Proc. Natl. Acad. Sci. USA*, 1996, **93**, 12053-12058.

27. RYAN, C.A., Protease inhibitors in plants - genes for improving defenses against insects and pathogens, *Annu. Rev. Phytopath.*, 1990, **28**, 425-449.

28. DUFFEY, S.S., STOUT, M.J., Antinutritive and toxic components of plant defense against insects, *Arch. Insect Biochem. Physiol.*, 1996, **32**, 3-37.

29. KESSLER, A., BALDWIN, I.T., Plant responses to insect herbivory: The emerging molecular analysis, *Ann. Rev. Plant Biol.*, 2002, **53**, 299-328.

30. WALLING, L.L., The myriad plant responses to herbivores, *J. Plant Growth Regul.*, 2000, **19**, 195-216.

31. CONSTABEL, C.P., A survey of herbivore-inducible defensive proteins and phytochemicals, *in:* Induced Plant Defenses Against Herbivores and Pathogens (A.A. Agrawal, Bent, E., Tuzun, S., eds.), APS Press, St. Paul. 1999, pp. 137-166

32. DUDAREVA, N., PICHERSKY, E., GERSHENZON, J., Biochemistry of plant volatiles, *Plant Physiol.*, 2004, **135**, 1893-1902.

33. KESSLER, A., BALDWIN, I.T., Defensive function of herbivore-induced plant volatile emissions in nature, *Science*, 2001, **291**, 2141-2144.

34. TAKABAYASHI, J., DICKE, M., Plant-carnivore mutualism through herbivore-induced carnivore attractants, *Trends Plant Sci.*, 1996, **1**, 109-113.

35. PARE, P.W., TUMLINSON, J.H., Plant volatiles as a defense against insect herbivores, *Plant Physiol.*, 1999, **121**, 325-331.

36. PICHERSKY, E., GERSHENZON, J., The formation and function of plant volatiles: perfumes for pollinator attraction and defense, *Curr. Op. Plant Biol.*, 2002, **5**, 237-243.

37. MCCONN, M., CREELMAN, R.A., BELL, E., MULLET, J.E., BROWSE, J., Jasmonate is essential for insect defense in *Arabidopsis*, *Proc. Natl. Acad. Sci. USA*, 1997, **94**, 5473-5477.

38. LI, C.Y., WILLIAMS, M.M., LOH, Y.T., LEE, G.I., HOWE, G.A., Resistance of cultivated tomato to cell content-feeding herbivores is regulated by the octadecanoid-signaling pathway, *Plant Physiol.*, 2002, **130**, 494-503.

39. HALITSCHKE, R., BALDWIN, I.T., Antisense LOX expression increases herbivore performance by decreasing defense responses and inhibiting growth-related transcriptional reorganization in *Nicotiana attenuata*, *Plant J.*, 2003, **36**, 794-807.

40. KESSLER, A., HALITSCHKE, R., BALDWIN, I.T., Silencing the jasmonate cascade: Induced plant defenses and insect populations, *Science*, 2004, **305**, 665-668.

41. HAVILL, N.P., RAFFA, K.F., Effects of elicitation treatment and genotypic variation on induced resistance in *Populus*: impacts on gypsy moth (Lepidoptera : Lymantriidae) development and feeding behavior, *Oecologia*, 1999, **120**, 295-303.

42. ROBISON, D.J., RAFFA, K.F., Effects of constitutive and inducible traits of hybrid poplars on forest tent caterpillar feeding and population ecology, *Forest Sci.*, 1997, **43**, 252-267.

43. GLYNN, C., HERMS, D.A., EGAWA, M., HANSEN, R., MATTSON, W.J., Effects of nutrient availability on biomass allocation as well as constitutive and rapid induced herbivore resistance in poplar, *Oikos*, 2003, **101**, 385-397.

44. RYAN, C.A., PEARCE, G., Systemins: A functionally defined family of peptide signal that regulate defensive genes in Solanaceae species, *Proc. Natl. Acad. Sci. USA*, 2003, **100**, 14577-14580.

45. HWANG, S.Y., LINDROTH, R.L., Clonal variation in foliar chemistry of aspen: Effects on gypsy moths and forest tent caterpillars, *Oecologia*, 1997, **111**, 99-108.

46. PIERPOINT, W.S., Salicylic acid and its derivatives in plants: medicines, metabolites and messenger molecules, *Adv. Bot. Res.*, 1994, **20**, 164-235.

47. RUUHOLA, T., JULKUNEN-TIITTO, R., VAINIOTALO, P., *In vitro* degradation of willow salicylates, *J. Chem. Ecol.*, 2003, **29**, 1083-1097.

48. CLAUSEN, T.P., REICHARDT, P.B., BRYANT, J.P., WERNER, R.A., POST, K., FRISBY, K., Chemical model for short-term induction in quaking aspen (*Populus tremuloides*) foliage against herbivores, *J. Chem. Ecol.*, 1989, **15**, 2335-2346.

49. RUUHOLA, T., TIKKANEN, O.P., TAHVANAINEN, J., Differences in host use efficiency of larvae of a generalist moth, *Operophtera brumata* on three chemically divergent *Salix* species, *J. Chem. Ecol.*, 2001, **27**, 1595-1615.

50. HARUTA, M., PEDERSEN, J.A., CONSTABEL, C.P., Polyphenol oxidase and herbivore defense in trembling aspen (*Populus tremuloides*): cDNA cloning, expression, and potential substrates, *Physiol. Plant.*, 2001, **112**, 552-558.

51. LINDROTH, R.L., KINNEY, K.K., Consequences of enriched atmospheric CO_2 and defoliation for foliar chemistry and gypsy moth performance, *J. Chem. Ecol.*, 1998, **24**, 1677-1695.

52. OSIER, T.L., LINDROTH, R.L., Effects of genotype, nutrient availability, and defoliation on aspen phytochemistry and insect performance, *J. Chem. Ecol.*, 2001, **27**, 1289-1313.

53. KAO, Y.Y., HARDING, S.A., TSAI, C.J., Differential expression of two distinct phenylalanine ammonia-lyase genes in condensed tannin-accumulating and lignifying cells of quaking aspen, *Plant Physiol.*, 2002, **130**, 796-807.

54. PORTER, L.J., HRSTICH, L.N., CHAN, B.G., The conversion of procyanidins and prodelphinidins to cyanidin and delphidin, *Phytochemistry*, 1986, **25**, 223-230.

55. MARLES, M.A.S., RAY, H., GRUBER, M.Y., New perspectives on proanthocyanidin biochemistry and molecular regulation, *Phytochemistry*, 2003, **64**, 367-383.

56. AYRES, M.P., CLAUSEN, T.P., MACLEAN, S.F., REDMAN, A.M., REICHARDT, P.B., Diversity of structure and antiherbivore activity in condensed tannins, *Ecology*, 1997, **78**, 1696-1712.

57. WINKEL-SHIRLEY, B., Flavonoid biosynthesis. A colorful model for genetics, biochemistry, cell biology, and biotechnology, *Plant Physiol.*, 2001, **126**, 485-493.

58. HAGERMAN, A.E., BUTLER, L.G., Tannins and lignins, *in:* Herbivores: Their Interaction with Secondary Metabolites (G.A. Rosenthal, M.R. Berenbaum, eds.), Academic Press, San Diego. 1991, pp. 355-387

59. BARBEHENN, R.V., MARTIN, M.M., Tannin sensitivity in larvae of *Malacosoma disstria* (Lepidoptera) - roles of the peritrophic envelope and midgut oxidation, *J. Chem. Ecol.*, 1994, **20**, 1985-2001.

60. APPEL, H.M., Phenolics in ecological interactions - the importance of oxidation, *J. Chem. Ecol.*, 1993, **19**, 1521-1552.

61. HEMMING, J.D.C., LINDROTH, R.L., Intraspecific variation in aspen phytochemistry - effects on performance of gypsy moths and forest tent caterpillars, *Oecologia*, 1995, **103**, 79-88.

62. PETERS, D.J., CONSTABEL, C.P., Molecular analysis of herbivore-induced condensed tannin synthesis: cloning and expression of dihydroflavonol reductase from trembling aspen (Populus tremuloides), *Plant J.*, 2002, **32**, 701-712.

63. TANNER, G.J., FRANCKI, K.T., ABRAHAMS, S., WATSON, J.M., LARKIN, P.J., ASHTON, A.R., Proanthocyanidin biosynthesis in plants - Purification of legume leucoanthocyanidin reductase and molecular cloning of its cDNA, *J. Biol. Chem.*, 2003, **278**, 31647-31656.

64. XIE, D.Y., SHARMA, S.B., PAIVA, N.L., FERREIRA, D., DIXON, R.A., Role of anthocyanidin reductase, encoded by BANYULS in plant flavonoid biosynthesis, *Science*, 2003, **299**, 396-399.

65. ABRAHAMS, S., LEE, E., WALKER, A.R., TANNER, G.J., LARKIN, P.J., ASHTON, A.R., The Arabidopsis TDS4 gene encodes leucoanthocyanidin dioxygenase (LDOX) and is essential for proanthocyanidin synthesis and vacuole development, *Plant J.*, 2003, **35**, 624-636.

66. TURLINGS, T.C.J., LOUGHRIN, J.H., MCCALL, P.J., ROSE, U.S.R., LEWIS, W.J., TUMLINSON, J.H., How caterpillar-damaged plants protect themselves by attracting parasitic wasps, *Proc. Natl. Acad. Sci. USA*, 1995, **92**, 4169-4174.

67. DE MORAES, C.M., LEWIS, W.J., PARE, P.W., ALBORN, H.T., TUMLINSON, J.H., Herbivore-infested plants selectively attract parasitoids, *Nature*, 1998, **393**, 570-573.

68. CHILCOTE, C.A., WITTER, J.A., MONTGOMERY, M.E., STOYENOFF, J.L., Intraclonal and interclonal variation in gypsy moth larval performance on bigtooth and trembling aspen, *Can. J. For. Res.*, 1992, **22**, 1676-1683.

69. HEMMING, J.D.C., LINDROTH, R.L., Effects of light and nutrient availability on aspen: Growth, phytochemistry, and insect performance, *J. Chem. Ecol.*, 1999, **25**, 1687-1714.

70. LINDROTH, R.L., OSIER, T.L., BARNHILL, H.R.H., WOOD, S.A., Effects of genotype and nutrient availability on phytochemistry of trembling aspen (Populus

tremuloides Michx.) during leaf senescence, *Biochem. System. Ecol.*, 2002, **30**, 297-307.

71. RICHARDSON, M., Seed storage proteins: the enzyme inhibitors, *Meth. Plant Biochem.*, 1991, **5**, 259-305.

72. LAWRENCE, S.D., NOVAK, N.G., A rapid method for the production and characterization of recombinant insecticidal proteins in plants, *Molec. Breed.*, 2001, **8**, 139-146.

73. HEIBGES, A., GLACZINSKI, H., BALLVORA, A., SALAMINI, F., GEBHARDT, C., Structural diversity and organization of three gene families for Kunitz-type enzyme inhibitors from potato tubers (Solanum tuberosum L.), *Molec. Gen. Genom.*, 2003, **269**, 526-534.

74. HEGEDUS, D., BALDWIN, D., O'GRADY, M., BRAUN, L., GLEDDIE, S., SHARPE, A., LYDIATE, D., ERLANDSON, M., Midgut proteases from *Mamestra configurata* (Lepidoptera: Noctuidae) larvae: Characterization, cDNA cloning, and expressed sequence tag analysis, *Arch. Insect Biochem. Physiol.*, 2003, **53**, 30-47.

75. HARUTA, M., MAJOR, I.T., CHRISTOPHER, M.E., PATTON, J.J., CONSTABEL, C.P., A Kunitz trypsin inhibitor gene family from trembling aspen (*Populus tremuloides* Michx.): cloning, functional expression, and induction by wounding and herbivory, *Plant Molec. Biol.*, 2001, **46**, 347-359.

76. BROADWAY, R.M., Dietary proteinase inhibitors alter complement of midgut proteases, *Arch.Insect Biochem.Physiol.*, 1996, **32**, 39-53.

77. JONGSMA, M.A., BAKKER, P.L., PETERS, J., BOSCH, D., STIEKEMA, W.J., Adaptation of *Spodoptera exigua* larvae to plant proteinase inhibitors by induction of gut proteinase activity insensitive to inhibition, *Proc. Natl. Acad. Sci. USA*, 1995, **92**, 8041-8045.

78. BOWN, D.P., WILKINSON, H.S., GATEHOUSE, J.A., Regulation of expression of genes encoding digestive proteases in the gut of a polyphagous lepidopteran larva in response to dietary protease inhibitors, *Physiol. Entomol.*, 2004, **29**, 278-290.

79. MIRANDA, M., CHRISTOPHER, M.E., CONSTABEL, C.P., The variable nature of herbivore defense: evidence for a rapidly diverging Kunitz trypsin inhibitor gene in *Populus*, *in:* Plant Adaptation: Molecular Genetics and Ecology (Q.C.B. Cronk, J. Whitton, R.H. Ree, I.E.P. Taylor, eds.), NRC Press, Ottawa. 2004, pp. 153-158

80. CONSTABEL, C.P., RYAN, C.A., A survey of wound- and methyl jasmonate-induced leaf polyphenol oxidase in crop plants, *Phytochemistry*, 1998, **47**, 507-511.

81. CONSTABEL, C.P., BERGEY, D.R., RYAN, C.A., Polyphenol oxidase as a component of the inducible defense response of tomato against herbivores, *in:* Phytochemical Diversity and Redundancy in Ecological Interactions (J.T. Romeo, J.A. Saunders, P. Barbosa, eds.), Plenum Press, New York. 1996, pp. 231-252

82. FELTON, G.W., DONATO, K., DELVECCHIO, R.J., DUFFEY, S.S., Activation of plant foliar oxidases by insect feeding reduces nutritive quality of foliage for noctuid herbivores, *J. Chem. Ecol.*, 1989, **15**, 2667-2694.

83. FELTON, G.W., DONATO, K.K., BROADWAY, R.M., DUFFEY, S.S., Impact of oxidized plant phenolics on the nutritional quality of dietary protein to a noctuid herbivore, S*podoptera exigua, J. Insect Physiol.*, 1992, **38**, 277-285.

84. WANG, J., CONSTABEL, C.P., Polyphenol oxidase overexpression in transgenic *Populus* enhances resistance to herbivory by forest tent caterpillar (*Malacosoma disstria*), *Planta*, 2004, **220**, 87-96.

85. DAVIS, J.M., CLARKE, H.R.G., BRADSHAW, H.D., GORDON, M.P., *Populus* chitinase genes - structure, organization, and similarity of translated sequences to herbaceous plant chitinases, *Plant Molec. Biol,*, 1991, **17**, 631-639.

86. CLARKE, H.R.G., LAWRENCE, S.D., FLASKERUD, J., KORHNAK, T.E., GORDON, M.P., DAVIS, J.M., Chitinase accumulates systemically in wounded poplar trees, *Physiol. Plant.*, 1998, **103**, 154-161.

87. COLLINGE, D.B., KRAGH, K.M., MIKKELSEN, J.D., NIELSEN, K.K., RASMUSSEN, U., VAD, K., Plant chitinases, *Plant J.*, 1993, **3**, 31-40.

88. BARBEHENN, R.V., Roles of peritrophic membranes in protecting herbivorous insects from ingested plant allelochemicals, *Arch Insect Biochem Physiol.*, 2001, **47**, 86-99.

89. BROADWAY, R.M., GONGORA, C., KAIN, W.C., SANDERSON, J.P., MONROY, J.A., BENNETT, K.C., WARNER, J.B., HOFFMANN, M.P., Novel chitinolytic enzymes with biological activity against herbivorous insects, *J. Chem. Ecol.*, 1998, **24**, 985-998.

90. DING, X.F., GOPALAKRISHNAN, B., JOHNSON, L.B., WHITE, F.F., WANG, X.R., MORGAN, T.D., KRAMER, K.J., MUTHUKRISHNAN, S., Insect resistance of transgenic tobacco expressing an insect chitinase gene, *Transgen. Res.*, 1998, **7**, 77-84.

91. SMITH, C.M., RODRIGUEZ-BUEY, M., KARLSSON, J., CAMPBELL, M.M., The response of the poplar transcriptome to wounding and subsequent infection by a viral pathogen, *New Phytol.*, 2004, **164**, 123-136.

92. DAVIS, J.M., EGELKROUT, E.E., COLEMAN, G.D., CHEN, T.H.H., HAISSIG, B.E., RIEMENSCHNEIDER, D.E., GORDON, M.P., A family of wound-induced genes in *Populus* shares common features with genes encoding vegetative storage proteins, *Plant Molec. Biol.*, 1993, **23**, 135-143.

93. LAWRENCE, S.D., GREENWOOD, J.S., KORHNAK, T.E., DAVIS, J.M., A vegetative storage protein homolog is expressed in the growing shoot apex of hybrid poplar, *Planta*, 1997, **203**, 237-244.

94. BERGER, S., BELL, E., SADKA, A., MULLET, J.E., Arabidopsis thaliana Atvsp is homologous to soybean Vspa and Vspb, genes encoding vegetative storage protein acid phosphatases, and is regulated similarly by methyl jasmonate, wounding, sugars, light and phosphate, *Plant Molec. Biol.*, 1995, **27**, 933-942.

95. HERTZBERG, M., ASPEBORG, H., SCHRADER, J., ANDERSSON, A., ERLANDSSON, R., BLOMQVIST, K., BHALERAO, R., UHLEN, M., TEERI, T.T., LUNDEBERG, J., SUNDBERG, B., NILSSON, P., SANDBERG, G., A transcriptional roadmap to wood formation, *Proc. Natl. Acad. Sci. USA*, 2001, **98**, 14732-14737.

96. COOKE, J.E.K., BROWN, K.A., WU, R., DAVIS, J.M., Gene expression associated with N-induced shifts in resource allocation in poplar, *Plant Cell Env.*, 2003, **26**, 757-770.

97. KOHLER, A., DELARUELLE C., MARTIN, D., ENCELOT N., MARTIN F., The poplar root transcriptome: analysis of 7000 expressed sequence tages, *FEBS Lett.*, 2003, **542**, 37-41.

98. ANDERSSON, A., KESKITALO, J., SJODIN, A., BHALERAO, R., STERKY, F., WISSEL, K., TANDRE, K., ASPEBORG, H., MOYLE, R., OHMIYA, Y., BRUNNER, A., GUSTAFSSON, P., KARLSSON, J., LUNDEBERG, J., NILSSON, O., SANDBERG, G., STRAUSS, S., SUNDBERG, B., UHLEN, M., JANSSON, S., NILSSON, P., A transcriptional timetable of autumn senescence, *Genome Biol.*, 2004, **5**.

99. STERKY, F., BHALERAO, R.R., UNNEBERG, P., SEGERMAN, B., NILSSON, P., BRUNNER, A.M., CHARBONNEL-CAMPAA, L., LINDVALL, J.J., TANDRE, K., STRAUSS, S.H., SUNDBERG, B., GUSTAFSSON, P., UHLEN, M., BHALERAO, R.P., NILSSON, O., SANDBERG, G., KARLSSON, J., LUNDEBERG, J., JANSSON, S., A *Populus* EST resource for plant functional genomics, *Proc. Natl. Acad. Sci. USA*, 2004, **101**, 13951-13956.

100. MEKHEDOV, S., DE ILARDUYA, O.M., OHLROGGE, J., Toward a functional catalog of the plant genome. A survey of genes for lipid biosynthesis, *Plant Physiol*, 2000, **122**, 389-401.

101. CHEONG, Y.H., CHANG, H.S., GUPTA, R., WANG, X., ZHU, T., LUAN, S., Transcriptional profiling reveals novel interactions between wounding, pathogen, abiotic stress, and hormonal responses in Arabidopsis, *Plant Physiol*, 2002, **129**, 661-677.

102. SCHENK, P.M., KAZAN, K., WILSON, I., ANDERSON, J.P., RICHMOND, T., SOMERVILLE, S.C., MANNERS, J.M., Coordinated plant defense responses in Arabidopsis revealed by microarray analysis, *Proc. Natl. Acad. Sci. USA*, 2000, **97**, 11655-11660.

103. PECHAN, T., COHEN, A., WILLIAMS, W.P., LUTHE, D.S., Insect feeding mobilizes a unique plant defense protease that disrupts the peritrophic matrix of caterpillars, *Proc. Natl. Acad. Sci. USA*, 2002, **99**, 13319-13323.

104. KONNO, K., HIRAYAMA, C., NAKAMURA, M., TATEISHI, K., TAMURA, Y., HATTORI, M., KOHNO, K., Papain protects papaya trees from herbivorous insects: Role of cysteine proteases in latex, *Plant J.*, 2004, **37**, 370-378.

105. VOELCKEL, C., BALDWIN, I.T., Herbivore-induced plant vaccination. Part II. Array-studies reveal the transience of herbivore-specific transcriptional imprints and a distinct imprint from stress combinations, *Plant J.*, 2004, **38**, 650-663.

106. HERMSMEIER, D., SCHITTKO, U., BALDWIN, I.T., Molecular interactions between the specialist herbivore *Manduca sexta* (Lepidoptera, Sphingidae) and its natural host Nicotiana attenuata. I. Large-scale changes in the accumulation of growth- and defense-related plant mRNAs, *Plant Physiol.*, 2001, **125**, 683-700.

107. GATEHOUSE, J.A., Plant resistance towards insect herbivores: A dynamic interaction, *New Phytol.*, 2002, **156**, 145-169.

108. CREELMAN, R.A., MULLET, J.E., Biosynthesis and action of jasmonates in plants, *Annu. Rev. Plant Physiol. Plant Molec. Biol.*, 1997, **48**, 355-381.

109. DAVIS, J.M., GORDON, M.P., SMIT, B.A., Assimilate movement dictates remote sites of wound-induced gene expression in poplar leaves, *Proc. Natl. Acad. Sci. USA.*, 1991, **88**, 2393-2396.

110. ARNOLD, T.M., SCHULTZ, J.C., Induced sink strength as a prerequisite for induced tannin biosynthesis in developing leaves of *Populus*, *Oecologia*, 2002, **130**, 585-593.

111. JONES, C.G., HOPPER, R.F., COLEMAN, J.S., KRISCHNIK, V.A., Control of systemically induced herbivore resistance by plant vascular architecture, *Oecologia*, 1983, **93**, 452-456.

112. SCHITTKO, U., BALDWIN, I.T., Constraints to herbivore-induced systemic responses: Bidirectional signaling along orthostichies in *Nicotiana attenuata, J. Chem. Ecol.*, 2003, **29**, 763-770.

113. LI, C.Y., LIU, G.H., XU, C.C., LEE, G.I., BAUER, P., LING, H.Q., GANAL, M.W., HOWE, G.A., The tomato Suppressor of prosystemin-mediated responses2 gene encodes a fatty acid desaturase required for the biosynthesis of jasmonic acid and the production of a systemic wound signal for defense gene expression, *Plant Cell*, 2003, **15**, 1646-1661.

114. DICKE, M., HILKER, M., Induced plant defences: From molecular biology to evolutionary ecology, *Basic Appl. Ecol.*, 2003, **4**, 3-14.

115. ALBORN, T., TURLINGS, T.C.J., JONES, T.H., STENHAGEN, G., LOUGHRIN, J.H., TUMLINSON, J.H., An elicitor of plant volatiles from beet armyworm oral secretion, *Science*, 1997, **276**, 945-949.

116. HALITSCHKE, R., SCHITTKO, U., POHNERT, G., BOLAND, W., BALDWIN, I.T., Molecular interactions between the specialist herbivore *Manduca sexta* (Lepidoptera, Sphingidae) and its natural host *Nicotiana attenuata*. III. Fatty acid-amino acid conjugates in herbivore oral secretions are necessary and sufficient for herbivore-specific plant responses, *Plant Physiol.*, 2001, **125**, 711-717.

117. SCHITTKO, U., HERMSMEIER, D., BALDWIN, I.T., Molecular interactions between the specialist herbivore *Manduca sexta* (Lepidoptera, Sphingidae) and its natural host *Nicotiana attenuata*. II. Accumulation of plant mRNAs in response to insect-derived cues, *Plant Physiol.*, 2001, **125**, 701-710.

Chapter Six

TROPICAL FORESTS AS SOURCES OF NATURAL INSECTICIDES

Murray B. Isman

Faculty of Agricultural Sciences
University of British Columbia
Vancouver, Canada V6T 1Z4

Address for correspondence, email: murray.isman@ubc.ca

INTRODUCTION

It is commonly held that tropical and subtropical plants are richer sources of defensive allelochemicals than their temperate zone counterparts because the pressure from insect herbivores, plant pathogens, and grazing mammals is substantially greater in tropical ecosystems.   Indeed, sources of the botanical insecticides rotenone (*Derris* and *Lonchocarpus*, Leguminosae) and nicotine (*Nicotiana*, Solanaceae) are tropical in origin, but the most widely used botanical, pyrethrum (from *Tanacetum cinerariorum*, Asteraceae), originates from and grows best in more temperate climates.   A number of factors together have provided the impetus for continued phytochemical exploration of tropical plants in pursuit of new botanical (naturally-sourced) insecticides or as agrochemical leads. These include: 1) the 'western discovery' of the profound insect control properties of extracts from the neem tree (*Azadirachta indica*, Meliaceae) and their commercialization as botanical insecticides in India, the U.S.A., and Germany; 2) the continuing belief that new pesticides and medicines with fewer detrimental health and environmental effects are yet to be discovered from plants; 3) rampant deforestation with resulting loss of biodiversity in Old and New World tropical rainforests; and 4) the hope that discovery of valuable non-timber forest products in tropical regions will produce competition for logging and justify forest conservation, particularly in developing countries.

Since the present subject was last reviewed over a decade ago,[1-5] numerous reports on the bioactivity of allelochemicals from tropical trees against insects have appeared, but with little commercial exploitation of the source species.   In this chapter, I review some recent work on tropical trees from the families Meliaceae, Annonaceae, and Myrtaceae, each of which shows promising bioactivity against insects, and may be, at the least, suitable for indigenous use as crop protectants, if not for harvest, refinement, and export for crop protection in industrialized countries.

SPECIES OF MELIACEAE AS SOURCES OF NATURAL INSECTICIDES

There can be no doubt that interest in the mahogany family (Meliaceae) was spurred by the development of natural insecticides from the Indian neem tree, *Azadirachta indica* A. Juss. This species and its commercial utility has been the subject of several international conferences, hundreds of research papers, and at least a dozen major volumes.[6-7] This interest was manifest in the collection and screening of more than 100 species from the family for insect bioactivity, with particular emphasis on the genera *Aglaia* and *Trichilia*. Even before this, attention had turned to members of the genus most closely related to neem, *Melia*, namely the Asian chinaberry tree *M. azedarach* L. and the East African *M. volkensii* Gurke.

Melia volkensii

This species grows at moderate elevations in Kenya and adjoining east African countries. A crude fruit extract was first recognized as an antifeedant to desert locusts over twenty years ago,[8] but extracts have subsequently been shown to be toxic to or interfere with growth of a wide range of insect pests.[9] At least eight limonoids have been isolated from the fruits; although the putative active ingredients from neem seeds, the azadirachtins (**1,2**), do not occur in *M. volkensii*, other limonoids such as salannin occur in both species. A major and unique constituent, volkensin (**3**), is an effective antifeedant, but none of the compounds isolated shows insect growth regulatory (IGR) activity in the Mexican bean beetle, *Epilachna varivestis* (Fig. 6.1).

In our hands, a fruit extract of *M. volkensii* proved to be a potent antifeedant and larval growth inhibitor to both the cabbage looper, *Trichoplusia ni* (Noctuidae) and the armyworm, *Pseudaletia unipuncta* (Noctuidae).[10] However, unlike the armyworm, the looper is capable of habituating to the extract, *i.e.* feeding deterrence diminishes following continuous exposure.[11]

This fast-growing species begins to set fruit in 4-5 years, and unlike neem, produces fruit almost year-round, making it suitable for small-scale regional use in east Africa. On the other hand, the complex chemistry and dearth of proper toxicological data make it highly unlikely to be developed for use in tightly regulated countries.

Melia azedarach

The chinaberry tree, native to eastern Asia, has been widely introduced and cultivated throughout the tropics and subtropics. Fruit extracts of *M. azedarach* have long been known to have insecticidal activities, in some cases comparable to that of neem.[12] Like *M. volkensii*, *M. azedarach* lacks azadirachtins, but fruits produce the chemically related meliacarpins (*e.g.* **4**) that have azadirachtin-like bioactivity in many types of insect.[13] Unlike neem, exploitation of chinaberry extracts for crop protection has been avoided owing to the presence of meliatoxins (*e.g.* **5**), limonoid constituents with demonstrated toxicity to mammals. However, fruit extracts from *M. azedarach* growing in Argentina lack toxicity in rats, and moreover, contain a unique limonoid, meliartenin (**6**), that has both antifeedant and insecticidal activities against a range of insects[14] (Fig. 6.1).

	R₁	R₂	R₃	R₄	
1	OTig	OAc	COOCH₃	OH	Azadirachtin
2	OH	OTig	COOCH₃	H	3-Tigloylazadirachtol ('Aza B')
8	OTig	OAc	H	OAc	Marrangin
9	OBenz	OAc	COOCH₃	OH	1-Benzoyl-1-detigloylazadirachtin
10	OH	OIsoval	COOCH₃	H	3-Isovaleroylazadirachtol

3 Volkensin

4 1-Cinnamoyl-3-feruoyl-11-hydroxymeliacarpin

R = MeCH₂CHMeCO

5 Meliatoxin A₁

6 Meliartenin

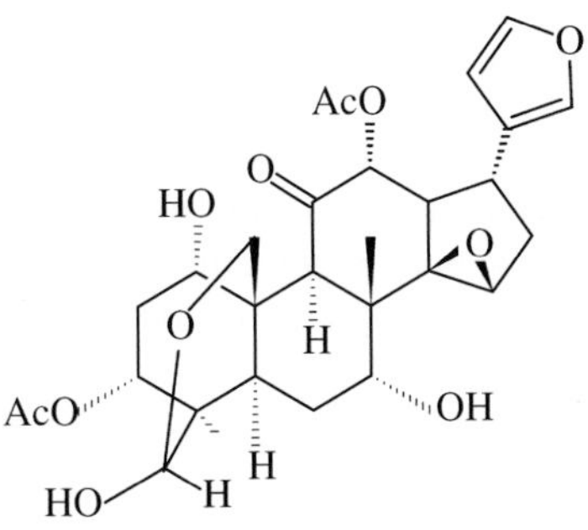

7 Toosendanin

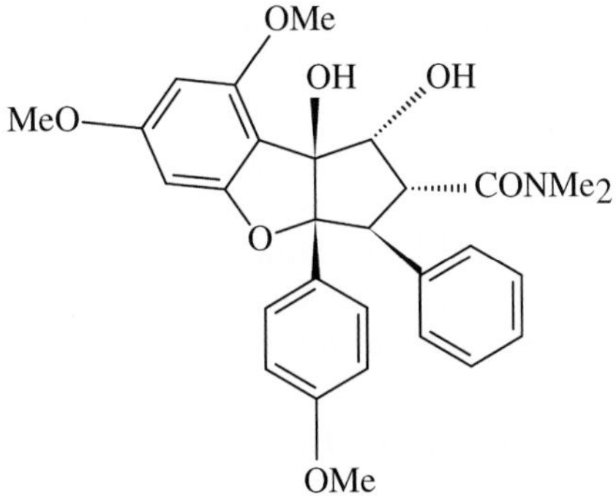

11 Rocaglamide

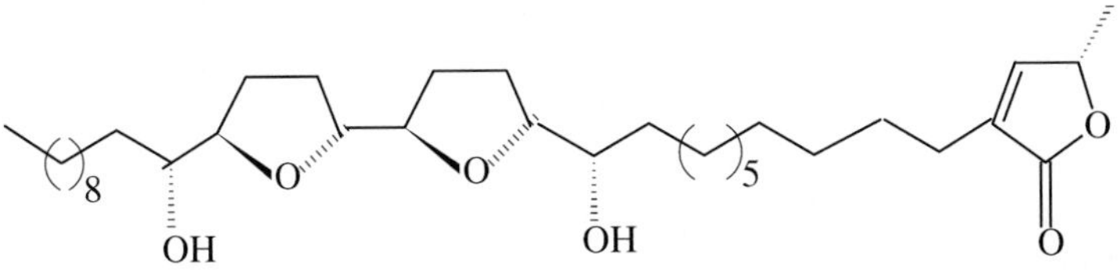

12 Squamocin-I

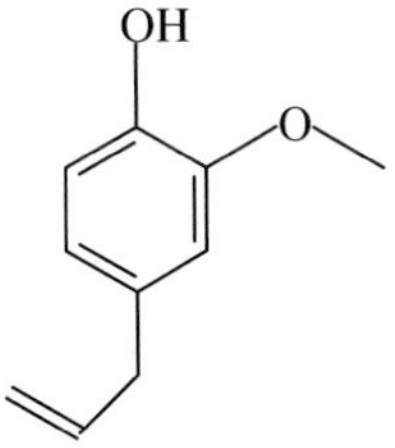

13 Eugenol 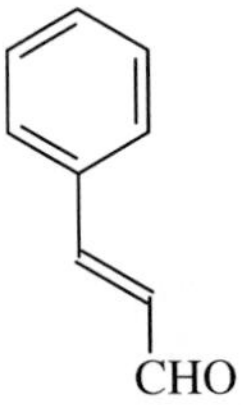   **14** Cinnamaldehyde 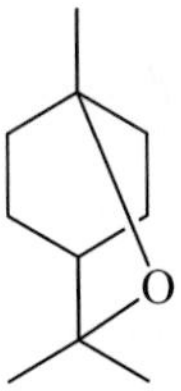   **15** 1,8-Cineole

Fig. 6.1: Chemical structures of natural insecticides in tropical plants.

One form of *M. azedarach* native to China, is recognized as a distinct species (*M. toosendan* Sieb. & Zucc.) by botanists in that country. A wide range of bioactive limonoids have been isolated from the seeds of this tree, including salannin, meliacarpins, and an analogue of volkensin.[15] The most notable constituent however is toosendanin (**7**), which occurs in the bark at concentrations as high as 0.5%. A semi-refined bark extract containing toosendanin has been commercialized in China as a botanical insecticide. The extract contains a series of analogues of toosendanin, all of which have antifeedant effects against insects.[5]

Azadirachta excelsa

Ermel et al.[16] first reported the insect growth regulatory activity of an extract from the seeds of *A. excelsa* (Jack), as well as the isolation of a novel limonoid they called marrangin (**8**), an analogue of azadirachtin. Other investigators have isolated additional limonoids from the seeds or fruits of this lowland rainforest species, but azadirachtin itself appears to be absent. Random screening of sawdust from Malaysian timber species led to the discovery of potent insecticidal activity from one species. Bioassay driven fractionation of the wood extract led to the isolation of azadirachtin B (**2**) and three other azadirachtin analogues, two of which were novel (**9,10**) (Fig. 6.1).[17] The wood was later determined to be from *A. excelsa*. Given that crude extracts of *A. excelsa* stemwood are comparable in efficacy to that of neem seed extracts, and *A. excelsa* has recently been extensively planted in Malaysia as a teak substitute, a patent has been issued on the use of wood extracts for insect control, and commercialization of a botanical insecticide based on this species appears imminent.

Trichilia americana

Earlier systematic screening of the Meliaceae revealed significant bioactivity in several species in the genus *Trichilia*.[1,18-19] Among these, the Costa Rican species *T. americana* (Sesse & Mocino) Pennington proved particularly inhibitory to the Asian armyworm *Spodoptera litura*.[19] Methanolic extracts of twigs of this species dramatically prolonged larval and pupal development and decreased pupal and adult weights at dietary concentrations of 10-75 ppm fresh weight. More detailed investigation of the bioactivity of the extract indicated that the effect was a consequence of feeding inhibition alone: neither topical administration nor abdominal injection of the extract resulted in any toxicity, and when armyworms were removed from diet with the extract and allowed to feed on untainted diet, normal growth ensued.[20] In a laboratory experiment, cabbage plants sprayed with a 0.5% methanolic solution of the twig extract were almost completely protected from herbivory by 4th instar armyworms for 24 hours (Fig. 6.2).[21]

Bioassay-driven fractionation of the extract was performed in an attempt to isolate the active principle(s) therein, but was unsuccessful. The most active fraction following preparative HPLC remained chemically complex, and the putative active constituents appeared difficult to resolve and present in very small quantities. Based on phytochemical investigations of *Trichilia* species by others, it is likely that the active principles are limonoids related to hirtin.[18] The genus, which contains more than 200 species with its center of diversity in Brazil, retains considerable scope for further phytochemical exploration and biological evaluation.

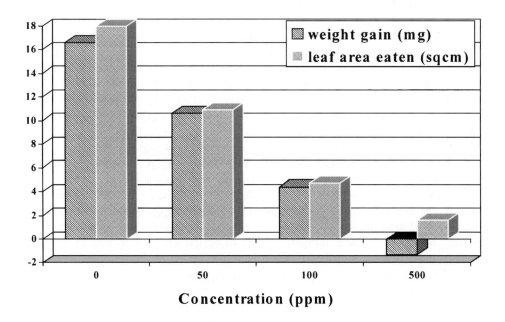

Fig. 6.2: Larval weight gain and leaf area eaten by 4th instar *Spodoptera litura* on cabbage plants sprayed with a crude twig extract from *Trichilia americana*. Adapted from reference 21.

Aglaia species

Interest in this large tropical genus was stimulated by the discovery of potent insecticidal activity in the frequently cultivated Asian species *A. odorata* Lour. Screening and phytochemical studies indicate that numerous other species in the genus are biologically active.[22-24] Bioactivity to insects can be attributed to a series of highly modified benzofurans related to rocaglamide (**11**) (Fig. 6.1); four such compounds were originally isolated from *A. odorata*.[24] In the decade since then, numerous compounds of the rocaglamide-type have been isolated from additional species of *Aglaia*, most of which inhibit insect growth or deter feeding.[25-26] Excepting azadirachtin, rocaglamide is the most potent natural insecticide isolated from higher plants to date. The mode of action of rocaglamide is inhibition of protein synthesis,[27] which accounts for its potent but slow action against insects.[28]

SPECIES OF ANNONACEAE AS SOURCES OF NATURAL INSECTICIDES

The Annonaceae, or custard apple family, is a large family of predominantly tropical trees and shrubs consisting of more than 2300 species. While certain species have enjoyed traditional use as vermifuges and for repelling lice, the broad spectrum insecticidal properties of twig extracts of the North American paw paw tree (*Asimina triloba* Dunal.) and seed extracts of the tropical sweetsop (*Annona squamosa* L.) and soursop (*A. muricata* L.) were only well documented in the past twenty years.[29-30] McLaughlin and colleagues in particular have isolated more than 100 acetogenins (*e.g.* **12**) (Fig. 6.1) – C-32 or C-34 linear fatty acids containing a 2-propanol unit to form a γ-lactone. These substances, found exclusively in the Annonaceae, are not only the insecticidal principles, but also potent anti-tumor agents. While McLaughlin has demonstrated both the efficacy of standardized extracts against pest insects and their relative safety to mammals, regulatory costs have prevented their commercialization to date.[30] Acetogenins are mitochondrial poisons, inhibiting cellular energy production through a mode of action identical to that of the well known botanical insecticide and fish poison, rotenone.[31]

Another approach to the utilization of these natural substances is the preparation of crude extracts of sweetsop and soursop seeds in developing countries for local use as crop protectants. For example, these species are widely cultivated in eastern Indonesia for the edible fruit and fruit juices they provide; in this case, the seeds are simply waste products that could be collected at minimal cost.

Annona squamosa

To determine the feasibility of the above-noted approach, seeds from soursop and sweetsop trees were collected in and around Ambon, Indonesia from 1996 to 1999. Collections of seeds from different locations and in different years were individually extracted in methanol, and their larval growth inhibitory properties were evaluated against the Asian armyworm *Spodoptera litura*. We observed around 8-fold variation in bioactivity among samples of sweetsop (*A. squamosa*) and around 6-fold variation among samples of soursop (*A. muricata*). However, sweetsop seed extracts were, on average, about 20 times more potent than those from soursop seeds.[32] Both geographic and temporal variations were significant (Fig. 6.3).

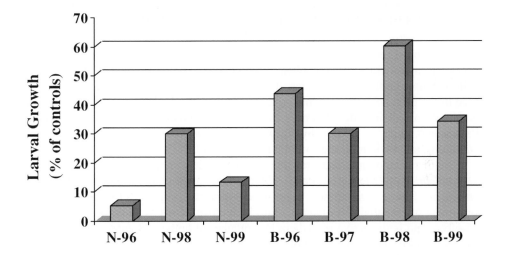

Fig. 6.3: Growth inhibitory effect of crude seed extracts of sweetsop (*Annona squamosa*) from different locations in Indonesian and years of collection on neonate *Spodoptera litura*. N = Namlea (1996-1999); B = Batugantung (1996-1999). All extracts tested at a dietary concentration of 250 ppm fwt. Adapted from reference 32.

Comparisons of the bioactivity of aqueous emulsions of methanolic seed extracts of *A. squamosa* with direct aqueous seed extracts indicated that the former are approximately 20 times more potent to 1st instar cabbage loopers (*Trichoplusia ni*) and 10 times more potent to 1st instar diamondback moth larvae (*Plutella xylostella*). Moreover, the latter insect is 13-25 times more susceptible than the former.[33]

Based on these results, we tested the efficacy of aqueous emulsions and direct aqueous extracts against diamondback moth larvae on cabbage plants in a greenhouse trial. Aqueous emulsions of seed extract as low as 0.5% concentration produced >80% larval mortality, and were superior to the use of 1% rotenone dust, a commercial botanical insecticide. An aqueous extract at 7.5% concentration produced >90% larval mortality, comparable to that obtained with 0.1% pyrethrum, another commercial botanical product.[34]   Overall, these results suggest that crude local preparations based on sweetsop seeds could be effective as low cost crop protectants where the seeds are readily available.

CLOVE (MYRTACEAE) AND OTHER ESSENTIAL OIL-BEARING PLANTS AS SOURCES OF NATURAL INSECTICIDES

The most recent group of botanical products that have seen some commercial success as insecticides are the plant essential oils. Though some of these have traditional uses dating back decades, if not longer, commercialization has only taken place in the past 7-8 years. Plants producing essential oils that have been exploited for insect control include a number of herbs, most notably from the mint family (Lamiaceae), such as garden thyme (*Thymus vulgaris* L.), rosemary (*Rosmarinus officinalis*), and various species of mint (*Mentha* spp.).[35-36]

Other important sources are tropical trees, notably clove (*Syzygium aromaticum* [L.] Merr. et Perry, Myrtaceae) and cinnamon (*Cinnamomum zealanicum* Blume, Lauraceae). The phenyl propene, eugenol (**13**), is the major constituent of essential oils obtained from both clove buds (90+%) and clove leaves (45-60%), and is the active ingredient in a number of home and garden insecticides. It is particularly useful in this application owing to its rapid knockdown action on flies and crawling insects, including cockroaches, but is toxic to a range of agricultural pests as well.[37]   Interestingly, eugenol is also widely used as an anaesthetic in fish research, and at high rates, as a broad spectrum natural herbicide. Essential oils prepared from cinnamon bark are rich in cinnamaldehyde (**14**), a substance with a number of useful biological actions. Products based on cinnamon oil have proven useful in the control of greenhouse pests and the fungal plant disease powdery mildew, but phytotoxicity has been a concern in commercial greenhouses growing ornamental plants. The long-standing cultivation of cloves and cinnamon in Southeast Asia and their worldwide use as culinary spices has led to their ready availability and relatively low cost, attractive features for their development and marketing as natural pesticides. Evaluation of essential oils prepared from related species of *Syzygium* and *Cinnamomum* for their pest control properties would appear to have merit. Finally, essential oils from *Eucalyptus* species (Myrtaceae), rich in the monoterpene 1,8-cineole (**15**), may have potential for exploitation as natural insecticides based on recent reports.[38] Evidence for this comes from the fact that the essential oil of rosemary (*Rosmarinus officinalis*, Lamiaceae) is the active ingredient in two botanical insecticides currently used in the United States (Hexacide[TM] and Ecotrol[TM]), and 1,8-cineole typically makes up approximately 50% of rosemary oil by weight (Fig. 6.1).

COMPARATIVE BIOACTIVITY OF SELECTED EXTRACTS AGAINST NOCTUID CATERPILLARS

Although all of the aforementioned plants have been demonstrated to produce extracts or oils with pronounced bioactivity against insects, few laboratories use the same insect species or employ the same bioassay protocols making direct comparisons of bioactivity impossible. Thus, my colleagues and I undertook a series of experiments in my laboratory to directly compare the efficacy of a number of botanical preparations against two important agricultural pests, the cabbage looper *Trichoplusia ni* and the true armyworm *Pseudaletia unipuncta*. To evaluate toxicity and growth inhibitory activities, neonate larvae of each species (n = 20) were placed on leaf discs (1.77 cm^2; cabbage for the looper, canary grass for the armyworm) treated with 10 μl (5 μl each side) of an aqueous emulsion of each botanical at concentrations of 0.25, 0.5, 1, or 2%. The botanicals tested in the experiments are listed in Table 61. Larvae were allowed to feed on the discs for 3 days, after which those surviving were placed individually on artificial diet and allowed to feed for a further 4 days. Larvae were weighed after the 7th day as a measure of larval growth. To measure feeding deterrence directly, 3rd instar larvae of both species were offered leaf discs treated with the test botanicals in a leaf disc choice test described elsewhere.[10]

Table 6.1: Botanical extracts tested for toxicity, larval growth inhibition and feeding deterrence in the cabbage looper and true armyworm.

Plant	Type of Extract	Active ingredient(s)
Annona squamosa (sweetsop)	Crude MeOH extract of seeds	Acetogenins (eg. squamocin-I); content unknown
Azadirachta indica (neem)	Refined extract of seeds	Azadirachtin (31%), azadirachtin B (6%)
Azadirachta excelsa (sentang)	Crude MeOH extract of stemwood	Azadirachtin analogues; content unknown
Melia toosendan (syn. *M. azedarach*)	Crude extract of bark	Toosendanin (3%) and limonoid analogues
Melia volkensii	Semi-refined extract of seeds	Mixture of limonoids; content unknown
Ryania speciosa	Dust containing powdered stemwood (50%)	Ryanodine (0.05%) and related alkaloids
Syzygium aromaticum (cloves)	Essential oil of leaves	60% eugenol
Trichilia americana	Crude MeOH extract of twigs	Unknown (limonoids?)

Extracts from both *Azadirachta* species and from *Melia volkensii* strongly inhibit early larval growth in both noctuid species, although the armyworm (*Pseudaletia*) is somewhat less susceptible (Table 6.2). In terms of insecticidal action, neem is most effective against the looper (*Trichoplusia*), whereas *M. volkensii* is most effective against the armworm. Extracts of *Ryania speciosa* wood were less active than the meliaceous species, but the level of active principle (ryanodine) is known to be low in this material (Table 6.1). The bark extract of *M. toosendan* was the least active in our bioassays, although this was expected in that the level of toosendan is rather low (3%), and this compound in purity is significantly less active than the azadirachtins found in neem (*A. indica*) and sentang (*A. excelsa*).

With respect to feeding deterrence in older larvae, the *M. volkensii* extract proved to be the most potent against the looper, whereas the neem extract was the most potent against the armyworm (Table 6.3). The sentang extract was somewhat less active against both species, but all of these were substantially more active than the other three extracts. *M. toosendan* and clove leaf oil (*Syzygium*) equally deterred both noctuid species, and were somewhat comparable in activity. *Ryania* was relatively ineffective as a feeding deterrent.

Table 6.2: Toxicity and growth inhibition by selected plant extracts in two species of noctuid caterpillars.

	Trichoplusia ni		*Pseudaletia unipuncta*	
	LC_{50} (%)*	EC_{50} (%)**	LC_{50} (%)*	EC_{50} (%)**
Annona squamosa	≪ 0.10	≪ 0.10	--	--
Azadirachta indica	≪ 0.10	≪ 0.10	0.83	≪ 0.10
Melia volkensii	0.15	≪ 0.10	0.61	0.24
Azadirachta excelsa	0.21	≪ 0.10	≫ 2.0	0.40
Trichilia americana	0.12	0.52	--	--
Ryania speciosa	0.35	0.65	0.88	≫ 2.0
Melia toosendan	≫ 2.0	--	~ 3.0	~ 2.3

*concentration causing 50% mortality, and **concentration causing 50% growth inhibition; both shown as % aqueous concentrations of extracts applied to leaf discs fed to neonate caterpillars for three days; larval weight and survival determined after seven days altogether

Table 6.3: Feeding deterrence by selected plant extracts in two species of noctuid caterpillars.

	Trichoplusia ni DC_{50} ($\mu g\ cm^{-2}$)*	*Pseudaletia unipuncta* DC_{50} ($\mu g\ cm^{-2}$)*
Azadirachta indica	21.9	0.6
Melia volkensii	5.8	10.8
Azadirachta excelsa	36.7	46.9
Ryania speciosa	725	400
Melia toosendan	288	249
Syzygium aromaticum	217	206

* concentration causing 50% feeding deterrence

SUMMARY

Numerous investigations worldwide have demonstrated that certain species of tropical trees produce phytochemicals with potent bioactivities against insects. A mere handful of these species have been exploited in the commercial development of botanical insecticides for crop protection, largely because bioactivity to target pests is only one of many criteria that must be satisfied for commercial development to be feasible. Other equally (if not more) important criteria are the availability of the needed starting biomass on a sustainable basis, modest cost of extraction and refinement, and stability of the active principles upon storage of the extract. Stringent regulatory requirements and the costs associated with meeting those requirements preclude access of many botanicals to the most lucrative markets in Europe and North America.

Among the botanicals discussed here, only neem has seen commercial success in North America, and that success has been limited. Yet the same factors that limit the introduction and success of botanical insecticides in highly industrialized countries should enhance their adoption in the developing countries where many of the source species grow. Many of the species discussed here are more suitable for local development and use in tropical countries where issues of standardization and stability are less critical than those of absolute cost. *Annona squamosa* is a good example of this; in rural Indonesia the starting material (seeds) is widely available at little cost, and a crude extract can be effective as a crop protectant. The same is probably true for *Melia volkensii* in its native East Africa.

At the other end of the availability spectrum, plant essential oils (viz. clove, cinnamon, eucalyptus) are currently exploited for other uses (in aromatherapy and as herbal remedies, or as flavorings) and, therefore, the infrastructure for their production is already in place, reducing their prices on the global market. Use of plant essential oils as pesticides at this point is in its infancy.

That so few botanical insecticides discovered or developed in the past 25 years have become established in the marketplace should not lessen our quest for additional botanical sources of phytochemicals with potential use in crop protection. Screening studies repeatedly confirm that the probability of finding additional species with useful bioactivity is high. We are surely more limited by scientific resources needed to characterize active principles and demonstrate their bioactivities than by the actual numbers of species in tropical forests suitable for use that remain to be investigated. It is my sincere hope that we have time to study these species before they are lost forever.

ACKNOWLEDGMENTS

I thank Dr. Yasmin Akhtar, Sylvia Altius, Veronica Robertson, and Ikkei Shikano for conducting bioassays as reported in Tables 6.2-3, and Nancy Brard for insect rearing. Supported by a Discovery grant from the Natural Sciences and Engineering Research Council of Canada (NSERC).

REFERENCES

1. ARNASON, J.T., MACKINNON, S., DURST, A., PHILOGENE, B.J.R., HASBUN, C., SANCHEZ, P., POVEDA, L., SAN ROMAN, L., ISMAN, M.B., SATASOOK, C., TOWERS, G.H.N., WIRIYACHITRA, P., MCLAUGHLIN, J.L., Insecticides in tropical plants with non-neurotoxic modes of action, *in*: Phytochemical Potential of Tropical Plants (K.R. Downum, J.T. Romeo, and J.A. Stafford, eds.), Plenum Press, New York. 1993, pp. 107-131.
2. MACKINNON, S., CHAURET, D., WANG, M., MATA, R., PEREDA-MIRANDA, R., JIMINEZ, A., BERNARD, C.B., KRISHNAMURTY, H.G., POVEDA, J., SANCHEZ-VINDAS, P.E., ARNASON, J.T., DURST, J., Botanicals from the Piperaceae and Meliaceae of the American neotropics: phytochemistry, *in*: Phytochemicals for Pest Control (P.A. Hedin, R.M. Hollingworth, E.P. Masler, J. Miyamoto, and D.G. Thompson, eds.), American Chemical Society, Washington. 1997, pp. 49-57.
3. ISMAN, M.B., GUNNING, P.J., SPOLLEN, K.M., Tropical timber species as sources of botanical insecticides, *in*: Phytochemicals for Pest Control (P.A. Hedin, R.M. Hollingworth, E.P. Masler, J. Miyamoto, and D.G. Thompson, eds.), American Chemical Society, Washington. 1997, pp. 27-37.

4. ISMAN, M.B., Leads and prospects for the development of new botanical insecticides, *in:* Reviews in Pesticide Toxicology, Volume 3 (R.M. Roe and R. J. Kuhr, eds.) Toxicology Communications Inc., Raleigh, 1995, pp. 1-20.

5. ISMAN, M.B., MATSUURA, H., MACKINNON, S., DURST, T., TOWERS, G.H.N., ARNASON, J.T., Phytochemistry of the Meliaceae: So many terpenoids, so few insecticides, *in:* Phytochemical Diversity and Redundancy in Ecological Interactions (J. T. Romeo, J. Saunders, and P. Barbosa, eds.), Plenum Press, New York. 1996, pp. 155-178.

6. SCHMUTTERER, H. (ed.), The Neem Tree, 2nd Edition. Neem Foundation, 2002, 893 p.

7. KOUL, O., WAHAB, S. (eds.), Neem: Today and in the New Millennium. Kluwer, Dordrecht, 2004, 276 p.

8. MWANGI, R.W., Locust antifeedant activity in fruits of *Melia volkensii, Entomol. Exp. Appl.*, 1982, **32,** 277-280.

9. REMBOLD, H., MWANGI, R.W., *Melia volkensii* Gurke, *in:* The Neem Tree, 2nd Edition (H. Schmutterer, ed.), Neem Foundation, Mumbai. 2002, pp. 827-832.

10. AKHTAR, Y., ISMAN, M.B., Comparative growth inhibitory and antifeedant effects of plant extracts and pure allelochemicals on four phytophagous insect species, *J. Appl. Ent.*, 2004, **128,** 32-38.

11. AKHTAR, Y., ISMAN, M.B., Generalization of a habituated feeding deterrent response to unrelated antifeedants following prolonged exposure in a generalist herbivore, *Trichoplusia ni, J. Chem. Ecol.*, 2004, **30,** 1333-1346.

12. ASCHER, K.R.S, SCHMUTTERER, H., MAZOR, M., ZEBITZ, C.P.W., NAQVI, S.N.H.., The Persian lilac or chinaberry tree: *Melia azedarach* L., *in:* The Neem Tree, 2nd Edition (H. Schmutterer, ed.), Neem Foundation, Mumbai. 2002, pp. 770-820.

13. KRAUS, W., Azadirachtin and other triterpenoids, *in:* The Neem Tree, 2nd Edition (H. Schmutterer, ed.), Neem Foundation, Mumbai. 2002, pp. 39-111.

14. CARPINELLA, M.C., DEFAGO, M.T., VALLADARES, G., PALACIOS, S.M., Antifeedant and insecticide properties of a limonoid from *Melia azedarach* (Meliaceae) with potential use for pest management, *J. Agric. Food Chem.,* 2003, **51,** 369-374.

15. XU, H.-H., CHIU, S.-F., *Melia toosendan* Sieb. & Zucc., *in:* The Neem Tree, 2nd Edition (H. Schmutterer, ed.), Neem Foundation, Mumbai. 2002, pp. 821-826.

16. ERMEL, K., KALINOWSKI, H.-O., SCHMUTTERER, H., Isolierung und charakterisierung von marrangin, einer neuen insektenmetamorphose-storenden substanz aus samenkornern des marrango-baumes *Azadirachta excelsa* (Jack), *J. Appl. Ent.*, 1991, **112,** 512-519.

17. SCHMUTTERER, H.,ERMEL, K., ISMAN, M.B., The tiam, sentang or marrango tree: *Azadirachta excelsa* (Jack), *in:* The Neem Tree, 2nd Edition (H. Schmutterer, ed.), Neem Foundation, Mumbai. 2002, pp. 760-769.

18. XIE, Y.S., ISMAN, M.B., GUNNING, P., MACKINNON, S., ARNASON, J.T., TAYLOR, D.R., SANCHEZ, P., HASBUN, C., TOWERS, G.H.N., Biological activity of extracts of *Trichilia* species and the limonoid hirtin against lepidopteran larvae, *Biochem. Syst. Ecol.*, 1994, **22,** 129-136.

19. WHEELER, D.A., ISMAN, M.B., SANCHEZ-VINDAS, P., ARNASON, J.T., Screening of Costa Rican *Trichilia* species for biological activity against the larvae of *Spodoptera litura* (Lepidoptera: Noctuidae), *Biochem. Syst. Ecol.*, 2001, **29**, 347-358.

20. WHEELER, D.A., ISMAN, M.B., Antifeedant and toxic activity of *Trichilia americana* extract against the larvae of *Spodoptera litura*, *Entomol. Exp. Appl.*, 2001, **98**, 9-16.

21. WHEELER, D.A., ISMAN, M.B., Effect of *Trichilia americana* extract on feeding behavior of Asian armyworm, *Spodoptera litura*, *J. Chem. Ecol.*, 2000, **26**, 2791-2800.

22. JANPRASERT, J., SATASOOK, C., SUKUMALANAND, P., CHAMPAGNE, D.E., ISMAN, M.B., WIRIYACHITRA, P., TOWERS, G.H.N., Rocaglamide, a natural benzofuran insecticide from *Aglaia odorata*, *Phytochemistry*, 1993, **32**, 67-69.

23. SATASOOK, C., ISMAN, M.B., ISHIBASHI, F., MEDBURY, S., WIRIYACHITRA, P., TOWERS, G.H.N., Insecticidal bioactivity of crude extracts of *Aglaia* species, *Biochem. Syst. Ecol.*, 1994, **22**, 121-127.

24. ISHIBASHI, F., SATASOOK, C., ISMAN, M.B., TOWERS, G.H.N., Insecticidal 1H-cyclopentatetrahydro[*b*]benzofurans from *Aglaia odorata*, *Phytochemistry*, 1993, **32**, 307-10.

25. SCHNEIDER, C., BOHNENSTENGEL, F.I., NUBROHO, B.W., WRAY, V., WITTE, L., HUNG, P.D., KIET, L.C., PROKSCH, P., Insecticidal rocaglamide derivatives from *Aglaia spectabilis* (Meliaceae), *Phytochemistry*, 2000, **54**, 731-736.

26. GREGER, H., PACHER, T., BREM, B., BACHER, M., HOFER, O., Insecticidal flavaglines and other compounds from Fijian *Aglaia* species, *Phytochemistry*, 2001, *57*, 57-64.

27. OHSE, T., OHBA, S., YAMAMOTO, T., KOYANO, T., UMEZAWA, K., Cyclopentabenzofuran lignan protein synthesis inhibitors from *Aglaia odorata*, *J. Nat. Prod.*, 1996, **59**, 650-652.

28. SATASOOK, C., ISMAN, M.B., WIRIYACHITRA, P., Toxicity of rocaglamide, an insecticidal natural product to the variegated cutworm, *Peridroma saucia* (Lepidoptera: Noctuidae), *Pestic. Sci.*, 1993, **36**, 53-58.

29. RUPPRECHT, J.K., HUI, Y.-H., MCLAUGHLIN, J.L., Annonaceous acetogenins: A review, *J. Nat. Prod.*, 1990, **53**, 237-278.

30. MCLAUGHLIN, J.L., ZENG, L., OBERLIES, N.H., ALFONSO, D., JOHNSON, J.A., CUMMINGS, B.A., Annonaceous acetogenins as new natural pesticides: recent progress, *in*: Phytochemicals for Pest Control (P.A. Hedin, R.M. Hollingworth, E.P. Masler, J. Miyamoto, and D.G. Thompson, eds.), American Chemical Society, Washington. 1997, pp. 117-133.

31. LONDERSHAUSEN, M., LEICHT, W., LIEB, F., MOESSCHLER, H., Molecular mode of action of annonins, *Pestic. Sci.*, 1991, **33**, 427-433.

32. LEATEMIA, J.A., ISMAN, M.B., Insecticidal activity of crude seed extracts of *Annona* spp., *Lansium domesticum* and *Sandoricum koetjape* against lepidopteran larvae, *Phytoparasitica*, 2004, **32**, 30-37.

33. LEATEMIA, J.A., ISMAN, M.B., Toxicity and antifeedant activity of crude seed extracts of *Annona squamosa* (Annonaceae) against lepidopteran pests and natural enemies, *Intl. J. Tropical Insect Sci.*, 2004, **24**, 1-10.

34. LEATEMIA, J.A., ISMAN, M.B., Efficacy of crude seed extracts of *Annona squamosa* against diamondback moth, *Plutella xylostella* L. in the greenhouse, *Intl. J. Pest Mgmt.*, 2004, **50,** 129-133.
35. ISMAN, M.B., Pesticides based on plant essential oils, *Pestic. Outlook*, 1999, **10,** 68-72.
36. ISMAN, M.B., Plant essential oils as green pesticides for pest and disease management, *in*: Agricultural Applications in Green Chemistry (W.M. Nelson, ed.), American Chemical Society, Washington. 2004, in press.
37. ISMAN, M.B., Plant essential oils for pest and disease management, *Crop Protection*, 2000, **19,** 603-608.
38. CHOI, W.-I., LEE, S.-G., PARK, H.-M., AHN, Y.-J., Toxicity of plant essential oils to *Tetranychus urticae* (Acari: Tetranychidae) and *Phytoseiulus persimilis* (Acari: Phytoseiidae), *J. Econ. Entomol.*, 2004, **97,** 553-558.

Chapter Seven

HIGH MOLECULAR WEIGHT PLANT POPLYPHENOLS (TANNINS): PROSPECTIVE FUNCTIONS

Takashi Yoshida,* Tsutomu Hatano, Hideyuki Ito

Faculty of Pharmaceutical Sciences,
Okayama University,
Tsushima, Okayama 700-8530,
Japan

*Author for correspondence: yoshida@pheasant.pharm.okayama-u.ac.jp

INTRODUCTION

Antioxidant plant polyphenols have received increasing attention of late owing to their multifunctional properties, which have been shown to be beneficial to human health. Antioxidant polyphenols are believed to reduce the risk of chronic diseases such as cancer, diabetes, and coronary heart disease. As each of these diseases is associated with long-term accumulation of cellular damage due to oxidative stress, the antioxidant and free radical scavenging activities of polyphenols are believed to underlie the beneficial actions of these substances in humans.

Antioxidant polyphenols include various compounds that range from relatively low molecular weight compounds, such as flavonoids, to higher molecular weight polyphenols; the latter are classified as tannins. In contrast to flavonoids, the biological functions of the individual tannin constituents of medicinal plants, beverages, and plant-based foods were unclear until the late 1970s. However, recent progress in the study of tannin structures, in various plants (particularly medicinal plants), has made possible the investigation of the diverse physiological functions of individual molecules with defined structures.[1,2]

Evidence has been accumulating from *in vitro* and *in vivo* studies that tannins with high molecular weights have stronger or more characteristic effects than low molecular weight polyphenols. These effects include the inhibition of enzymes such as reverse transcriptase,[3] DNA topoisomerase-II,[4] and protein kinase C,[5] as well as lipid peroxidation.[6,7] Host-mediated antitumorigenic,[8] antiviral,[9-11] and anti-iflammatory[12] effects have also been reported. More recently, some tannins have been found that prevent cancer (*e.g.*, polyphenols from green tea), exhibit antibacterial activities against *Helicobacter pylori*,[13] and have synergistic effects with β-lactam antibiotics against methicillin-resistant *Staphylococcus aureus* (MRSA).[14-16] The aforementioned actions of tannins often depend on the structure or type (hydrolyzable versus condensed tannins). For example, some dehydroellagitannins, such as geraniin, exhibit immunomodulatory, antileishmanial activity against intracellular amastigotes of *Leishmania donovani* that reside within macrophage-like cells.[17,18] Moreover, several high molecular weight (>1500) oligomeric hydrolyzable tannins are stronger inhibitors of poly(ADP-ribose) glycohydrolase (the metabolism of which is associated with gene activation, *i.e.*, DNA repair, replication, and transcription) than are monomeric tannins, condensed tannins, and related low molecular weight polyphenols.[19,20] Consequently, the physiological effects of the high molecular weight class of polyphenols have been increasingly investigated.

This chapter provides an outline of the structural diversity of tannins, including condensed tannins and hydrolyzable tannins, after which selected biological functions of tannins, discovered primarily in our laboratory, are described in more detail.

STRUCTURAL DIVERSITY OF PLANT POLYPHENOLS (TANNINS)

Condensed tannins comprise one of two large groups of tannins, and are probably the most ubiquitous compounds in the plant kingdom. They are found in a wide range of plants, ranging from ferns and gymnosperms to angiosperms. Today, condensed tannins are designated as proanthocyanidins, because this type of polyphenol is degraded oxidatively to form reddish anthocyanindin pigments upon heating in acidic media.

Proanthocyanidins are composed of flavan-3-ols that are linked through single C-C bonds, usually C4→C6' or C4→C8' (B-type) or doubly linked with an additional bond at C2→O→C7' (A-type) (Fig. 7.1). The structural diversity of proanthocyanidins arises from differences both in the degree of polymerization and the pattern of hydroxylation of the constituent flavan-3-ol units, as well as from the different positions and configurations of interflavan linkages.[1] Galloylation or glycosidation of flavan-3-ol units at C-3 confers much structural complexity. Polyphenols from grape-seed[21] and cranberry (*Vaccnium macrocarpon* fruits)[22] are examples of 3-*O*-galloylated procyanidin oligomers and A-type proanthocyanidins oligomers, respectively. These polyphenols have attracted considerable interest owing to the ameliorative effects of red wine in people with cardiovascular disorders and the use of cranberry juice to treat urinary tract infections.

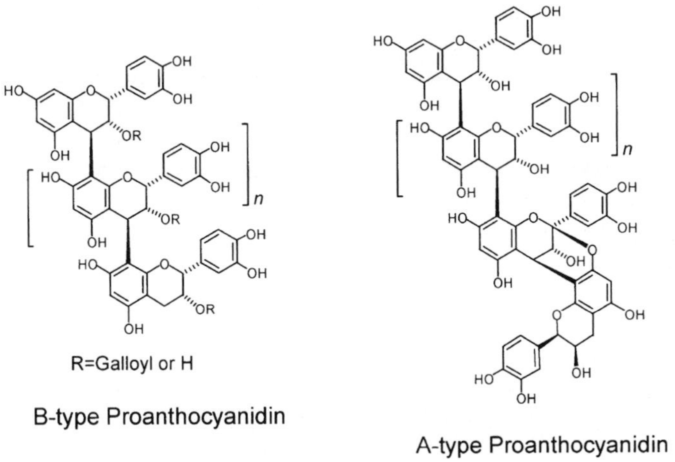

R=Galloyl or H

B-type Proanthocyanidin

A-type Proanthocyanidin

Fig. 7.1: Structures of galloylated procyanidin B oligomer (red wine) and cranberry tannin.

The other large group of tannins consists of hydrolyzable tannins (Fig. 7.2, 7.3). In contrast to the ubiquitous proanthocyanidins, hydrolyzable tannins are found in relatively few species of plants and occur only in dicotyledonous angiosperms (although there are a few exceptions).[2] Hydrolyzable tannins are essentially polyesters of glucose and either gallic acid or the dimeric metabolite of gallic acid, hexahydroxydiphenic acid; these compounds are referred to as gallotannins and

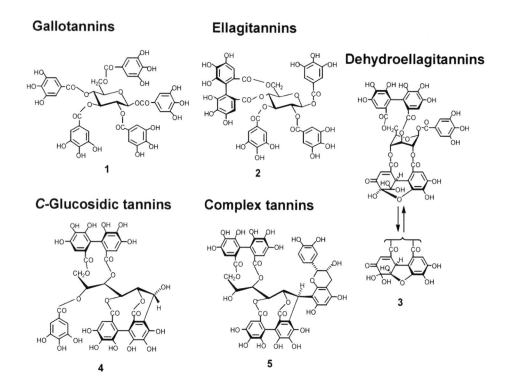

Fig. 7.2: Subgroups of hydrolyzable tannins: gallotannin, ellagitannin, dehydroellagitannin, *C*-glucosidic ellagitannin and complex tannin.

ellagitannins, respectively [*e.g.*, pentagalloylglucose (**1**) and tellimagrandin II (**2**)]. Besides several possible combinations of binding sites with respect to the galloyl and hexahydrodiphenoyl (HHDP) groups within the glucose core, further oxidative modifications of the HHDP group within the ellagitannin molecule lead to enormous structural diversity.[1,2]

Ellagitannins are subdivided into two groups: dehydroellagitannins, which have a dehydrohexahydroxydiphenoyl group [*e.g.*, geraniin (**3**)[23,24]], and *C*-glucosidic ellagitannins, which have an open-chain glucose residue [*e.g.*, casuarinin (**4**)[25]]. Among the *C*-glucosidic ellagitannins are the so-called 'complex tannins,' within which a *C*-glucosidic ellagitannin and a flavan unit are linked via a C-C bond between C-1 of the open-chain glucose and C-8 or C-6 of the flavan [*e.g.*, camelliatannin A (**5**)[26]].

Among the more than 500 hydrolyzable tannins that have been described to date, almost 40% are oligomeric ellagitannins, which are metabolites that are produced biogenetically through intermolecular C-O oxidative coupling(s) between the galloyl or HHDP group of one monomeric tannin and the HHDP or galloyl group of another monomeric tannin. The structural variety of oligomeric ellagitannins has been enhanced by multiplication of the number of monomer units, by variations in the linking modes, and by the extent of oligomerization among the monomers.

In the putative oligomerization modes, there is preferential C-O coupling to form a dehydrodigalloyl (DHDG) or valoneoyl group as the linking unit of monomers, as exemplified by agrimoniin (**6**)[27] from *Agrimonia pilosa* (Rosaceae) and cornusiin A (**7**)[28] from *Cornus officinalis* (Cornaceae). It is noteworthy that the linking modes often seem to be specific to a particular plant family. For example, tamaricaceous trees produce characteristic oligomers, such as hirtellins A (**8**)[29] and T_1 (**9**),[30] that have DHDG linking unit(s) in which the C-O coupling occurs between the galloyl groups at C-2 of one monomer and C-1 of another monomer; this is unlike the coupling that characterizes **6**. Onagraceous and euphorbiaceous plants are examples of families in which plants produce oligomers in which the valoneoyl group is the linking unit, such as oentheins B (**10**)[31] from *Oenothera erythrosepala* and euphorbin A (**11**)[32] from *Euphorbia hirta*. *C*-Glucosidic dimers are produced through intermolecular C-C coupling between a C-1 of the *C*-glucosidic tannin monomer and a galloyl group of another molecule, as exemplified by casuglaunin A (**12**) from *Elaeagnus umbellata*.[33]

6

7

8

10

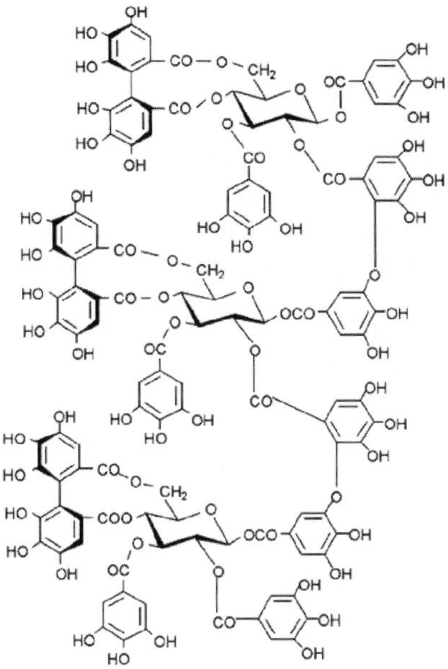

9

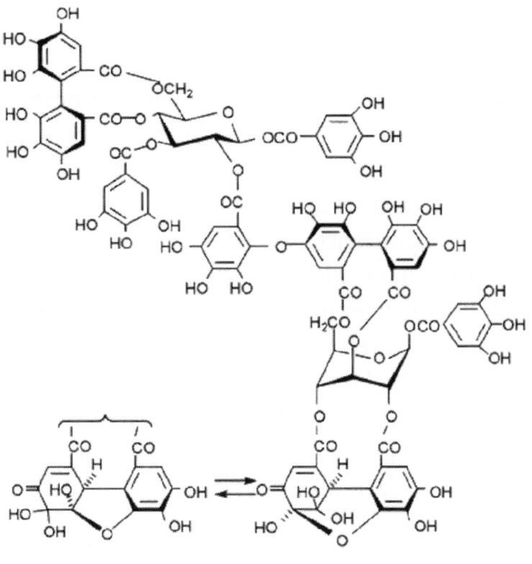

11

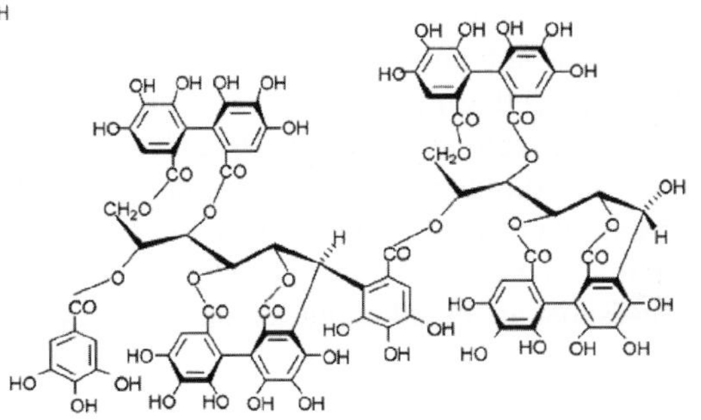

12

Fig. 7.3: Examples of hydrolyzable tannin oligomers characteristic of plant family.

SELECTED BIOLOGICAL PROPERTIES OF TANNINS

Inhibition of Polygalacturonase

Tannins protect plants against herbivores such as pathogenic microbes, insects, and animals.[1,34] Such protection is generally attributed to interactions between tannins and proteins, in addition to antifungal and antiviral activity. Some fungi that are pathogenic to plants, such as *Phytophthora infestans* (potato blight),[35] secrete polygalacturonase while attacking host plant cell walls to establish an infection. Tannins have long been known to be inhibitors of polygalacturonase; however, the effects of structurally defined tannins have not been investigated thoroughly.

The effect of monomeric and oligomeric ellagitannins isolated from various medicinal plants on polygalacturonase from *Aspergillu niger* (Sigma) has been assessed. The inhibitory effect was evaluated by quantifying the liberation of reducing sugars from substrate by pectic acid in the presence and absence of tannins.[36] Oenotheins B (**10**) and A (**13**), as well as camellliin B (**14**), each of which has a macrocyclic structure (Fig. 7.4), exhibited remarkable inhibition of pectic acid with an IC_{50} of 1.2–18.5 μg/ml[37,38]. By contrast, the inhibitory actions of the constituent monomers of each of these compounds, namely tellimagrandin I (**15**) (Fig. 7.4), was less potent (IC_{50} = 50.5 μg/ml).[39] However, the aforementioned oligomers did not inhibit amylase or cellulase (other hydrolytic enzymes that dissolve the plant cell wall), even at concentrations as high as 100 μg/ml, which suggested that their activity is specific to polygalacturonase.[39] In another study, the maceration of radish induced by polygalacturonase was found to be suppressed in a dose-dependent manner by oenothein A (**13**).[39] Therefore, it would appear that ellagitannin oligomers are promising candidates for providing plants with protection against diseases that are based on the action of polygalacturonase.

Antioxidants

Antioxidant Polyphenols of Walnuts

Walnuts are the seeds of *Juglans regia* L. (Juglandaceae) and are highly nutritious. They are rich in oil that is composed of unsaturated fatty acids that are susceptible to oxidation, such as linoleic acid and oleic acid. Despite the low content of α-tocopherol (an antioxidant) in walnuts, as compared to other types of nut,[40] walnuts are fairly stable when preserved; this implies that an unknown antioxidant(s) within walnuts inhibits lipid autoxidation. Indeed, an aqueous ethanol extract from commercially available walnuts showed marked superoxide dismutase (SOD)-like

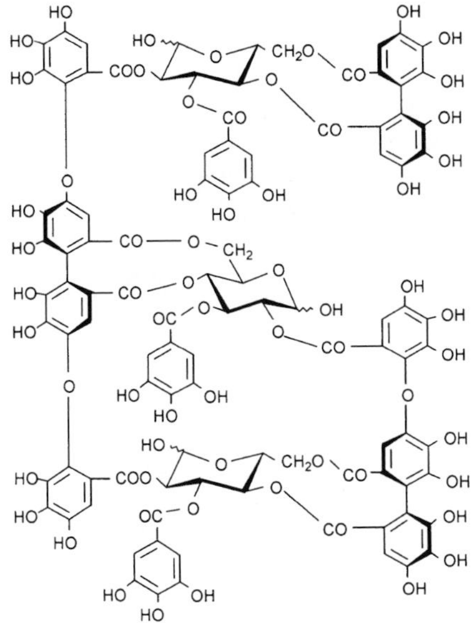

Fig. 7.4: Inhibitors against polygalacturonase.

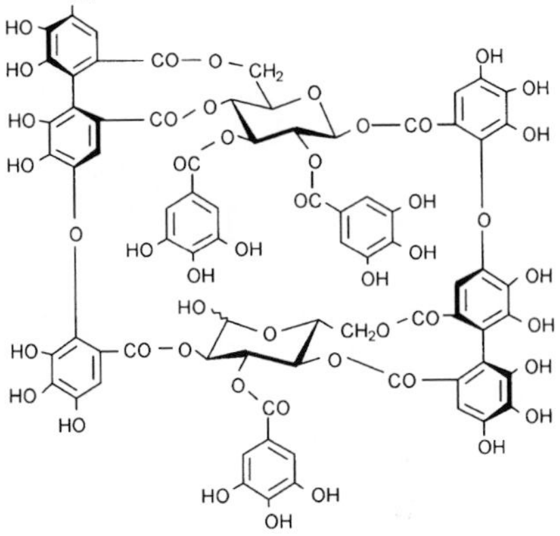

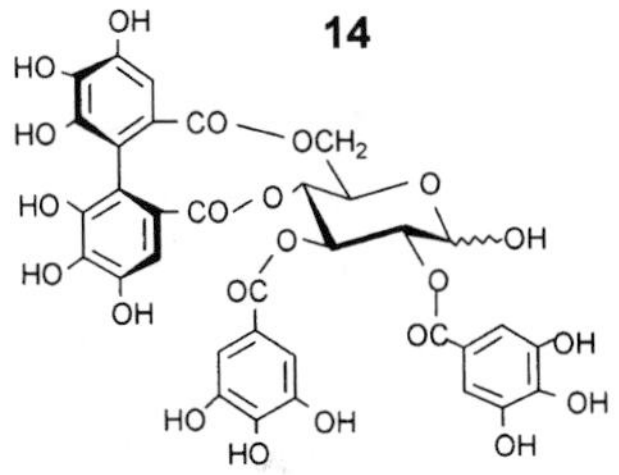

activity, and assay-guided fractionation of the *n*-butanol-soluble portion of these extracts resulted in the isolation of three novel hydrolyzable tannins, namely glansrins A (**16**), B (**17**), and C (**18**) (Fig. 7.5), along with adenosine, adenine, and several known ellagitannins (see Table 7.1). The novel tannins were characterized as ellagitannins that contain a rare tergalloyl-based acyl group at O-4/O-6 of the glucose core.[41]

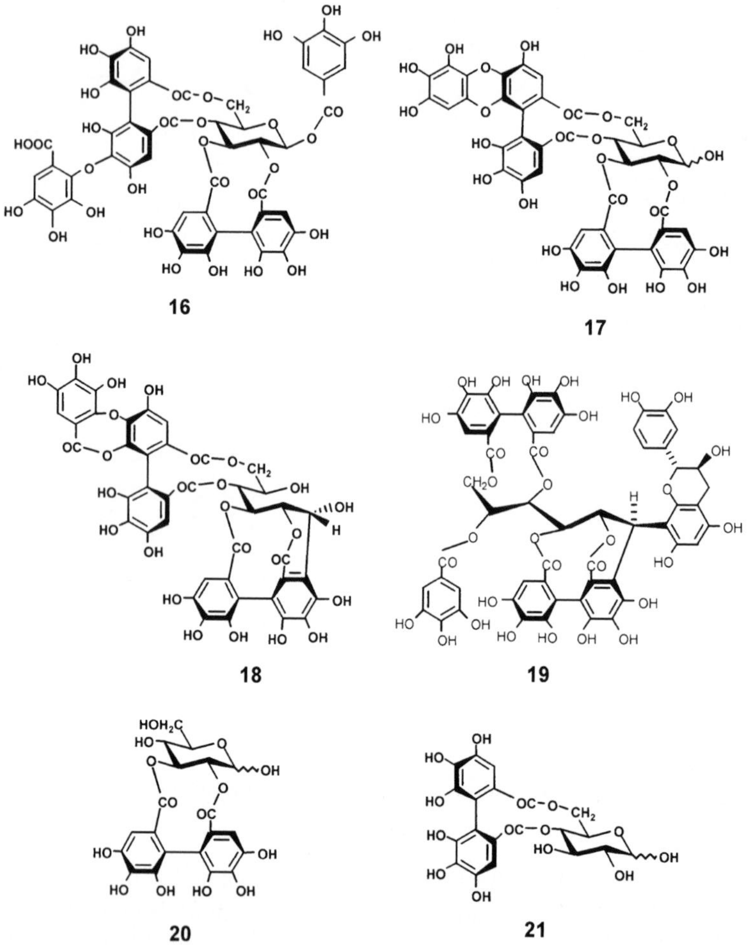

Fig. 7.5: Structures of new antioxidant polyphenols, glansrins A (**16**), B (**17**), C (**18**) and stenophyllanin A (**19**) from walnuts.

Table 7.1: SOD-like activity of walnut polyphenols.

Compound	SOD-like activity EC_{50} (μM)
Glansrin A (**16**)	190
Glansrin B (**17**)	41.9
Glansrin C (**18**)	21.4
Stenophyllanin A (**19**)	35.6
2,3-HHDP-D-glucose (**20**)	166
Pedunculagin (**32**)	63.7
Tellimagrandin I (**15**)	53.4
Tellimagrandin II (**2**)	94.8
1,2-di-*O*-galloyl-4,6-HHDP-β-D-glucose	76.3
Rugosin C	45.3
Casuarinin (**4**)	57.7
Adenosine	>1000
Adenine	695
Gallic acid	31.7
L-Ascorbic acid	34.6

The SOD-like activity of the constituents of the extract from walnuts was estimated by suppression of superoxide anion radicals (O_2^-) generated by the xanthine–xanthine oxidase (XOD) system.[6] The results are shown in Table 7.1. Glansrin C (**18**) and stenophyllanin A (**19**) (Fig 7.5) exhibited a strong SOD-like effect with an IC_{50} of 21.4 μM and 35.6 μM, respectively; these effects were comparable to those of the positive controls, L-ascorbic acid (IC_{50} = 34.6 μM) and gallic acid (IC_{50} = 31.7 μM). The IC_{50} of the activity of the remaining polyphenols in the walnut extract was on the order of 10^{-5} M, except for glansrin A (**16**) (IC_{50} = 190 μM) and 2,3-HHDP-D-glucopyranose (IC_{50} = 166 μM). By contrast, all of these polyphenols exhibited a greater degree of scavenging of 1,1-diphenyl-2-picrylhydrazyl (DPPH) (IC_{50} values between 10^{-7} and 10^{-6} M), as compared to the scavenging activity of L-ascorbic acid and gallic acid (IC_{50} = 6.24 and 5.88 μM, respectively). Although the relationship between the structure and activity of these polyphenols is not clear, ellagitannins that contain both a galloyl and an HHDP group in addition to a valoneoyl and/or its isomeric group have a tendency to exhibit more potent radical scavenging of DPPH than do polyphenols that contain only an HHDP group (*e.g.*, **20** and **21**, Fig. 7.5) (data not shown).

The *in vivo* antioxidative effects of polyphenol-rich walnut extract (WPF) were evaluated in type 2 diabetic mice.[42] Diabetes causes oxidative stress, and urinary 8-hydroxydeoxyguanine (8-OHdG) levels can be used as a biomarker of oxidative stress *in vivo*.[43] Oral administration of WPF (200 mg/kg) to diabetic mice for 4 weeks significantly reduced urinary 8-OHdG levels, which suggests that walnut polyphenols could be used as an antioxidative food to suppress diabetes-induced oxidative stress.[42]

Antioxidant Polyphenols of Cacao Liquor

Cacao liquor is prepared by fermenting and roasting cacao beans (*Theobroma cacao* L.). There are many reports on the antioxidant and related biological activities of polyphenol-rich fractions of cacao liquor.[44] In addition to the previously reported flavan monomers and oligomers that are found in fresh and fermented cacao beans,[45,46] we isolated a novel *C*-glucosidic flavan and *O*-glycosides of dimeric and trimeric A-type proanthocyanidins (**22-25**) from cacao liquor.[47] (Fig. 7.6, Table 7.2).

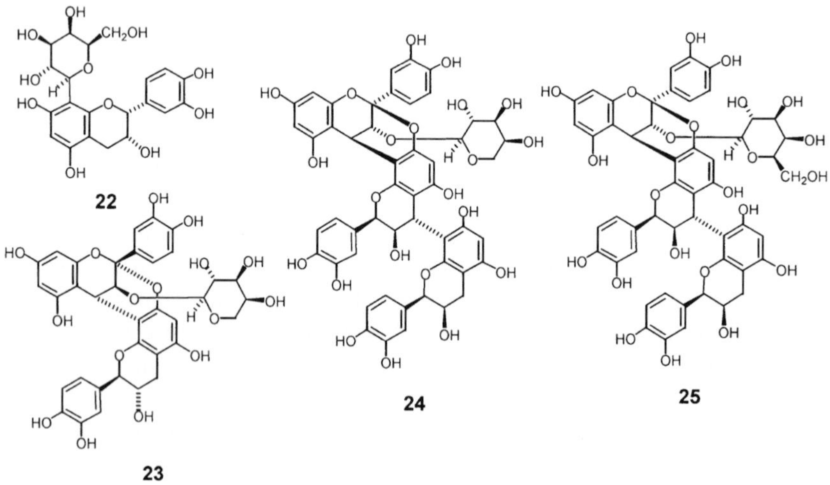

Fig. 7.6: Structures of cacao polyphenols.

Table 7.2: Inhibitory effect of cacao polyphenols on lipid peroxidation in rat liver microsomes and autoxidation of linoleic acid initiated by the addition of V-70

Compound	IC_{50} (μg/ml)	
	Lipid Peroxidation	Linoleic acid autoxidation
(–)-Epicatechin	29	0.62
3T-O-α-L-Arabinopyranosylcinnamtannin B$_1$ (**24**)	28	1.9
3T-O-β-D-Galactopyranosylcinnamtannin B$_1$ (**25**)	38	2.0
3T-O-α-L-Arabinopyranosyl-*ent*-epicatechin-(2α $\rightarrow$7, $4\alpha$$\rightarrow$8)-epicatechin (**23**)	21	2.2
Procyanidin B5	12	2.3
Cinnamtannin A$_2$	25	2.7
Procyanidin C1	68	5.3
(–)-Epicatechin 8-C-galactopyranoside (**22**)	>100	9.5
dl-α-Tocopherol	5.6	7.7

Some of the isolated polyphenols that are listed in Table 7.2 had a potent inhibitory effect on the autoxidation of linoleic acid by V-70 (2,2'-azobis(4-methoxy-2,4-dimethyl valeronitrile)); the IC_{50} of this inhibitory effect was 0.62–9.5 μg/ml. Except for procyanidin C-1, the potency of these polyphenols was greater than the inhibitory effect of α-tocopherol (IC_{50} = 7.7 μg/ml). The ranking of the polyphenols, according to the strength of the inhibitory activity of each, was comparable to the ability of these substances to scavenge DPPH (Table 7.2). Therefore, the antioxidant effects of cacao polyphenols are likely to be attributable to the scavenging of free radicals.

Antioxidant Polyphenols of Quercus *spp.*

Various tannin-rich plants have been used traditionally in many countries to treat gastritis and gastric ulcers. Based on the reports that extracts of *Quercus* spp. protect against ethanol-induced gastric lesions in mice when given orally before treatment with ethanol,[48] we investigated the tannin constituents of the leaves of Algerian *Q. coccifera* and *Q. suber* and isolated five novel ellagitannin oligomers, namely cocciferins D$_1$ (**26**), D$_2$ (**27**), D$_3$ (**28**), T$_1$ (**29**), and T$_2$ (**30**), together with many known hydrolyzable tannins including castalagin (**31**), pedunculagin (**32**), and phillyraeoidin A (**33**) (Fig 7.7).[49] Pretreatment of mice with a purified extract (50 mg/kg) within which the major tannins were present resulted in significant protection against ethanol-induced gastric lesions (Fig. 7.8). Among the aforementioned substances, castalagin (**31**) produced the greatest degree of gastroprotection.[50] There is a close relationship between ulcer formation and lipid peroxidation within the stomach.[51] In fact, the major ellagitannins of *Quercus* spp. strongly inhibited (by 55–65 % at a concentration of 10 µg/ml) lipid autoperoxidation in homogenated rabbit brain,[50] which suggested that the gastroprotective properties of the *Quercus* extract (and of the tannins isolated from the extract) might be related partly to strong antioxidant activity.

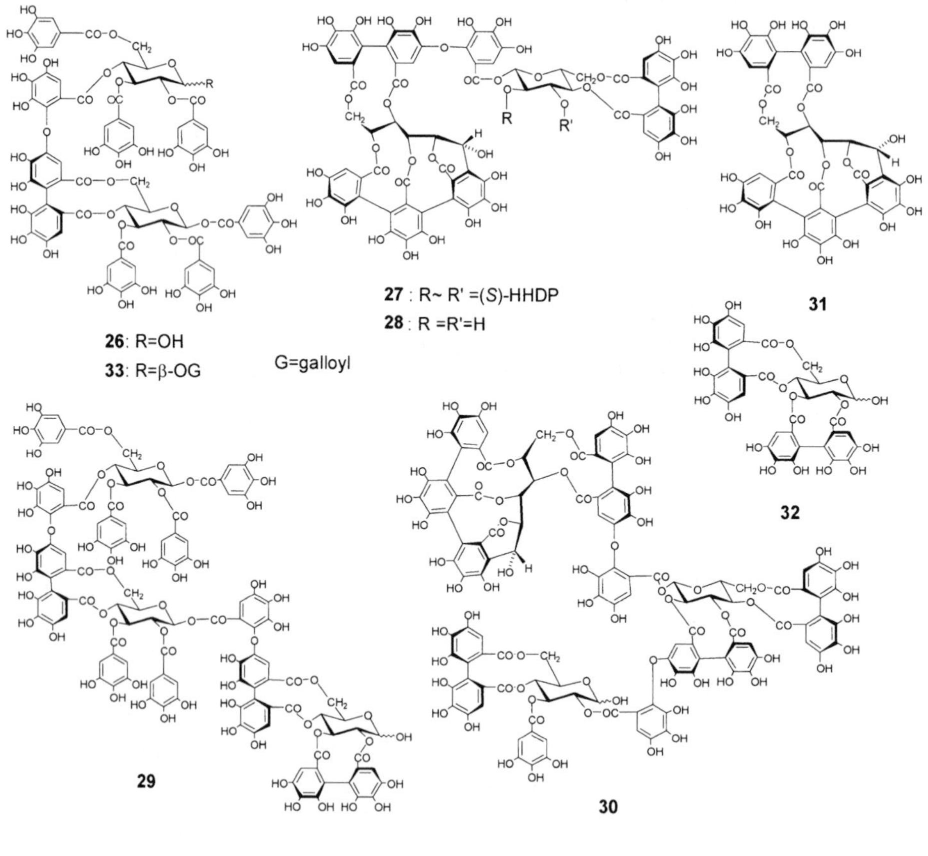

Fig. 7.7: Ellagitannin monomers and oligomers from *Qeurcus coccifera* and *Q. suber.*

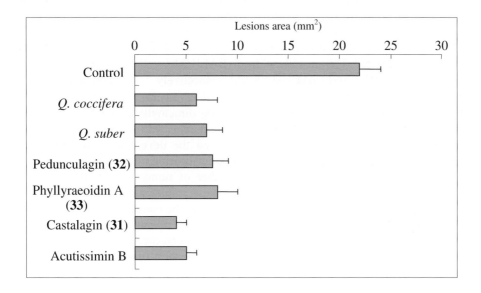

Fig. 7.8: Effect of extracts of *Q. suber* and *Q. coccifera*, and of pedunculagin (**32**), phillyraeoidin A (**33**), castalagin (**31**) and acutissimin B on ethanol-induced gastric lesions in the mouse. Test solutions (50 mg/kg) were given by orogastric instillation 1 h before 40% ethanol was administered. Results are expressed as the mean ± SEM (n = 8-10). * $P < 0.05$ versus the control.

Chemoprevention of Cancer

Chemoprevention is regarded as a promising strategy for cancer prevention. Based on the successful prevention of cancer by green tea[52] [which includes the antioxidant polyphenol, (-)-epigallocatechin gallate (EGCG) (**34**)], natural products and synthetic compounds that suppress the reversible development of tumors in humans have been sought out extensively in recent years. In mechanistic studies of antitumorigenic effects, tumor necrosis factor-α (TNF-α) and other cytokines were found to play an important role as endogenous tumor promoters, as these substances stimulate the development of tumors and the progression of initiated and premalignant cells.[53] Fujiki et al.[54] recently developed a useful method of screening for agents that prevent cancer by quantifying the degree of inhibition of TNF-α release from BALB/3T3 cells treated with okadaic acid, which promotes the development of tumors. This assay was applied to several plant extracts, among which the leaf extract of *Acer nikoense* (Aceraceae) was found to be an effective inhibitor of TNF-α release (IC$_{50}$ = 260 µg/ml).[55] The active components within the leaf extract were revealed to be the ellagitannins geraniin (3) and corilagin (37) (Fig. 7.9), which inhibited TNF-α release from BALB/3T3 cells in a dose-dependent manner, with an IC$_{50}$ of 43 µM and 76 µM, respectively.[55] The potencies of these ellagitannins were slightly lower than that of EGCG (34) (IC$_{50}$ = 26 µM). In a two-stage carcinogenesis assay on mouse skin, using dimethylbenz(a)anthracene as an initiator and okadaic acid as a promoter, treatment with geraniin prior to the application of okadaic acid significantly delayed the development of tumors, as indicated by a reduction in the percentage of tumor-bearing mice from 80% (control) to 40%, and a reduction in the average number of tumors per mouse from 3.8 (control) to 1.1 after 20 weeks (Fig. 7.10).[55] These results suggest that geraniin (3) and corilagin (37) might be used to prevent cancer in the manner of EGCG (34). It is noteworthy that geraniin (3) is found in many species of Euphorbiaceae, Rosaceae, Simaroubaceae, Cericididaceae, Geraniaceae, and Aceraceae.[2] Therefore, plants that contain 3 are also likely candidates for the identification of additional agents that might prevent cancer.

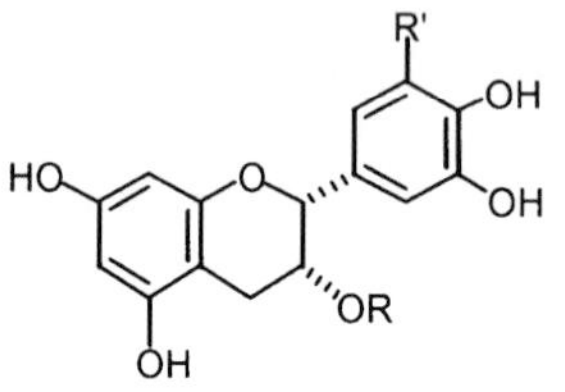

34: R=G, R'=OH

35: R=G, R'=H

36: R=R'=H   G=Galloyl

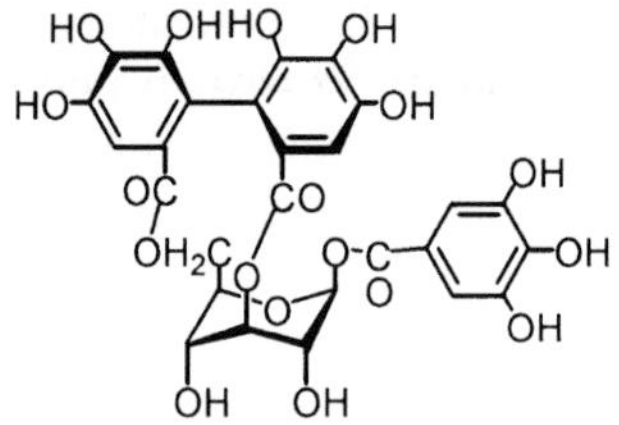

37

Fig. 7.9: Structures of green tea polyphenols (34-36) and corilagin (37).

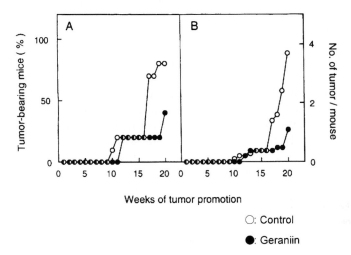

Fig. 7.10: Suppressive effect of geraniin (**3**) against two-stage carcinogenesis on mouse skin. (a) Average percentage of tumor-bearing mice. (b) Average numbers of tumors per mouse. ○ control (DMBA plus okadaic acid); ● DMBA plus okadaic acid plus **3**.

Antimicrobial Activity

Tannins and related polyphenols exhibit moderate antiviral activity against herpes simplex HSV-1,[9,10] human immunodeficiency virus,[11] and some influenza viruses.[56] Recently, hydrolyzable and condensed tannins have been shown to exhibit potential antiprotozoal activity,[19,20] as indicated by the observation that these substances stimulated the immune system after macrophages had been infected with intracellular amastigotes of *L. donovoni*. However, although tannin-rich extracts have long been known to exhibit significant antimicrobial activity, chemically defined tannins (irrespective of whether they are hydrolyzable or condensed) show weak or negligible antimicrobial activity against most fungi and pathogenic intestinal bacteria.[57] Nevertheless, we found that hydrolyzable tannins have remarkable antibacterial activity against *H. pylori*, a Gram-negative, helical rod bacterium that is found in human gastric epithelium.[13] In addition, some polyphenolics diminish the resistance of methicillin-resistant *S. aureus* (MRSA) to β-lactam antibiotics. [14-16,58]

Antibacterial Activity Against H. pylori

    H. pylori is a major etiological factor in gastroduodenal disorders such as chronic gastritis, peptic ulcer, and gastric cancer. Therefore, the treatment and prevention of these diseases would be facilitated by its eradication.[59] At present, triple therapies that comprise two antibiotics (clarithromycin and amphotericin B) and a proton pump inhibitor are used to eradicate *H. pylori*.[59] However, strains that are resistant to antibiotics have appeared. In addition, antibiotic treatment is associated with serious side effects such as nausea, vomiting, and diarrhea. Therefore, the discovery of novel antibacterial agents that are highly effective and safe is badly needed for the treatment of *H. pylori* infection.
    While exploring a new function of plant polyphenols, we examined the antibacterial activity of tannins and related polyphenols against *H. pylori in vitro.*[13] The antibacterial potency of different polyphenols that consist of hydrolyzable tannin monomers, oligomers, proanthocyanidins, and flavonoids was evaluated using standard strains (NCTC11638 and ATCC43504) and clinical isolates of *H. pylori*. Table 7.3 shows some of the results of this study. Procyanidin dimers and a trimer tested against the NCTC11638 strain showed weak antibacterial activity (minimum inhibitory concentration (MIC) = 50–100 μg/ml), whereas a procyanidin polymer (average dodecamer) showed no antibacterial activity at a concentration of 100 μg/ml. It was noticeable that a trihydroxyphenyl residue enhanced antibacterial activity, as evidenced by a comparison of the efficacy of various flavan-3-ol analogues; a green tea-derived tannin (EGCG (**34**)), containing two trihydroxybenzene rings, exhibited stronger antibacterial activity (MIC = 25 μg/ml) than (-)-epicatechin gallate (**35**) and (-)-epicatechin (**36**) (MIC = 50 μg/ml and > 100 μg/ml, respectively). All of the hydrolyzable tannins that were tested, especially monomeric hydrolyzable tannins (MICs 6.25-12.5 μg/ml), exhibited potent antibacterial activity. Although oligomerization of hydrolyzable tannins reduced the antibacterial activity somewhat, antibacterial activity against *H. pylori* would appear to be a characteristic of hydrolyzable tannins rather than of other types of polyphenol. None of the polyphenols that were tested exhibited any antibacterial activity against *E. coli*, which is a nonpathogenic bacterium that is normally found within the human intestinal tract, and the polyphenols were not cytotoxic to normal cells (MKN-28) derived from human gastric epithelium. Therefore, hydrolyzable tannins are potentially promising agents that could be used to suppress *H. pylori*, without affecting gastric epithelial cells and nonpathogenic intestinal bacteria.

Table 7.3: Antibacterial activity of polyphenols against *H. pylori* strains

Compound	MIC (μg/ml)[a]	
	H. pylori	
	NCTC11638	ATCC43504
Hydrolyzable tannins		
Monomers		
Penta-*O*-galloyl-β-D-glucose (**1**)	12.5	12.5
Strictinin	6.25	25
Pedunculagin (**32**)	12.5	12.5
Tellimagrandin I (**15**)	12.5	12.5
Tellimagrandin II (**2**)	12.5	12.5
Corilagin (**37**)	12.5	12.5
Geraniin (**3**)	12.5	25
Casuarinin (**4**)	25	25
Dimers		
Agrimoniin (**6**)	25	25
Nobotanin B	12.5	12.5
Rugosin D	25	50
Oenothein B (**10**)	25	12.5
Trimers		
Oenothein A (**13**)	50	50
Catechins		
(-)-Epicatechin (**36**)	>100	>100
(-)-Epicatechin gallate (**35**)	50	>100
(-)-Epigallocatechin gallate (**34**)	25	50
Proanthocyanidins		
Procyanidin B-1	100	100
Procyanidin B-3	100	>100
Procyanidin B-4	100	>100
Procyanidin B-5	50	100
Procyanidin C-1	100	100
Procyanidin polymer	>100	>100
Other polyphenols		
Chlorogenic acid	>100	>100
Isorhamnetin 3-*O*-rutinoside	>100	>100
Iridin	>100	>100
Piceatannol-4'-*O*-(6"-galloyl)-glucoside	25	50
Terpenoids		
3-*O*-Caffeoylbetulinic acid	25	50
Betulinic acid	>100	>100
Glycyrrhizin	>100	>100

[a]MICs were determined by the agar dilution method with PPLO agar supplemented with 0.2% DMCD.

Synergic Effects of Tannins and Antibiotics on MRSA

Methicillin-resistant *S. aureus* (MRSA) is a major cause of hospital-acquired (nosocomial) infections in many countries. MRSA infections are very difficult to cure because MRSA is resistant not only to β-lactams (methicillin, oxacillin, etc.), but also to most of other antibacterial agents. The *mecA* gene, which encodes the low-affinity penicillin-binding protein PBP2', as well as several other genes including *fem, fmt, llm,* and *sigB,* have been recognized as being important for resistance to MRSA infection.[60-64] However, the multi-drug resistance of MRSA is not yet fully understood. Given that the discovery of new antibiotics constitutes a difficult process, the development of novel drugs that are effective and safe, or of alternative therapies, is urgently required.

A promising strategy for overcoming MRSA resistance was suggested by the finding that the green tea polyphenols, **34** and **35**, caused a remarkable reduction of the MIC of β-lactam (oxacillin) against MRSA. Specifically, (-)-epicatechin gallate (**35**) at 25 μg/ml lowered the MIC of oxacillin against MRSA 250–500-fold, while EGCG (**34**) at 50 μg/ml caused an 8–120-fold reduction in the MIC.[14] As EGCG (**34**) was found to be unstable even in solutions of neutral pH, we investigated the oxidation products of **34,** and characterized several.[65] Among them, theasinensin A (**38**) (Fig. 7.11), the major product, showed noticeable suppressive effects on the MRSA resistance against β-lactam and aminoglycoside antibiotics. The effect on the viable cell numbers of MRSA by **38** in the presence of oxacillin (a β-lactam antibiotic) was bactericidal within 10 h, where the decrease of bacterial cell amount was shown (Fig. 7.11).[65]

The aforementioned findings mean that these polyphenols restored the effectiveness of β-lactams against MRSA. In a subsequent study of compounds that have similar actions against MRSA, extracts of rose red flowers (*Rosa canina* L.) and *Arctostaphylos uva-ursi* reduced the MICs of oxacillin.[15,16] Tellimagrandin I (**15**) (from the *R. canina* extract) and corilagin (**37**) (from the *A. uva-ursi* extract) were identified as the active components. Tellimagrandin I (**15**) alone had a weak antibacterial action against MRSA (MIC = 128 μg/ml), but the addition of 50 μg/ml to the medium markedly reduced the MIC of oxacillin against MRSA (128–512-fold).[15] Similarly, corilagin (**37**), at a concentration 16 μg/ml, reduced the MICs of various β-lactams (oxacillin, imipenem, cefmetazole, etc.) 100–2,000-fold.[16]

The aforementioned observations indicate that tannins administered together with antibiotics produce synergistic bactericidal effects. There are three possible mechanisms that would account for the synergistic action of tannins and β-lactam: (1) inhibition of PBP2' activity; (2) inhibition of the production of PBP2'; and (3)

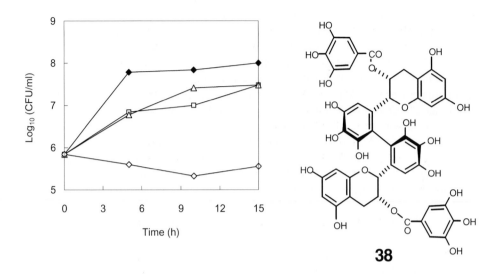

◆ : Control

□ : Oxacillin (5 μg/mL)

△ : Theasinensin A (32 μg/mL)

◇ : Oxacillin (5 μg/mL) + Theasinensin A (32 μg/mL)

Fig. 7.11: Time course for the change in the viable cell numbers of MRSA OM584 strain upon incubation with theasinensin A (**38**) and/or oxacillin. ◆ control; □ oxacillin (5 μg/ml); △ theasinensin A (**38**) (32 μg/ml); ◇ oxacillin (5 μg/ml) plus **38** (32 μg/ml).

inhibition of β-lactamase. Although the mechanism remains to be elucidated, there is evidence that tannins do not inhibit the production of PBP2', but do inhibit PBP2' activity that causes damage to the cell wall and membranes, and thus increases the permeability of the cell to drugs.[66]

SUMMARY

The marked progress that has been made in phytochemical studies of plant polyphenols, over several decades, has revealed numerous novel tannins and related compounds from a wide array of plant species. These findings have not only encouraged studies of the biosynthetic properties[67,68] and chemical synthesis[69,70] of these substances, but have also stimulated attempts to evaluate the physiological functions of chemically defined tannins. Newly discovered biological functions with respect to polyphenols from medicinal plants and edible fruits and beverages have led to claims that these substances have the potential to be used in the treatment of various chronic diseases that are related to oxidative stress and microbial infections. In the future, studies of plant polyphenols, based on the absorption and metabolism of ingested tannins and related polyphenols, will focus on the challenge of obtaining a better understanding of the bioavailability of these substances in humans.

ACKNOWLEDGEMENTS

We are indebted to Professors H. Fujiki, Toskushima Bunri University, K. Hirai, Jichi Medical School, T. Tsuchiya, Okayama University, and Dr. T. Tsuji, Sagami Chemical Research Center for their collaborations performing biological assays. Financial support (in part) by a Grant-in-Aid for Scientific Research from the Ministry of Science, Education, Sports and Culture of Japan is gratefully acknowledged.

REFERENCES

1.   HASLAM, E., Practical Polyphenolics - From Structure to Molecular Recognition and Physiological Action. Cambridge University Press, 1998.
2.   OKUDA, T., YOSHIDA, T., HATANO, T., Hydrolyzable tannins and related polyphenols, *Fortschritte der Chemie organischer Naturstoffe*, 1995, **66**, 1-117.
3.   KAKIUCHI, N., HATTORI, M., NAMBA, T., NISHIZAWA, M., YAMAGISHI, T., OKUDA, T., Inhibitory effect of tannins on reverse transcriptase from RNA tumor virus, *J. Nat. Prod.*, 1985, **48**, 614-621.
4.   KURAMOCHI-MOTEGI, A., KURAMOCHI, H., KOBAYASHI, F., EKIMOTO, H., TAKAHASHI, K., KADOTA, S., TAKAMORI, Y., KIKUCHI, T., Woodfruticosin

(woodfordin C), a new inhibitor of DNA topoisomerase II. Experimental antitumor activity, *Biochem. Pharmacol.*, 1992, **44**, 1961-1965.

5. GINSBURG, I., MITRA, R. S., GIBBS, D. F., VARANI, J., KOHEN, R., Killing of endothelial cells and release of arachidonic acid. Synergistic effects among hydrogen peroxide, membrane-damaging agents, cationic substances, and proteinases and their modulation by inhibitors, *Inflammation*, 1993, **17**, 295-319.

6. OKUDA, T., KIMURA, Y., YOSHIDA, T., HATANO, T., OKUDA, H., ARICHI, S., Studies on the activities of tannins and related compounds from medicinal plants and drugs. I. Inhibitory effects on lipid peroxidation in mitochondria and microsomes of liver, *Chem. Pharm. Bull.*, 1983, **31**, 1625-1631.

7. HATANO, T., EDAMATSU, R., HIRAMATSU, M., MORI, K., FUJITA, Y., YASUHARA, T., YOSHIDA, T., OKUDA, T., Effects of interaction of tannins with co-existing substances. VI. Effect of Tannins and Related Polyphenols on Superoxide Anion Radical, and on 1, 1-diphenyl-2-picrylhydrazyl radical, *Chem. Pharm. Bull.*, 1989, **37**, 2016-2021.

8. MIYAMOTO, K., NOMURA, M., MURAYAMA, T., FURUKAWA, T., HATANO, T., YOSHIDA, T., KOSHIURA, R., OKUDA, T., Antitumor activities of ellagitannins against sarcoma-180 in mice, *Biol. Pharm. Bull.*, 1993, **16**, 379-387.

9. FUKUCHI, K., SAKAGAMI, H., OKUDA, T., HATANO, T., TANUMA, S., KITAJIMA, K., INOUE, Y., INOUE, S., ICHIKAWA, S., NONOYAMA, M., KONNO, K., Inhibition of herpes simplex virus infection by tannins and related compounds, *Antiviral Res.*, 1989, **11**, 285-297.

10. TAKECHI, M., TANAKA, Y., TAKEHARA, M., NONAKA, G.-I., NISHIOKA, I., Structure and antiherpetic activity among the tannins, *Phytochemistry*, 1985, **24**, 2245-2250.

11. NAKASHIMA, H., MURAKAMI, T., YAMAMOTO, N., SAKAGAMI, H., TANUMA, S., HATANO, T., YOSHIDA, T., OKUDA, T., Inhibition of human immunodeficiency viral replication by tannins and related compounds, *Antiviral Res.*, 1992, **18**, 91-103.

12. KANOH, R., HATANO, T., ITO, H., YOSHIDA, T., AKAGI, M., Effects of tannins and related polyphenols on superoxide-induced histamine release from rat peritoneal mast cells, *Phytomedicine*, 2000, 7, 297-302.

13. FUNATOGAWA, K., HAYASHI, S., SHIMOMURA, H., YOSHIDA, T., HATANO, T., ITO, H., HIRAI, Y., Antibacterial activity of hydrolyzable tannins derived from medicinal plants against *Helicobacter pylori*, *Microbiol. Immunol.*, 2004, **48**, 251-261.

14. SHIOTA, S., SHIMIZU, M., MIZUSHIMA, T., ITO, H., HATANO, T., YOSHIDA, T., TSUCHIYA, T., Marked reduction in the minimum inhibitory concentration (MIC) of beta-lactams in methicillin-resistant *Staphylococcus aureus* produced by epicatechin gallate, an ingredient of green tea (*Camellia sinensis*), *Biol. Pharm. Bull.*, 1999, **22**, 1388-1390.

15. SHIOTA, S., SHIMIZU, M., MIZUSHIMA, T., ITO, H., HATANO, T., YOSHIDA, T., TSUCHIYA, T., Restoration of effectiveness of beta-lactams on methicillin-resistant *Staphylococcus aureus* by tellimagrandin I from rose red, *FEMS Microbiol. Lett.*, 2000, **185**, 135-138.

16. SHIMIZU, M., SHIOTA, S., MIZUSHIMA, T., ITO, H., HATANO, T., YOSHIDA, T., TSUCHIYA, T., Marked potentiation of activity of beta-lactams against methicillin-resistant *Staphylococcus aureus* by corilagin, *Antimicrob. Agents Chemother.*, 2001, **45**, 3198-3201.

17. KOLODZIEJ, H., KAYSER, O., KIDERLEN, A. F., ITO, H., HATANO, T., YOSHIDA, T., FOO, L. Y., Proanthocyanidins and related compounds: Antileishmanial activity and modulatory effects on nitric oxide and tumor necrosis factor-alpha-release in the murine macrophage-like cell line RAW 264.7, *Biol. Pharm. Bull.*, 2001, **24**, 1016-1021.

18. KOLODZIEJ, H., KAYSER, O., KIDERLEN, A. F., ITO, H., HATANO, T., YOSHIDA, T., FOO, L. Y., Antileishmanial activity of hydrolyzable tannins and their modulatory effects on nitric oxide and tumour necrosis factor-alpha release in macrophages *in vitro*, *Planta Med.*, 2001, **67**, 825-832.

19. AOKI, K., NISHIMURA, K., ABE, H., MARUTA, H., SAKAGAMI, H., HATANO, T., OKUDA, T., YOSHIDA, T., TSAI, Y. J., UCHIUMI, F., TANUMA, S., Novel inhibitors of poly(ADP-ribose) glycohydrolase, *Biochim. Biophys. Acta*, 1993, **1158**, 251-256.

20. AOKI, K., MARUTA, H., UCHIUMI, F., HATANO, T., YOSHIDA, T., TANUMA, S., A macrocircular ellagitannin, oenothein B, suppresses mouse mammary tumor gene expression via inhibition of poly(ADP-ribose) glycohydrolase. *Biochem. Biophys. Res. Commun.*, 1995, **210**, 329-337.

21. SILVA, J. M. R. d., RIGAUD, J., CHEYNIER, V., CHEMINAT, A., MOUTOU, M., Procyanidin dimers and trimers from grape seeds, *Phytochemistry*, 1991, **30**, 1259-1264.

22. FOO, L. Y., LU, Y., HOWELL, A. B., VORSA, N., The structure of cranberry proanthocyanidins which inhibit adherence of uropathogenic *P*-fimbriated *Escherichia coli in vitro*, *Phytochemistry*, 2000, **54**, 173-181.

23. OKUDA, T., YOSHIDA, T., HATANO, T., Constituents of *Geranium thunbergii* Sieb. *et* Zucc. Part 12. Hydrated stereostructure and equilibration of geraniin, *J. Chem. Soc. Perkin Trans. 1*, 1982, 9-14.

24. LUGER, P., WEBER, M., KASHINO, S., AMAKURA, Y., YOSHIDA, T., OKUDA, T., BEURSKENS, G., DAUTER, Z., Structure of the tannin geraniin based on conventional X-ray data at 295 K and on synchrotron data at 293 and 120 K. *Acta Crystallographica*, 1998, **B54**, 687-694.

25. OKUDA, T., YOSHIDA, T., ASHIDA, M., YAZAKI, K., Tannins of *Casuarina* and *Stachyurus* Species. Part 1. Structures of pedunculagin, casuarictin, strictinin, casuarinin, casuariin, and stachyurin. *J. Chem. Soc. Perkin Trans. 1*, 1983, 1765-1772.

26. HATANO, T., SHIDA, S., HAN, L., OKUDA, T., Tannins of theaceous plants. III. Camelliatannins A and B, two new complex tannins from *Camellia japonica* L., *Chem. Pharm. Bull.*, 1991, **39**, 876-880.

27. OKUDA, T., YOSHIDA, T., KUWAHARA, M., MEMON, M. U., SHINGU, T., Agrimoniin and potentillin, ellagitannin dimer and monomer having α-glucose core, *J. Chem. Soc., Chem. Commun.*, 1982, 163-164.

28. HATANO, T., OGAWA, N., KIRA, R., YASUHARA, T., OKUDA, T., Tannins of cornaceous plants. I. Cornusiins A, B and C, dimeric, monomeric and trimeric

hydrolyzable tannins from *Cornus officinalis*, and orientation of valoneoyl group in related tannins, *Chem. Pharm. Bull.*, 1989, **37**, 2083-2090.

29. YOSHIDA, T., AHMED, A. F., MEMON, M. U., OKUDA, T., Tannins of tamaricaceous plants. II. New monomeric and dimeric hydrolyzable tannins from *Reaumuria hirtella* and *Tamarix pakistanica*, *Chem. Pharma. Bull.*, 1994, **39**, 2849-2854.

30. AHMED, A. F., YOSHIDA, T., OKUDA, T., Tannins of tamaricaceous plants. V. New dimeric, trimeric and tetrameric ellagitannins from *Reaumuria hirtella*, *Chem. Pharma. Bull.*, 1994, **42**, 246-253.

31. HATANO, T., YASUHARA, T., MATSUDA, M., YAZAKI, K., YOSHIDA, T., OKUDA, T., Oenothein B, a dimeric, hydrolysable tannin with macrocyclic structure, and accompanying tannins from *Oenothera erythrosepala*, *J. Chem. Soc. Perkin Trans. 1*, **1990**, 2735-2743.

32. YOSHIDA, T., CHEN, L., SHINGU, T., OKUDA, T., Tannins and related polyphenols of euphorbiaceous plants. IV. Euphorbins A and B, novel dimeric dehydroellagitannins from *Euphorbia hirta* L., *Chem. Pharm. Bull.*, 1988, **36**, 2940-2949.

33. ITO, H., MIKI, K., YOSHIDA, T., Elaeagnatins A-G, *C*-glucosidic ellagitannins from *Elaeagnus umbellata*, *Chem. Pharm. Bull.*, 1999, **47**, 536-542.

34. SCHULTZ, J. C., HUNTER, M. D., APPEL, H. M., Antimicrobial activity of polyphenols mediates plant-herbivore interaction, *in* Plant polyphenols: Synthesis, Properties, Significance (R. W. Hemingway, P. E. Laks, eds.), Plenum Press, 1992, pp. 621-637.

35. TAYLOR, J. L., FRITZEMEIER, K. H., HAUSER, I., KOMBRINK, E., ROHWER, F., SCHRODER, M., STRITTMATTER, G., HAHLBROCK, K., Structural analysis and activation by fungal infection of a gene encoding a pathogenesis-related protein in potato, *Mol. Plant Microbe. Interact.*, 1990, **3**, 72-77.

36. NELSON, N., A photometric adaptation of the somogyi method for the determination of glucose, *J. Biol. Chem.*, 1944, **153**, 373-380.

37. YOSHIDA, T., CHOU, T., MATSUDA, M., YASUHARA, T., YAZAKI, K., HATANO, T., OKUDA, T., Woodfordin D and oenothein A, trimeric hydrolyzable tannins of macro-ring structure with anti-tumor activity, *Chem. Pharm. Bull.*, 1991, **39**, 1157-1162.

38. YOSHIDA, T., CHOU, T., MARUYAMA, Y., OKUDA, T., Tannins of theaceous plants. II. Camellins A and B, two new dimeric hydrolyzable tannins from flower buds of *Camellia japonica* L. and *Camellia sasanqua* THUNB, *Chem. Pharm. Bull.*, 1990, **38**, 2681-2686.

39. TSUJI, T., SEITAI, U., SHU, T., YOSHIDA, T. 1996. Polygalacturonase inhibitors isolation from plant. Jpn. Kokai Tokkyo Koho. Japan. p. 7.

40. KAGAWA, Y., Standard Tables of Food Composition in Japan. Kagawa Nutrition University Press, 2001.

41. FUKUDA, T., ITO, H., YOSHIDA, T., Antioxidative polyphenols from walnuts (*Juglans regia* L.), *Phytochemistry*, 2003, **63**, 795-801.

42. FUKUDA, T., ITO, H., YOSHIDA, T., Effect of the walnut polyphenol fraction on oxidative stress in type 2 diabetes mice, *BioFactors*, 2004, **21**, 251-253.

43. MIYAKE, Y., YAMAMOTO, K., TSUJIHARA, N., OSAWA, T., Protective effects of lemon flavonoids on oxidative stress in diabetic rats, *Lipids*, 1998, **33**, 689-695.

44. OSAKABE, N., NATSUME, M., ADACHI, T., YAMAGISHI, M., HIRANO, R., TAKIZAWA, T., ITAKURA, H., KONDO, K., Effects of cacao liquor polyphenols on the susceptibility of low-density lipoprotein to oxidation in hypercholesterolemic rabbits, *J. Atheroscler. Thromb.*, 2000, 7, 164-168 and references cited therein.

45. PORTER, L. J., MA, Z., CHAN, B. G., Cacao procyanidins: major flavanoids and identification of some minor metabolites, *Phytochemistry*, 1991, **30**, 1657-1663.

46. THOMPSON, R. S. J., D.; HASLAM, E.; TANNER, R. J. N., Plant proanthocyanidins. I. Introduction. Isolation, structure, and distribution in nature of plant procyanidins, *J. Chem. Soc., Perkin Trans 1*, 1972, 1387-1399.

47. HATANO, T., MIYATAKE, H., NATSUME, M., OSAKABE, N., TAKIZAWA, T., ITO, H., YOSHIDA, T., Proanthocyanidin glycosides and related polyphenols from cacao liquor and their antioxidant effects, *Phytochemistry*, 2002, **59**, 749-758.

48. KHENNOUF, S., GHARZOULI, K., AMIRA, S., GHARZOULI, A., Effects of *Quercus ilex* L. and *Punica granatum* L. polyphenols against ethanol-induced gastric damage in rats, *Pharmazie*, 1999, **54**, 75-76.

49. ITO, H., YAMAGUCHI, K., KIM, T. H., KHENNOUF, S., GHARZOULI, K., YOSHIDA, T., Dimeric and trimeric hydrolyzable tannins from *Quercus coccifera* and *Quercus suber*, *J. Nat. Prod.*, 2002, **65**, 339-345.

50. KHENNOUF, S., BENABDALLAH, H., GHARZOULI, K., AMIRA, S., ITO, H., KIM, T. H., YOSHIDA, T., GHARZOULI, A., Effect of tannins from *Quercus suber* and *Quercus coccifera* leaves on ethanol-induced gastric lesions in mice, *J. Agric. Food Chem.*, 2003, **51**, 1469-1473.

51. NISHIDA, K., OHTA, Y., KOBAYASHI, T., ISHIGURO, I., Involvement of the xanthine-xanthine oxidase system and neutrophils in the development of acute gastric mucosal lesions in rats with water immersion restraint stress, *Digestion*, 1997, **58**, 340-351.

52. YOSHIZAWA, S., HORIUCHI, H., FUJIKI, H., YOSHIDA, T., OKUDA, T., SUGIMURA, T., Antitumor promoting activities of (-)-epigallocatechin gallate, the main constituent of "Tannin" in green tea, *Phytotherapy Res.*, 1987, **1**, 44-47.

53. SUGANUMA, M., OKABE, S., MARINO, M. W., SAKAI, A., SUEOKA, E., FUJIKI, H., Essential role of tumor necrosis factor alpha (TNF-α) in tumor promotion as revealed by TNF-α–deficient mice, *Cancer Res.*, 1999, **59**, 4516-4518.

54. KOMORI, A., YATSUNAMI, J., SUGANUMA, M., OKABE, S., ABE, S., SAKAI, A., SASAKI, K., FUJIKI, H., Tumor necrosis factor acts as a tumor promoter in BALB/3T3 cell transformation, *Cancer Res.*, 1993, **53**, 1982-1985.

55. OKABE, S., SUGANUMA, M., IMAYOSHI, Y., TANIGUCHI, S., YOSHIDA, T., FUJIKI, H., New TNF-alpha releasing inhibitors, geraniin and corilagin, in leaves of *Acer nikoense*, megusurino-ki, *Biol. Pharm. Bull.*, 2001, **24**, 1145-1148.

56. MANTANI, N., ANDOH, T., KAWAMATA, H., TERASAWA, K., OCHIAI, H., Inhibitory effect of *Ephedrae herba*, an oriental traditional medicine, on the growth of influenza A/PR/8 virus in MDCK cells, *Antiviral Res.*, 1999, **44**, 193-200.

57. KOLODZIEJ, H., KAYSER, O., LATTE, K. P., KIDERLEN, A. F., Enhancement of antimicrobial activity of tannins and related compounds by immune modulatory

effects, *in* Plant Polyphenols: Chemistry, Biology, Pharmacology, Ecology (G. G. Gross, R. W. Hemingway, T. Yoshida, eds.), Kluwer and Plenum, 1999, pp. 575-594.

58. HATANO, T., KUSUDA, M., HORI, M., SHIOTA, S., TSUCHIYA, T., YOSHIDA, T., Theasinensin A, a tea polyphenol formed from (-)-epigallocatechin gallate, suppresses antibiotic resistance of methicillin-resistant *Staphylococcus aureus*, *Planta Med.*, 2003, **69,** 984-989.

59. ULMER, H. J., BECKERLING, A., GATZ, G., Recent use of proton pump inhibitor-based triple therapies for the eradication of *H pylori*: a broad data review, *Helicobacter*, 2003, **8,** 95-104.

60. UBUKATA, K., NONOGUCHI, R., MATSUHASHI, M., KONNO, M., Expression and inducibility in *Staphylococcus aureus* of the mecA gene, which encodes a methicillin-resistant *S. aureus*-specific penicillin-binding protein, *J. Bacteriol.*, 1989, **171,** 2882-2885.

61. BERGER-BACHI, B., BARBERIS-MAINO, L., STRASSLE, A., KAYSER, F. H., FemA, a host-mediated factor essential for methicillin resistance in *Staphylococcus aureus*: molecular cloning and characterization, *Mol. Gen. Genet.*, 1989, **219,** 263-269.

62. KOMATSUZAWA, H., SUGAI, M., OHTA, K., FUJIWARA, T., NAKASHIMA, S., SUZUKI, J., LEE, C. Y., SUGINAKA, H., Cloning and characterization of the fmt gene which affects the methicillin resistance level and autolysis in the presence of triton X-100 in methicillin-resistant *Staphylococcus aureus*, *Antimicrob. Agents Chemother.*, 1997, **41,** 2355-2361.

63. MAKI, H., YAMAGUCHI, T., MURAKAMI, K., Cloning and characterization of a gene affecting the methicillin resistance level and the autolysis rate in *Staphylococcus aureus*, *J. Bacteriol.*, 1994, **176,** 4993-5000.

64. WU, S., DE LENCASTRE, H., TOMASZ, A., Sigma-B, a putative operon encoding alternate sigma factor of *Staphylococcus aureus* RNA polymerase: molecular cloning and DNA sequencing, *J. Bacteriol.*, 1996, **178,** 6036-6042.

65. HATANO, T., KUSUDA, M., HORI, M., SHIOTA, S., TSUCHIYA, T., YOSHIDA, T. Theasinensin A, a tea polyphenol formed from (-)-epigallocatechin gallate, suppresses antibiotic resistance of methicillin-resistant *Staphylococcus aureus*, *Planta Medica*, 2003, **69,** 984-989.

66. ZHAO, W. H., HU, Z. Q., OKUBO, S., HARA, Y., SHIMAMURA, T., Mechanism of synergy between epigallocatechin gallate and beta-lactams against methicillin-resistant *Staphylococcus aureus*, *Antimicrob. Agents Chemother.*, 2001, **45,** 1737-1742.

67. NIEMETZ, R., GROSS, G. G., Biosynthesis and biodegradation of complex gallotannins, *in* Plant Polyphenols: Chemistry, Biology, Pharmacology, Ecology (G. G. Gross, R. W. Hemingway, T. Yoshida, eds.), Kluwer and Plenum, 1999, pp. 63-82.

68. NIEMETZ, R., GROSS, G. G., Ellagitannin biosynthesis: laccase-catalyzed dimerization of tellimagrandin II to cornusiin E in *Tellima grandiflora*, *Phytochemistry*, 2003, **64,** 1197-1201.

69. KHAMBABAEE, K., VAN REE, T., Strategies for the synthesis of ellagitannins, *Synthesis*, 2001, **11,** 1585-1610.

70. QUIDEAU, S., FELDMAN, K. S., Ellagitannin chemistry, *Chem. Rev.*, 1996, **96,** 475-503.

Chapter Eight

ODOR PERCEPTION AND THE VARIABILITY IN NATURAL ODOR SCENES

Geraldine A. Wright* and Mitchell G.A. Thomson

Mathematical Biosciences Institute
Ohio State University
231 W. 18th Ave., Columbus, OH 43210

Author for correspondence, email: wright.571@osu.edu

INTRODUCTION

Naturally occurring olfactory stimuli, such as floral perfumes or animal pheromones, are typically complex combinations of many chemical compounds.[1-3] Odor compounds emitted by a single odor source may have diverse chemical structures,[4] and each compound present may differ quantitatively from the others by several orders of magnitude in concentration. [1,5] Odors also vary spatially and temporally as a function of the turbulent nature of the air medium into which they disperse.[6,7] Most animals use odor signaling for several functions that are important to their fitness, including: attracting mates, identifying kin, finding food, and avoiding predators. Because many odors are ecologically important stimuli, the olfactory system must solve the problem of identifying odor stimuli and discriminating them from other odors, even when they vary significantly in concentration and precise composition from one odor-producing object to the next. How the olfactory system produces a reliable representation of odors is not yet entirely known.[2]

This review draws on other areas of sensory science to suggest how one might quantify the statistical structure of naturally occurring olfactory stimuli. We identify likely sources of variation in ecologically valid olfactory scenes, and we discuss how this variation might influence the way in which animals recognize and discriminate among salient odors. We focus mainly on odor signals that are learned rather than those that produce innate behaviors.

ODORS AS NATURAL STIMULI

Odorant molecules are typically small organic molecules with a molecular weight of 26-300.[8] An odor consists of millions of volatile molecules emitted into the fluid media of air or water. The vapor pressure of a compound is a measure that directly relates to its volatility; odorants with high vapor pressures in ambient conditions volatilize readily.[9,10] Inanimate odor sources passively emit odors; animate odor sources may passively or actively emit them. The intensity and chemical nature of both passive and active emission are a function of the properties of the compounds present, the temperature, and the atmospheric pressure. Examples of inanimate passive emission would include an inanimate object such as a glass of wine emitting hundreds of compounds from the fluid surface into the "headspace" in the immediately vicinity of the fluid.[11] Animate passive emission often occurs as a function of a metabolic process.[12] For example, the odors typically associated with rotting meat are produced and emitted by microbes as a byproduct of metabolism.[13] Mammals emit CO_2 and lactic acid as a byproduct of metabolism; CO_2 and lactic acid may be used by insects, such as mosquitos, to locate mammalian hosts.[14,15] In active emission, odors are synthesized and emitted by an animate signaler. Examples of this are: odor synthesis by flowers of plants; [4,16,17] pheromone production by animals for sexual or conspecific signaling; synomone production by plants that subsequently attracts parasitoids when a plant is attacked by insect herbivores.[3,18,19] In the case of flowers, odorant compounds are produced in the cells of petals or other floral parts [20-23] and released through epidermal cells or specialized glands.[3] Pheromones are often produced in specific glandular tissues specialized for their production,[24,25] but they may also arise from other scent sources such as body tissues.[26] Volatile compounds, such as ethylene and methyl jasmonate, are produced by plants during herbivorous attack and have been shown to act as signals to other plants for the up-regulation of the production of defensive compounds, and to act as cues for parasitoids that attack herbivores.[18,27]

Given the remarkable variety of both odor chemistry and olfactory contexts, it is reasonable to suggest that characterizing the statistical structure of an animal's olfactory environment would help us appreciate how animals encode and use olfactory signals efficiently. This is a relatively new departure for olfactory science, and we might borrow here from the other sensory sciences where attempts to characterize systematic characterization are becoming highly developed. Vision scientists have been the driving force behind much of this research: the philosophy of placing vision within a "natural" context grew out of work in the 1960s and 1970s by neuroscientists and psychologists such as Barlow (1961),[28,29] Gibson (1982),[30] and Mackay (1986).[31] A wealth of information about the statistics of natural visual scenes has appeared in the scientific literature. A good example is provided by Field's (1987) landmark study: *"Relations between the statistics of natural images and the response profiles of cortical cells."*[32] This study showed that there are remarkable statistical regularities in ensembles of completely heterogeneous natural visual scenes. He also showed that the cellular physiology of the visual cortex is consistent with a coding system that takes advantage of these regularities to encode scenes efficiently. The notion of a match between the statistics of visual scenes and the processing in the human visual pathway has since been extended to include a wide variety of scene descriptors, such as orientation, color, spatial and temporal frequency.

For such studies to be possible, there must exist (i) a physical metrical space suitable for a general characterization of the physical relationships among different natural stimuli; (ii) measurable statistical regularities (non-randomness) in the stimuli when expressed in this metrical space; (iii) a perceptual metrical space suitable for a general characterization of the sensory discriminability of the natural stimuli. To take examples from vision, one might characterize the visual input physically by recording points of light as a function of time, wavelength, and space. Thus, a generalized, high-dimensional physical metrical space (a visual 'hypercube') exists for visual stimuli, though vision scientists work typically with down-projections from this space (*e.g.*, a static grayscale scene, which records only brightness as a function of space; or a color movie, which records only 3-dimensional colorimetric data as a function of space and time). Analyses of the statistics of natural visual scenes within these metrical subspaces have shown that, though remarkably heterogeneous visually, they are far from being statistically random. In fact, each natural visual stimulus is a sample drawn from a specific, highly complex distribution. If natural stimuli can be shown to have a definitive underlying distribution, one may start to make inferences about how these statistical regularities have influenced how the sensory system has evolved.[28,29,33,34] In addition, one can benchmark the sensory system by comparing its empirically observed performance with that of a so-called "ideal-observer model" (a theoretical model whose properties are optimized for the measured scene statistics): does the sensory system do as well as it could?

Following the above approach within the olfactory domain is likely to prove difficult. Taking the three requirements listed above, there are good reasons to expect that condition (ii) would be true in olfaction. A later section of this review will imply that there are probably a small, finite number of sources of variability in natural olfactory scenes, and it seems likely that the physical metrical space of olfactory stimuli would be just as structured (*i.e.*, nonrandom, with a characterizable statistical distribution) as that of visual or auditory stimuli. The key problems here are surely conditions (i) and (iii) above: providing definitive physical and perceptual characterizations of the olfactory scene. An important point here is that the scope of the physical metrical space must be at least as great, and preferably much greater, than that of the perceptual metrical space. In a sense, it is only by determining which physical characteristics our sensory systems *ignore* that we can answer those questions that help place perception in some sort of ecological context (*e.g.*, "why is human color vision merely three-dimensional when other animals have been shown to have many more color receptors?"[35]). In the olfactory domain, then, the physical metric space must encompass *at least* those properties of the stimulus known to be perceptually important. It is known, for example, that molecular shape is an important feature affecting odor perception.[36,37] Thus, an attempt to derive a chemical metric that could then be related to the perceptual qualities of odors must operate at least at the level of structural formulae. The difficulty in elaborating such a metric is clear, even for limited families of compounds; a completely general, structural-chemical metric would have to provide an index of similarity for diverse organic molecules. Indeed, the diversity and complexity of organic molecular structures makes it near-impossible to speculate even on how many *dimensions* such a metrical space should have.

Considering this, it should not be surprising that attempts to characterize the statistical features of naturally occurring odorants have been thus far limited to the detection and characterization of odor compounds and estimations of their relative concentrations. Further progress, however, can be made in spite of the difficulties outlined above — particularly the fact that we cannot determine the shape of the physical metrical space for characterizing odor chemistry — by considering the natural sources of variation in odor emission. Just as we can, in vision science, investigate the causes of surface color without concerning the dimensions or shape of the chromatic visual representation, we can look at the different types of variability (spatial, temporal, compositional, etc.) observed either across ensembles or within individual natural odor scenes and then use this natural variability to place odor scenes, and the recognition of odors in context. The next section discusses what these sources of variability might be.

PHYSICS OF ODOR PRODUCTION: SOURCES OF VARIATION

Spatiotemporal Plume Structure

Advances have been made in the last 20 years on characterizing the spatial distribution of odor plumes.[7,38-41] As odors are emitted in air, they disperse as complex, buoyant plumes.[42] The extent to which they disperse is a function of the momentum of the plume through the turbulent boundary layer of the fluid medium.[39] Dispersion produces a gradient of concentration that varies as a function of the distance from the source,[43] and depends on the prevailing environmental conditions such as the temperature and the air/wind speed.[42,44-46] Plumes of highly volatile compounds additionally have complex structures that arise from the distribution of the turbulent kinetic energy of the fluid medium in which they travel.[6,9,42,47] This structure is characterized by filamentous regions of odorant where the concentration varies dramatically within a local spatial region (Fig. 8.1).[7,37] Images taken of the spatial structure of odor plumes at specific time points during emission reveal that the structure of an odor plume changes as a function of the odor molecules' momentum when carried by a fluid medium such as air or water.[7,48] The contrast between odor concentrations in filaments tends to be greatest at the boundary of the odor plume than it is in the center of the plume near the source.[6]

Temporal Fluctuations in Active Emission

The temporal pattern of odor emission can also show marked periodicity on either a short timescale (milliseconds) or a longer one (hours or days). Scent production may occur on a short timescale such that odor quanta are produced and released discretely. For example, some species of female moths have been reported to produce odor "puffs" rather than steady odor production.[49] Odors used for defense, such as the spray secretions of skunks or bombardier beetles, may last only a few seconds.[50,51] On longer timescales, the scent emission of pheromones for some organisms may follow a temporal pattern that correlates with diurnal phases;[25,52-54] or with specific periods in the reproductive cycles of individuals.[55-58] Several studies of flowering plants have shown that floral scents are emitted for hours during those

WRIGHT and THOMSON

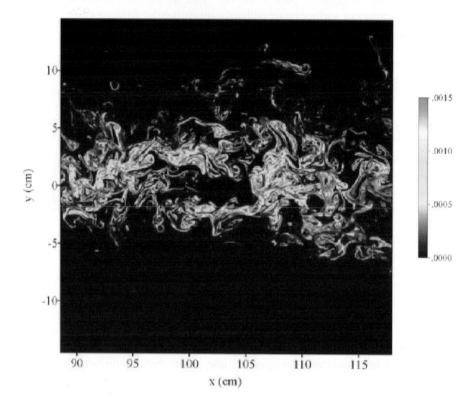

Fig. 8.1: An image of an odor plume taken using planar, laser-induced fluorescence. This image reveals the instantaneous scalar structure of the plume. The image was captured from the outer layer of the momentum boundary layer of the plume. It is a horizontal image spanning a lateral and streamwise range; it reveals the spatial patterns at a given vertical location. The color scale indicates the concentration of the odor in the plume; concentrations are normalized by the source concentration C_0 and color coded as shown in the legend. From Crimaldi et al., Journal of Turbulence, 2002, The relationship between mean and instantaneous structure in turbulent passive scalar plumes, vol. 3, pp. 1-24. Reproduced with the permission of the authors and Taylor and Francis Ltd. (www.tandf.co.uk/journals).

epochs in a diurnal cycle that correlate with the appearance of their pollinators.[59-62] Raguso et al. (2003) showed that moth-pollinated plants show a 2-10 fold increase in scent production during the early nighttime hours when their pollinators fly.[62]

In situations where these temporal periodicities are constant, they could be studied using the standard statistical technique of power spectral analysis,[63] which would illustrate the variance in the rate of emission as a function of temporal frequency in cycles per time interval. Variation in such a diurnal cycle of emission can, however, occur when the environmental conditions change, since floral scent emission is also influenced by environmental factors such as light and temperature.[59] Statistically, such cases would be characterized by a change in the variance of the original periodicity due to the environmental factors, and techniques borrowed from higher-order statistical analysis would be appropriate here.[64]

Qualitative and Quantitative Variation in Odor Sources: Floral Scent as an Example

In addition to the spatial and temporal structure of odor emissions, another significant source of variability in a natural olfactory scene is likely to be composition of the volatile compounds across a population of odor sources.[65] Odors produced by similar sources (*e.g.*, flowers of the same species) may differ with respect to the number of compounds present, the types of compounds, the quantity of each compound, and the overall odor intensity (sum of the odors in the mixture, see below).[65,66] Each of these types of inter-odor differences may alter an odor's perceptual qualities, and, therefore, could also alter the way in which animals respond to them.

In the following subsections we will use flowers as an example of an odor source. Plants produce flowers with visual and olfactory displays that attract animal pollinators.[67] Floral signals both attract new pollinators and help them learn to associate the floral cues with reward so that they will visit other similar flowers and perform pollination. Odors are important signals used for identification and discrimination among flowers; insect pollinators, such as moths, will not feed from flowers unless they have the appropriate odor stimulus.[68] Floral signals may also advertise when flowers have been pollinated by changing their scent composition after pollination.[69,70] It has also been observed that specific suites of compounds may be more attractive to a particular "pollinator guild," such as moth pollinators,[71] and that natural selection by pollinators may select the production of odors these pollinators recognize.[1] The two subsections that follow describe some of the qualitative and quantitative inter-odor differences observed among volatile compounds found in flowers and discuss how to quantify these differences.

Qualitative Variation

    Floral scents are often complex blends of several compounds.[1,4,65] These are mainly fatty-acid derivatives, benzenoids, phenylpropanoids, isoprenoids, nitrogen- and sulfur-containing compounds.[4] Knudsen et al.'s (1993) comprehensive review of floral odors reported floral scents with as few as three compounds and as many as several hundred.[4] A study of moth-pollinated flowers from the Nyctaginaceae reported as many as 77 compounds in floral scent and as few as one.[5] Studies that have focussed on identifying the site of volatile production in flowers report that they are produced mostly in the petals.[16] Volatiles are both synthesized and released by the epidermal cells or by special glandular trichomes on the petal surface.[72] Other studies have shown that the odor of pollen is also part of the odor signal.[21,73]

    Variation in the emission of specific volatile compounds is likely to arise from differences in gene expression of enzymes that produce volatiles and their substrates.[74] Qualitative interspecific differences in the production of scent compounds are generally greater than intraspecific differences.[5,54,65,75,76] Natural selection may lead to differences in scent between species that are advantageous to the plant, as specific pollinators may be less likely to generalize to other plant species and, therefore, pollinate more efficiently. Grison et al. (1999) observed that most fig species produce 4-6 odorant compounds.[77] They observed that one species had 22 compounds; another had a completely different suite of compounds. Both of these traits, an increase in number and a suite of completely different compounds, would change the scent and make it less likely for fig wasps to confuse the scent with the scent of another fig tree. However, natural selection towards the production of volatiles preferred by particular pollinators could also lead to unrelated plant species producing the same suite of compounds.[4,67,78] In some environments, it may be to the plant's advantage to be indistinguishable from other species of plants, so that it increases its pollination rate by generalist pollinators when plants of the same species are less common.[79,80]

    In general, examinations of intraspecific floral scent indicate considerable variability in scent composition. Most information about qualitative variation within species comes from studies of domestically cultivated flowering plants. For a given species, it is possible that all plants produce flowers that have the same number and type of compounds. This was shown to be the case for the scents of 4 snapdragon cultivars, in which 8 volatile compounds occurred in all cultivars.[81] A study by Kim et al. (2000) reported 41 compounds in the scent of roses including 8 different classes of compounds (alcohols, aldehydes, alkanes, monoterpenes, sesquiterpenes, ethers, esters, and ketones).[82] They examined 3 cultivars and found that each had approximately 30 compounds, but only 19 of those (63%) were found in all 3 cultivars. Another study of 4 cultivars of lavender by Kim and Lee (2002) reported that each produced approximately 28 compounds, with only 14 in common (50%) (Fig. 8.2 and 8.3).[83]

Variability in scent composition has been measured in some wild populations of flowering plants as well.[5,84-86] In a study of 20 species of moth pollinated flowers, Levin et al. (2001) found that variability in scent composition was different for each species.[5] They also noted that plants with more compounds in their scent may also have more variability in the types of scent compounds across the population.[5] Knudsen's (2002) study of 5 populations of *Geonoma macrostachys* from western Amazon reported that of 108 compounds found across all the samples, only 28 compounds were common to all; the remaining 70 were not consistently found in every sample.[85] The number of compounds found in the scent also varied among populations in a range from 39-95 compounds. Thus, even within a species, a substantial amount of qualitative variation may exist.

Quantitative Variation

Odor Intensity

The most variable aspect of odor is odor quantity. Even for individual flowers, the amount of scent produced can vary widely as a function of time of day, development stage, and environmental conditions.[16] The amount of scent produced by odor-emitting objects varies both in terms of the individual scent compounds and also in terms of the overall scent intensity. For our purposes, a rough measure of scent intensity is defined as the sum of the concentrations of each of the individual compounds. Considerations of the statistical features of naturally occurring odor objects and the way that animals perceive these objects should also include quantitative information. The overall intensity of floral odors has been reported to range over 4-5 orders of magnitude in concentration.[5,78,85] Differences in intensity may occur among species of plants; scent intensity is not dependent upon the number of compounds present in the odor, as complex scents may still have low intensities if all odor compounds are present at small concentrations. Scent intensity also varies for the same flower throughout its development,[87] and it may also vary as a function of the flower's diurnal cycle.[62] Raguso et al. (2003) reported up to a 10 fold increase in scent intensity over a diurnal cycle for moth pollinated flowers of *Nicotiana alata*.[62]

Correlations among Odorant Concentrations

Most naturally occurring odors are composed of several odorants; each compound is probably present at a different concentration (see Fig. 8.2). The concentration of individual odorants, in fact, can differ by orders of magnitude in concentration. Based on 6 studies of floral scent including the scents of 43 different

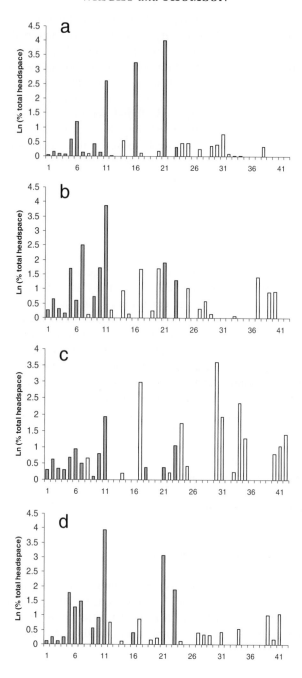

Fig. 8.2: Lavender cultivars have only 14 volatile compounds in common out of 43 compounds detected by gas chromatography (dark bars; see Fig. 8.3 for structures of each compound). Y-axis represents the natural log transformed percentage of each volatile in the floral headspace. Four cultivars are represented a) French; b) Fringed; c) Hidcote; d) Sweet. Each bar represents a different odor. Reproduced from Kim and Lee, Journal of Chromatography A, 2002, vol. 982, pp. 31-47, with the permission of the authors and Elsevier.

species, we estimated that the average floral scent has 31 compounds.[75,78,82,83,88,89] We also estimated the frequency of the compounds of different concentrations in the average floral scent. Approximately 17 compounds (54%) are present at low concentrations and are each less than 1% of the total scent output. Eight compounds (27%) each produce 1-5% of the total scent; 4 compounds (12%) each produce 5-20% of the total scent, and 2 of the compounds (6%) each produce over 20% of the scent. Analysis of correlations among compounds found in the scents of flowers of a population of plants can be accomplished by using statistical data-reduction techniques (*e.g.*, principal components analysis (PCA), characteristic vector analysis, factor analysis) designed to reduce many variables to a smaller subset of components, each of which represents linear correlations among the variables. A study by Ayasse et al. (2000) used PCA to examine correlations in 106 compounds extracted from the orchid, *Ophyrs sphegodes*.[90] They found that groups of compounds belonging to the same chemical class, such as n-alkenes, tended to be represented by the same principal component. Another study by Wright et al., (2005) showed that the first principal component computed from a dataset describing the volatile profile of the scents of snapdragon flowers arose largely as a result of a high, positive correlation between two compounds (*cis* and *trans*-methylcinnamate) produced by the same enzymatic pathway.[81]

Ratios of Odorant Concentrations

As a scent is composed of several compounds with different concentrations, the ratio of the concentrations may be a useful way of identifying the similarity of odor producing objects. These ratios are more likely to reflect the activity of the biosynthetic pathways producing odor than the differences in the vapor pressure of the compounds;[59,91] correlations among the presence and concentration of individual compounds could reflect common biosynthesis pathways.[27] For example, compounds produced by the same enzymatic pathway may be correlated.[54,62,81] Compounds

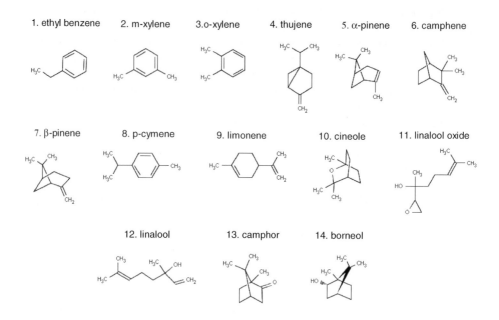

Fig. 8.3: The 2-d structures of the 14 volatile compounds in common among lavender cultivars, reported by Kim and Lee (*J. Chrom. A* 982, 31-47, 2002).[83] The compounds are numbered in order according to their retention time on an SPB-5 column from a gas chromatograph. Structures were drawn using ChemIDPlus at the NIH website (http://chem.sis.nlm.nih.gov/chemidplus/).

produced from the same substrate could also be correlated,[74] as their appearance is dependent upon the amount of substrate available. When the substrate is present and plentiful, each compound may be present in high amounts; if it is limiting, then the concentrations of each may decrease.

As observed for the qualitative aspects of scent, the ratios of the concentration of compounds in scents may show greater inter-species than intra-species variation. This is true both for floral scents and for other, naturally occurring odors. The amount that scent production varies over a specific time interval may vary widely from one individual to the next.[53,92,93] Additionally, some odorants found in scents may simply be intrinsically less variable across a population than others.[53,81,92] This variability can be characterized by measuring the coefficient of variation for each compound in a complex scent across a sampled population of odor-emitting objects.[12,81,90] The coefficient of variation is simply the standard deviation expressed as a percentage of the mean.[94] It provides a standardized way of comparing variation among populations that may exhibit great differences in their ranges of measurement.[94] Interestingly, Ayasse et al. (2000) reported less variability in the concentrations of the "bioactive" compounds in a sexually deceptive orchid than in the concentrations of the orchid's nonactive compounds.[90] The bioactive compounds are similar to the sex pheromone produced by the wasps and attract male wasps. The male wasps attempt to copulate with the flower and pollinate the flower. It is possible in this case that selection pressure is greatest on the odors mimicking the pheromones of the pollinating wasps, specifically such that a lower variation in their production is selected for.

One means of examining differences in variability in ratios of odors is by using methods of classification such as discriminant analysis or cluster analysis. Discriminant analysis is used to identify the maximal co-linearity among a set of variables and then to produce a function representing these linear relationships that maximally separates pre-defined populations or categories.[95,96] If the pre-determined categories are not distinguishable using the entered variables, then the discriminant functions produced will not be statistically significant. Thus, by using discriminant analysis, it is possible to compare the ratios of odor compounds among different species or different subpopulations.[96] When the variation in the ratios within a species or subpopulation is smaller than between species, significant differences will be observed.

This was recently shown in a study of 4 cultivars of snapdragons where even subtle differences in scent were found to be significantly different by discriminant analysis.[81] The analysis by Wright et al. (2005) reported 3 discriminant functions that split the cultivars into 4 significantly different groups.[81] Standardized coefficients are reported in Table 8.1; the magnitude of these coefficients reflects the importance of a specific odorant to the classification of each cultivar by the discriminant functions in the presence of the other odorants.[96] Each discriminant function represents a split of one cultivar from the others. The sign of the unstandardized coefficient reflects which cultivar was split from the group. This type of analysis provides information both about differences in scent profiles of putatively different odor producing objects and about which aspects of the scent are used to differentiate them.

Table 8.1: Discriminant analysis of the volatile compounds of the scents of 4 snapdragon cultivars. The standardized coefficients in bold indicate the volatiles that contributed the most to classification of the snapdragon varieties by each of the functions. The sign and magnitude of the unstandardized coefficients (in bold) indicate which cultivar was best separated from the others by the function. The order of the functions indicates the distance in similarity between the snapdragon cultivars. (From Wright et al., 2005,[81] reproduced with permission of Springer)

	Discriminant Function		
	1	2	3
Volatile compound	**Standardized Coefficients**		
Myrcene	**0.976**	0.417	0.049
E-β-ocimene	-0.306	**-0.757**	-0.222
MethylBenzoate	-0.260	**-1.12**	0.473
Acetophenone	-0.308	0.486	**0.632**
Linalool	0.244	0.370	0.410
Dimethoxytoluene	0.101	0.663	**-0.867**
C-methylcinnamate	-0.098	-0.268	-0.313
T-methylcinnamate	-0.173	0.547	0.241

Cultivar	**Unstandardized coefficients**		
PH	**-2.94**	0.088	0.021
MTP	2.15	**-1.68**	0.045
PP	3.59	0.501	**-0.442**
PW	3.01	0.591	0.811

Temporal Fluctuations in Ratios

The ratios of odorants in scent, however, may also change throughout a diurnal cycle of emission, as the increase in scent production during specific time intervals may not be the same for all compounds. Individual compounds may not have the same coefficient of variation.[81,90] One study by Helsper et al. (1998) showed that diurnal emission rates were different for each compound in the scent of single rose flowers (Fig. 8.4a).[59] Differences in the temporal structure of the emission of individual compounds changed the ratios of the concentrations of each of the compounds such that a different odor "profile" was produced depending on the sampling point (Fig. 8.4b). Additionally, if the intensity of the scent is calculated for this same study, the overall scent production was at a maximum 6 h into the light half of the day cycle. Thus, in a diurnal cycle, the ratios of the concentrations of the compounds may vary dramatically. The extent to which differences in scent emission are observed, however, is likely to be a function of the duration of the time interval of odor sampling.

ODOR PERCEPTION AND THE OLFACTORY SYSTEM

The olfactory system must allow animals to detect and discriminate among a vast array of possible odors. Interestingly, the olfactory systems of diverse animals exhibit common features, such as first-order processing at receptor neurons in the olfactory epithelium, followed by second-order processing at the glomeruli in the olfactory bulb (mammals) or antennal lobe (insects). It is likely that these structural features have evolved independently. The morphological identity of these olfactory systems may thus be a result of common evolutionary constraints,[97,98] which might in turn reflect statistical consistencies among naturally occurring odors.[34] Sensory transduction of odors occurs when odorant molecules come into contact with G-protein-coupled receptors present in the dendrites of receptor neurons. These receptor neurons may be "tuned" to respond to a specific range of odorant molecules; the extent of tuning may be proportional to the binding affinity of odorant molecules with that odorant receptor.[99-102] Depending on the animal, receptor neurons express one to several odorant receptors.[97,102] The receptor neurons are bipolar with dendrites expressing G-proteins for odorant binding and with axons that converge in the antennal lobe or the olfactory bulb.[102] The antennal lobe and olfactory bulb are composed of highly interconnected circuits of excitatory and inhibitory neurons that form glomeruli.[2] For animals expressing only one odorant receptor, axons from receptor neurons that express the same odorant receptor converge onto the same glomerulus or onto a small set of neighboring glomeruli.[100,103] The neurons in the antennal lobe relay information about an odor's identity via the action potentials of projection neurons to the higher centers of the brain.[2] The higher centers of the brain integrate information from other sensory modalities to form associations between sensory representations.[104,105]

Learning, Generalization and Discrimination

Animals show both innate and learned responses to odors in their environment. Examples of innate responses include the anemotactic behavior of those male moths that fly upwind towards a pheromone-emitting female[106] or the defensive responses of honeybees to their alarm pheromone.[107] Many animals also *learn* to associate odor with important events such as the presence of food. The

a

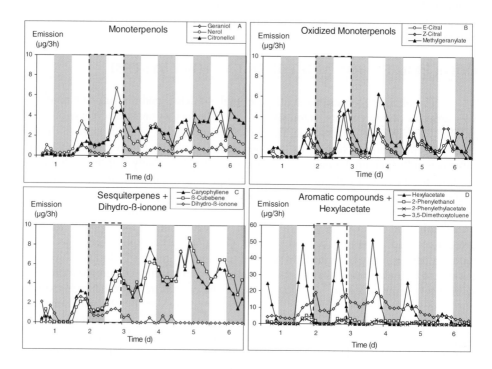

Fig. 8.4: The ratios of the concentrations of odor volatiles change as a function of the time period of sampling. a) Diurnal emission of individual odor volatiles by flowers of Rosa hybrida exposed to a 12-h photoperiod; shaded and non-shaded areas correspond to periods of darkness and light. Each point represents the amount of a given volatile at a 3 h sampling period. b) Representation of the odor volatiles for each 3 h sampling period over a 24 h period (shown in a dotted box on (a)). The odors are represented on the x-axis as: 1=geraniol, 2=nerol, 3=citronellol, 4=e-citral, 5=z-citral, 6=methylgeranylate, 7=caryophyllene, 8=β-cubebene, 9=dihydro-B-ionone, 10=hexylacetate, 11=2-phenylethanol, 12=2-phenylethyl-acetate, 13=3,5 dimethyoxytoluene. Reproduced from Helsper et al., Planta, 1998, Circadian rhythmicity in emission of volatile compounds by flowers of *Rosa hybrida* L. Cv. Honesty, vol. 207, iss. 1, pp. 88-95, with the permission of the authors and Springer.

b

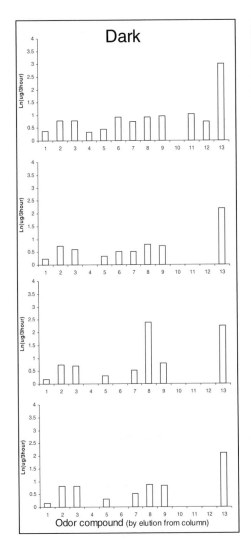

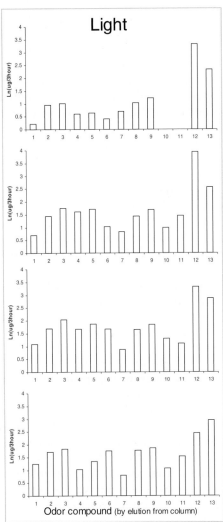

following sections discuss how animals perceive odors and how the statistics of natural odors may affect what they perceive. We focus mainly on odor signals that are learned by animals.

Studies of learned behavior via associative conditioning with odors have shown that learning strongly influences the ability of animals to perceive and identify odors.[108] Animals learn by associating a "conditioning" stimulus or a context with an important event.[109] Learning is defined as an animal's use of information about a conditioning stimulus to predict another stimulus or event.[110] By necessity, animals must take information about a conditioning stimulus from the first experience and generalize this information to the next experience because no conditioning stimulus is experienced in exactly the same way every time the animal encounters it.[109,111] Therefore, during learning, animals are forced also to generalize from one conditioning stimulus to the next. When perceiving a conditioning stimulus, an animal's responses are mediated by a comparison between previous experiences and the current stimulus. Generalization of a conditioned response, such as a learned association of an odor with food, occurs when animals perceive similarities among stimuli from one experience to the next.[110,111] Animals often face situations where they must generalize from an odor they have learned to another odor stimulus. One example of this is foraging by honeybees. Honeybees use pollen and nectar from flowers as their primary food source.[112] They use the odor of flowers to identify a good floral resource and forage on it exclusively to increase their foraging efficiency. When foraging, a honeybee learns the odor of a flower and compares its scent with the odors produced by new flowers to decide whether to forage on a new flower or not. It may use several features of the scent, including the types of compounds and their concentrations, to generalize what it learned to a new flower.[81,113]

In addition to generalizing from one odor experience to the next, animals must also discriminate among odors. Nursing lambs use the odor of their mother to identify her in a field of other ewes.[114,115] Honeybees also use the odors of flowers to discriminate rewarding flowers from unrewarding flowers; they may also place scent marks on flowers to discriminate a flower they just visited from a new flower of the same type.[116] The guard caste of honeybees patrols the entrance of the hive; they use the scent of entering workers to determine whether the worker is a hivemate or an intruder from another hive. In this case, a guard must generalize the scent of a nestmate to an entering worker that may have been exposed to other scents[117] yet discriminate nestmates from intruders attempting to rob the hive.[118,119]

Several variables influence both generalization and discrimination, including: the perceptual similarity of stimuli, the order of an animal's experience with stimuli, the amount of experience it has had with the stimuli, and its motivational state to attend to differences in the stimuli. The specific features of stimuli used to generalize from one stimulus to another may be different from the features used to discriminate among stimuli. Generalization from one stimulus to the next may arise because an animal is unable to perceive differences among stimuli; it also arises when an animal uses features of stimuli to classify stimuli with similar meanings.[111] The ability to discriminate is related to an animal's ability to detect differences among scents (for example, differences among the types of compounds or their concentrations), but discriminative *behavior* is also strongly affected by motivational factors; for example, the cost of making a mistake when discriminating among stimuli.[110]

Perception of Odor Quality: Odor Similarity and Odor Space

One aim of studying olfaction has been to gain an understanding of how the physical features of odor molecules correlate with their perceptual qualities.[120] Unfortunately, the perceptual qualities do not often follow an easily ordered metric that can be simply related to molecular structures. We might, therefore, speak of a dual problem in attempting to relate odor perception to the features of the corresponding molecules: no obvious metric is available to describe either the space of odor perceptions or the space of odor chemistry.

Some studies have examined the ability of animals to discriminate among odor stimuli and have attempted to correlate failures of discrimination with structural similarity among odor molecules.[121] These studies may be used to form indices of the perceptual similarity of odors, and this may provide information about the way that a limited variety of chemical structures (*e.g.*, alcohol groups or double-bonded oxygen moieties (ketones)) are used as features by the olfactory system[120,122] Recent behavioral experiments using aliphatic alcohols, ketones, and aldehydes have shown that the perceptual qualities of odors are correlated with molecular features such as carbon-chain length.[121,123-125] For example, Daly et al. (2001) found that discrimination of aliphatic alcohols and ketones by the hawkmoth, *Manduca sexta*, was greatest for compounds of different functional groups (alcohol vs. ketone); moreover, among odorants that belonged to the same functional group, compounds with the greatest differences in carbon-chain length (*e.g.*, alcohols with different carbon backbones) were easiest for the hawkmoth to discriminate.[124] Thus, it is tempting to conclude that compounds closely related in respect of their chemical formulae may also be difficult to distinguish, whereas compounds with substantially different chemical formulae are likely to be easy to distinguish. As we have already discussed, however, counterexamples to such a hypothesis exist in the form of chiral enantiomers of odor molecules: enantiomers of the same molecule sometimes exhibit substantially different perceptual qualities.[36,37,122] Thus, an attempt to derive a chemical metric that could then be related to the perceptual qualities of odors must operate at least at the level of structural formulae, since chemical formulae do not take chirality into account. In view of the metrical problems outlined earlier, then, it seems that for the moment we will have to accept that attempts to relate odor chemistry to olfactory physiology must restrict themselves to piecewise analyses of small, isolated areas of the global psychophysical odor space.

The study of the perception of compounds present in odor mixtures is also fraught with the sorts of nonlinear "contextual effects" with which vision and auditory scientists are all too familiar. At present, most studies use human subjects to examine the way that odor compounds affect perception in complex mixtures; such studies show that perceived odor similarity between two complex blends is likely to be mediated both by the types of compounds found in the mixtures and the relative number of compounds that the two mixtures have in common.[126,127] Perception of similarity is also be affected by the presence of specific volatiles in common that overshadow other volatiles, either because their concentration is greater or they are easier to perceive;[126-128] this is similar to the "masking" effects seen in other sensory systems. A complex mixture may also produce a percept whose qualities are quite independent of the qualities of the individual compounds,[129,130] making it difficult to relate the sensory properties of a mixture back to each of the components, especially in mixtures containing several compounds.[126,127,131] Additionally, not all of the odorants in a mixture may contribute equally to a scent's perceptual qualities; some may not contribute at all.[128] Both types of complication could be interpreted as a violation of linearity (the response to A plus B is not the response to A plus the response to B) analogous to that which is seen in the later stages of processing in both auditory and visual systems.[132]

A final problem is the potential for confusing odor discriminability with the discriminability of odor categories. Some studies have addressed the problem of finding satisfactory metrics to express the relation between odor structure and the way that humans perceive them by using descriptive language to classify odors (for a review see: Wise et al. (2000[120])). For example, in the classification of Zwaardemaker (1925), odors considered to be "nutritive" were classified in some of the following categories: etherous, floral and balsamic, aromatic, and ambrosaic[133] (reported from Wise et al., 2000[120]). Returning once again to a color-vision analogy, being asked to visually discriminate two slightly different shades of red is clearly an entirely different task from being asked whether a given color is red or orange; the former is a perceptual discrimination, whereas the latter is a category discrimination. Imposing categories upon the psychological odor space may help produce a perceptual metric, but it is a metric of categories, not a metric of odors, and it is conceivable that the relationship between the two might turn out to be as complex as it is in color vision.

Odor Concentration and Odor Perception

As mentioned above, the concentration of an odor stimulus is a key physical feature of naturally occurring odors. The concentration of an odor may vary widely within an animal's experiences; between salient odor objects, concentration may be a defining feature used to identify important odors. The concentration may also convey information about an animal's distance from the source of emission. Odor learning and discrimination are affected by concentration. Odors of low concentration are more difficult to learn.[134-136] Discrimination among odors also increases as a function of odor concentration (Fig. 8.5).[136-139] This has been shown both for the discrimination of monomolecular compounds[136,138] and for discrimination among

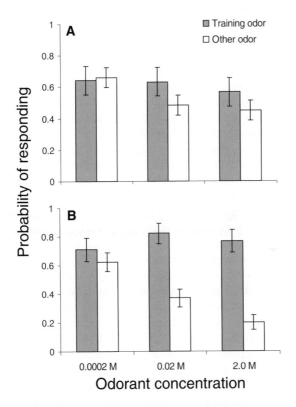

Fig. 8.5: Discrimination increases with concentration for dissimilar odors (b) but not for similar odors (a). Conditioning and testing were performed at the same concentration for the similar (**a**; n = 87 animals) or dissimilar (**b**; n = 88 animals) test odors. Generalization response levels to novel odors (open columns) are shown next to the response to the conditioned odor (shaded column) at a given training/testing concentration. Reproduced from Wright and Smith, Different thresholds for detection and discrimination of odors in the honey bee (*Apis mellifera*). Chemical Senses, 2004, vol. 29, pp. 127-135 with permission of the Oxford University Press.

mixtures.[81,113] That odors are harder to discriminate at low concentrations suggests two possibilities: (i) as in other sensory systems, detectability is a function of the signal-to-noise ratio, such that a higher-strength signal is more reliably detected and thus easier to discriminate; (ii) that odors may not be perceptually invariant as a function of concentration. These two possibilities are not mutually exclusive, andindeed both may well turn out to be true. Variation in the perceptual qualities of odors as a function of concentration has been noted for many years[139-143] but is difficult to test exhaustively. Recent studies suggest that samples of the same odor at different concentrations are discriminated reliably by animals[134,144] and that odors may not be perceptually invariant as a function of concentration.[139] Additionally, for several insects with innate responses to odor, odor may be attractive at one concentration and repulsive at a higher concentration.[145,146] Erbilgin et al. (2003) recently showed that pine beetles (*Ips pini*) demonstrate an apparently highly nonlinear response to their aggregation pheromone, such that low concentrations of pheromone are not attractive, mid-range concentrations are highly attractive, and high concentrations are repulsive.[146] Dethier (1976) also showed a similar nonlinear response to odor in blowflies: he observed that blowflies were attracted to low concentrations of iso-valeraldehyde and repelled by high concentrations.[145]

Behavioral responses that vary as a function of concentration may indicate that the odor concentration is the feature that triggers the innate response for some animals. In other words, the animal may know something *a priori* about the statistics of the natural distribution of odorant concentration, and reject those concentrations that are in some sense ecologically valid. For this to be true, it must be the case that the odor in question simply does not remain perceptually invariant as a function of concentration: the animal perceives high-concentration odorant as a different substance altogether to low-concentration odorant. The existence of concentration-dependent variation in odor quality has profound implications for our understanding of olfactory coding and its relationship to the statistics of natural odor scenes. As thousands of different odor molecules exist, there may be evolutionary trade-offs between coding for a diverse set of odors and coding for an extensive range of odors. In particular, failure of invariance may suggest that the need to encode diverse odor molecules has exerted a greater pressure on the olfactory system than the need to identify a specific odor over a large range of concentrations. This may have arisen because the concentration range of naturally occurring odors is not large for a specific population of odor scenes. In this case, it may be important to maintain odor identity within the ecologically relevant range; if the animal is not likely to encounter an odor outside of this range, then the ability to render its perceptual qualities invariantly may be less important. Further research into the relationship between the concentration of natural odor scenes and the way that animals use odor scenes is necessary to test this hypothesis.

Odor Perception and Spatiotemporal Plume Structure

One additional aspect of odor concentration and its meaning to animals may be related to the spatial statistics of odor-concentration distribution. The concentration may convey information about the distance of the animal from the odor source.[40,41,106,147-151] In spite of studies that show that animals will perform upwind anemotaxis towards an odor source, it has yet to be shown definitively that behavior is simply concentration-driven: animals may not understand innately that low concentrations of an odor necessarily mean that they are far away from the source of emission.[152] Instead, other aspects of the signal, such as the frequency of encounter of (high-concentration) filaments of an odor plume or other —possibly higher-order — spatial statistics of the plume, may be the stimulus driving anemotaxis behavior.[40,41,150,151] The way that spatial statistics of odor plumes are used by animals may also depend on the animal and its environment.[44] A recent study by Keller and Weissburg (2004) showed that the chemosensation of the blue crab (*Callinectes sapidus*) used large-scale variation (pulses of 1-3 sec) in concentration in odor plumes to detect odor sources,[153] rather than using either fine-scale variation (less than 1 second) or acting as a flux detector.[154]

Perception of Ratios of Compounds in Complex Mixtures

In complex odor mixtures, the ratios of odorant concentrations may be important features that animals use to discriminate. This has been shown to be especially true of pheromone blends, where the ratio of the concentration of each pheromone compound in the blend may affect several behaviors, from anemotaxis to contact with the source of emission.[92,155] Variability of emission of the ratios of odorants in pheromone blends can occur within individuals[91] and across populations[92,156] or between species.[157,158] The strength of the effects of variation on behavior appears to be dependent upon the species involved, however.[92] In non-pheromonal odors, the ratios of odor compounds in scent may also be useful for discriminating among odor-emitting objects. Differences attributed to scent as a function of the ratios of odorant compounds may be more subtle than perceptual differences that occur from the subtraction or addition of different odorant compounds to an odor mixture. As these differences may be hard to detect, the extent to which animals use differences in the ratios of compounds may be governed by the cost of making a mistake between scents with different ratios.

Depending on the complexity of the blend, there may also be a limit to which these ratios actually affect the perceptual qualities of an odor. Other variables, such as intensity, may also change the effects of ratios on an odor's perceptual qualities.[126] There is evidence that both vertebrates and invertebrates can make fine discriminations based on the ratios of odorant concentrations. A study by Osada et al. (2003) showed that the ratios of the odor compounds in mouse urine changed as a function of age, and that mice could learn to discriminate among these ratios.[159] They found 38 odor compounds; eight were different between adult mice and aged mice. Five were greater in aged mice, and 3 were smaller in concentration. As the concentrations of all eight compounds were not greater in the aged mice, the mice were using the ratio rather than a change in the intensity of the scent. Honeybees are also able to use the ratios of compounds found in floral scents to make subtle discriminations. Wright et al. (2005) showed that honeybees could learn to discriminate the scents of 4 snapdragon cultivars.[81] Each cultivar had 8 compounds present; the cultivars were significantly different in the ratios of scent they emitted when classified by a discriminant analysis (see section above Table 8 1). The ability of honeybees to discriminate among the scents correlated with the differences in the ratios among the cultivars. Additionally, scent discrimination was also affected by scent intensity, such that discrimination using the ratios was more difficult when the scent was less intense.

Odor Perception and Temporal Variation

Animals experience odors sequentially. In particular, this is true of foraging animals that rely on scent for the identification of food items. For example, honeybees foraging on flowers go from one flower to the next; they may visit tens or hundreds of flowers on a single foraging trip.[160] Other animals, whether they are searching for mates using pheromones[92] or identifying nestmates,[118] are also likely to encounter variation among odor signals. Variability in the signal itself may, therefore, become an important feature that animals can use while learning to identify salient odors. As might be expected, the sequence of odor experience has been shown to affect which features an animal uses to identify a previously experienced scent.[108] Of particular importance is the presence of common features that occur from one olfactory stimulus to the next. Studies of the behavioral mechanism known as olfactory blocking have shown that an odorant that is common throughout conditioning will become a dominant perceptual feature of subsequently experienced binary mixtures.[161-164] In this case, the presence of an odorant that has the least variability throughout conditioning also acquires the most salience.

A recent study by Wright and Smith (2004a) showed that multiple features of odor mixtures influence which features are used by honeybees to identify previously experienced odors.[165] In particular, the study observed that both variation in the concentration of individual odorants in a mixture and the overall intensity of an odor mixture affected what honeybees learned about odors during conditioning. They conditioned honeybees with mixtures of odorants where the concentration of one odorant remained the same throughout conditioning and the concentration of the other odorants varied. Generalization was greatest from the conditioned mixture to the non-varying odorant. However, this occurred only for odors composed of structurally dissimilar odorants. Generalization was also affected both by the overall amount of variability from one conditioning trial to the next and the average odor intensity experienced during conditioning. Generalization increased when variability in the mixture increased (Fig. 8.6). Honeybees conditioned with mixtures with an average high concentration generalized less to low-concentration odorants. Thus, the second-order statistics of a population of natural odor scenes have been shown to be detectable features used by honeybees to identify salient odors.

Source-to-source odor variation in a population of odor-producing objects may be an important statistical feature affecting olfactory perception and behavior. Another recent study showed that variability in the major-histocompatability-complex (MHC) odor signal and the amount of scent marking by individuals affected scent use by female mice when they were selecting mates.[166] They showed that a

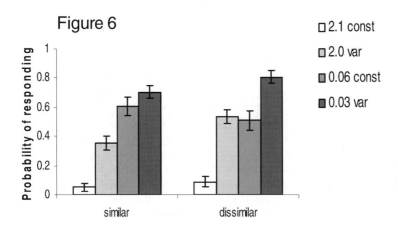

Fig. 8.6: Generalization from one odor stimulus to another increases as a function of variation in odors. Generalization to low concentration odorants is also affected by the intensity of the conditioning mixture. When honeybees were conditioned with high intensity mixtures (2.0 and 2.1 M) they responded less than honeybees conditioned with low intensity mixtures (0.03 and 0.06 M). Honeybees conditioned with highly variable mixtures (*var*) responded with a higher probability than honeybees conditioned with no variation present in the mixture (*const*). Reproduced from Wright and Smith, Proceedings of the Royal Society B, 2004, vol. 271, iss. 1535, pp. 147-152, with the permission of the Royal Society of London.

female mouse determined her genetic similarity to other mice through information about the variability in the MHC complexes and the marking rates of the males she encountered. Variability in the MHC was a function of the population of males she encountered. When variability in the MHC complex was small, a female would use marking rate as a means of choosing a male. In this case, variation in the MHC was used in a context-dependent manner. It is likely that as variation in natural odor scenes becomes easier to measure and more common to report that future studies will also observe that it affects feature recognition by other animals

CONCLUSIONS

The study of the statistics of naturally occurring odor scenes is likely to yield rich information about the way that organisms produce and use scent in their interactions. One approach to studying odor perception and recognition may be via detailed studies that examine both the statistics of natural odor scenes and the way in which animals use these features to generalize among odor scenes. Studies of this kind may show that in addition to second-order features (*i.e.*, variation), the higher-order features of odor scenes are also important features used for odor recognition. As has been shown in the visual and auditory sciences, relating olfactory physiology to the statistics of natural scenes may also lead to insight into the way in which odors are represented by the nervous system. This is true not just at the level of the olfactory periphery but also higher up in the nervous system. Studies of odor coding in the insect antennal lobe and the olfactory bulbs of fish and mammals indicate that higher-order coding at these levels contributes to the way that the brain organizes information about odors,[2] and this higher-order neural coding may provide a means of reliably producing a neural representation in the face of variation in natural stimuli. Future work that examines both the statistics of odors and the physiology of these higher-order odor representations may reveal that the overall morphology of the olfactory system is adapted for the detection and discrimination of specifically those odors that have ecological signification for a given animal.

ACKNOWLEDGEMENTS

The authors would like to thank Brian H. Smith for years of support at the Ohio State University. We owe special gratitude to the authors that allowed us to use their figures (John Crimaldi, Johannes Helsper, Dong-Sun Lee). We also would like to thank the organizers of the International Society for Chemical Ecology for inviting G.A.W. to participate in the main symposium. G.A.W. was funded in part by a National Institute of Health grant awarded to Brian Smith (NCRR 9 R01 RR1466); G.A.W. and M.G.A.T were both funded in part by a National Science Foundation grant awarded to the Mathematical Biosciences Institute (agreement 0112050).

REFERENCES

1. DOBSON, H.E.M., Floral volatiles in insect biology. *in*: Insect-plant Interactions (E.A. Bernays, ed,), CRC Press, Boca Raton, FL. 1994, pp. 47-81.
2. LAURENT, G., Olfactory network dynamics and the coding of multidimensional signals. *Nat. Rev. Neurosci.*, 2002, **3**, 884-895.
3. PICHERSKY, E., GERSHENZON, J., The formation and function of plant volatiles: perfumes for pollinator attraction and defense. *Curr. Opin. Plant. Biol.*, 2002, **5**, 237-243.
4. KNUDSEN, J.T., TOLLSTEN, L., BERGSTROM, L.G., Floral scents – a checklist of volatile compounds isolated by headspace techniques. *Phytochemistry*, 1993, **33**, 253-280.
5. LEVIN, R.A., RAGUSO, R.A., MCDADE, L.A., Fragrance chemistry and pollinator affinities in Nyctaginaceae. *Phytochemistry.* 2001, **58**, 429-440.
6. MURLIS, J., Odor plumes and the signal they provide. *in*: Insect Pheromone Research. (R.T. Carde and A.K. Minks, eds,) Chapman and Hall, London, UK. 1997, pp. 221-231.
7. CRIMALDI, J.P., WILEY, M.B., KOSEFF, J.R., The relationship between mean and instantaneous structure in turbulent passive scalar plumes. *J. Turbul.*, 2002, **3**, 1-24.
8. MORI, K., NAGAO, H., SASAKI, Y.F., Computation of molecular information in mammalian olfactory system. *Network: Comp. Neur. Sys.*, 1998, **9**, 79-102.
9. BRADY, J.E., HUMISTON, G.E., General Chemistry: Principles and Structure. 4th ed. John Wiley and Sons, New York, NY. 1986.
10. LETCHER, T.M., NAICKER, P.K. Determination of vapor pressures using gas chromatography. *J. Chrom. A*, 2004, **1037**, 107-114.
11. NOBLE, A.C., EBELER, S.E., Use of multivariate statistics in understanding wine flavor. *Food Rev. Intl.*, 2002, **18**, 1-21.
12. COLLINS, S.B., PEREZ-CAMARGO, G., GETTINBY, G., BUTTERWICK, R.F., BATT, R.M., GIFFARD, C.J., Development of a technique for the in vivo assessment of flatulence in dogs. *Amer. J. Vet. Res.*, 2001, **62**, 1014-1019.
13. WHITFIELD, F.B., 1998. Microbiology of food taints. *Intl. J. Food Sci. Tech.*, 1998, **33**, 31-51.
14. DEKKER, T., STEIB, B., CARDE, R.T., GEIER, M., L-lactic acid: A human-signifying host cue for the anthropophilic mosquito, *Anopheles gambiae. Med. Vet. Entomol.*, 2002, **16**, 91-98.
15. BARROZO, R.B., LAZZARI, C.R., The response of the blood-sucking bug *Triatoma infestans* to carbon dioxide and other host odours. *Chem. Sens.*, 2004, **29**, 319-329.
16. DUDAREVA, N., PICHERSKY, E., Biochemical and molecular genetic aspects of floral scents. *Plant. Phys.*, 2000, **122**, 627-633.
17. PIECHULLA, B., POTT, M.B., Plant scents - mediators of inter- and intraorganismic communication. *Planta*, 2003, **217**, 687-689.

18. TURLINGS, T. C. J., LOUGHRIN, J. H., RÖSE, U., MCCALL, P. J., LEWIS, W. J., TUMLINSON, J. H., How caterpillar-damaged plants protect themselves by attracting parasitic wasps. *Proc. Natl. Acad. Sci. USA*, 1995, **92**, 4169-4174.

19. BALDWIN, I.T., KESSLER, A., HALITSCHKE, R., Volatile signaling in plant-plant-herbivore interactions: what is real?, *Curr. Opin. Plant Biol.*, 2002, **5**, 351-354.

20. BERGSTROM, G., DOBSON, H.E.M., GROTH I., Spatial fragrance patterns within the flowers of *Ranunculus-acris* (Ranunculaceae). *Plant Syst. Evol.*, 1995, **195**, 221-242.

21. COOK, S.M., BARTLET, E., MURRAY, D.A., WILLIAMS, I.H., The role of pollen odour in the attraction of pollen beetles to oilseed rape flowers. *Ent. exp. app.*, 2002, **104**, 43-50.

22. GOODWIN, S.M., KOLOSOVA, N., KISH, C.M., WOOD, K.V., DUDAREVA, N., JENKS, M.A., Cuticle characteristics and volatile emissions of petals in *Antirrhinum majus*. *Physiol. Plant,* 2003, **117**, 435-443.

23. SHALIT, M., GUTERMAN, I., VOLPIN, H., BAR, E., TAMARI, T., MENDA, N., ADAM, Z., ZAMIR, D., VAINSTEIN, A., WEISS, D., PICHERSKY, E., LEWINSOHN, E., Volatile ester formation in roses. Identification of an acetyl-coenzyme A. Geraniol/citronellol acetyltransferase in developing rose petals. *Plant Phys.*, 2003, **131**, 1868-1876.

24. SRENG, L., Cockroach mating behaviors, sex-pheromones, and abdominal glands (Dictyoptera, Blaberidae). *J. Insect Behav.*, 1993, **6**, 715-735.

25. FOSTER, S.P., Periodicity of sex pheromone biosynthesis, release and degradation in the lightbrown apple moth, *Epiphyas postvittana* (Walker). *Arch. Insect Biochem. Physiol.*, 2000, **43**, 125-136.

26. BRENNAN, P.A., KEVERNE, E.B., Something in the air? New insights into mammalian pheromones. *Curr. Biol.*, 2004, **14**, R81-R89.

27. DEGEN, T., DILLMANN, C., MARION-POLL, F., TURLINGS,. T.C.J., High genetic variability of herbivore-induced volatile emission within a broad range of maize inbred lines. *Plant Physiol.*, 2004, **135**, 1928-1938.

28. BARLOW, H.B., The coding of sensory messages. *in*: Current Problems in Animal Behaviour. (W.H. Thorpe, O.L. Zangwill, eds,), Cambridge Univ. Press, Cambridge, UK. 1961a, pp 331-360.

29. BARLOW, H.B., Possible processes underlying the transformation of sensory messages. *in*: Sensory Communication (W. Rosenblith, ed,), MIT Press, Cambridge, MA. 1961b, pp. 217.

30. GIBSON J.J., The Perception of the Visual World. Houghton Mifflin, Boston, MA. 1966

31. MACKAY, D. M., 1986. Vision - the capture of optical variation. *in*: Visual Neuroscience, (J.D. Pettigrew, K.T. Sandison, W.R. Levick, eds,), Cambridge Univ. Press, Cambridge, UK. 1986, pp 365-373.

32. FIELD, D.J., Relations between the statistics of natural images and the response profiles of cortical cells. *J. Opt. Soc. Amer.* 1987, **4**, 2379-2394.

33. ATTNEAVE, F., Informational aspects of visual perception, *Psychol. Rev.,* 1954, **61**, 183-13.

34. SIMONCELLI, E.P., OLSHAUSEN, B.A., Natural image statistics and neural representation. *Annu. Rev. Neurosci.*, 2001, **24**, 1193-1216.
35. MARSHALL, J., OBERWINKLER, J., The colourful world of the mantis shrimp. Nature, **401**, 873-874.
36. RUBIN, B. D. KATZ, L.C., Optical imaging of odorant representations in the mammalian olfactory bulb. *Neuron*, 1999, **23**, 499-511.
37. LINSTER, C., JOHNSON, B.A., YUE, E., MORSE, A., XU, Z., HINGCO, E.E., CHOI, Y.J., CHOI, M., MESSIHA, A., LEON, M., Perceptual correlates of neural representations evoked by odorant enantiomers. *J. Neurosci* , 2001, **21**, 9837-9843.
38. MURLIS, J., WILLIS, M.A., CARDE, R.T., Spatial and temporal structures of pheromone plumes in fields and forests. *Physiol. Entomol.*, 2000, **25**, 211-222.
39. CRIMALDI, J.P., KOSEFF, J.R., High-resolution measurements of the spatial and temporal scalar structure of a turbulent plume. *Exp. Fluids*, 2001, **31**, 90-102.
40. WEISSBURG M.J., DUSENBERY, D.B., Behavioral observations and computer simulations of blue crab movement to a chemical source in a controlled turbulent flow. *J. Exp. Biol.,* 2002, **205**, 3387-3398.
41. WEISSBURG, M.J., DUSENBERY, D.B., ISHIDA, H. JANATA, J. KELLER, T., ROBERTS, P.J.W. WEBSTER, D.R., A multidisciplinary study of spatial and temporal scales containing information in turbulent chemical plume tracking. *Environ. Fluid Mech.,* 2002, **2**, 65–94.
42. MURLIS, J., ELKINTON, J.S., CARDE, R.T., Odor plumes and how insects use them. *Annu. Rev. Entomol.*, 1992, **37**, 505-532.
43. JONES, C. D., On the structure of instantaneous plumes in the atmosphere. *J. Hazard Mat.*, 1983, **7**, 87-112.
44. FINELLI, C. M., PENTCHEFF, N.D., ZIMMER, R.K.. WETHEY, D.S., Physical constraints on ecological processes: a field test of odor mediated foraging. *Ecol.*, 2000, **81**, 784–797.
45. MOORE, P.A., GRILLS, J.L., SCHNEIDER, R.W.S., Habitat-specific signal structure for olfaction: An example from artificial streams. *J. Chem. Ecol.*, 2000, **26** (2), 565-584.
46. TOMBA, A.M., KELLER, T.A., MOORE, P.A., 2001. Foraging in complex odor landscapes: chemical orientation strategies during stimulation by conflicting chemical cues. *J. N. Amer. Benthol. Soc.*, 2001, **20**, 211-222.
47. MYLNE, K.R., MASON, P.J., Concentration fluctuation measurements in a dispersing plume at a range of up to 1000-m. *Quart. J. Royal Meteorol. Soc.*, 1991, **117**, 177-206.
48. GRASSO, F. W., BASIL, J.A., How lobsters, crayfishes, and crabs find important sources of odor: current perspectives and future directions. *Curr. Opin. Neurobiol.*, 2002, **12**, 721-727.
49. EISNER, T. For Love of Insects. Belknap Press of Harvard University Press: Cambridge, MA. 2003.
50. LARIVIERE S., MESSIER, F., Aposematic behaviour in the striped skunk, *Mephitis mephitis, Ethol.*, 1996, **102**, 986-992.

51. EISNER, T, ANESHANSLEY, D.J., YACK, J., ATTYGALLE, A.B., EISNER, M., Spray mechanism of crepidogastrine bombardier beetles (Carabidae; Crepidogastrini). *Chemoecol.*, 2001, **11**, 209-219.

52. KIM, J.Y., HASEGAWA, M., LEAL, W.S., Individual variation in pheromone emission and termination patterns in female *Anomala cuprea. Chemoecol.*, 2002, **12**, 121-124.

53. MIKLAS, N., RENOU, M., MALOSSE, I., MALOSSE, C., Repeatability of pheromone blend composition in individual males of the southern green stink bug, *Nezara viridula. J. Chem. Ecol.*, 2000, **26**, 2473-2485.

54. BARKMAN, T.J., BEAMAN, J.H., GAGE, D.A., Floral fragrance variation in Cypripedium: Implications for evolutionary and ecological studies, *Phytochem.* 1997, **44**, 875-882.

55. MA, W., KLEMM, W.R., Variations of equine urinary volatile compounds during the oestrous cycle. *Vet. Res. Commun.*, 1997, **21**, 437-446.

56. REKWOT, P.I., OGWU, D., OYEDIPE, E.O., SEKONI, V.O., The role of pheromones and biostimulation in animal reproduction. *Anim. Reprod. Sci.*, 2001, **65**, 157-170.

57. SWAISGOOD, R.R., LINDBURG, D.G., ZHANG, H., Discrimination of oestrous status in giant pandas (*Ailuropoda melanoleuca*) via chemical cues in urine. *J. Zool.*, 2002, **257**, 381-386.

58. KEVERNE, E.B., Vomeronasal/accessory olfactory system and pheromonal recognition. *Chem. Sens.*, 1998, **23**, 491-494.

59. HELSPER, J.P.F.G., DAVIES, J.A., BOUWMEESTER, H.J., KROL, A.F., VAN KAMPEN, M.H., Circadian rhythmicity in emission of volatile compounds by flowers of *Rosa hybrida* L. Cv. Honesty. *Planta*, 1998, **207**, 88-95.

60. KOLOSOVA, N., SHERMAN, D., KARLSON, D., DUDAREVA, N., Cellular and subcellular localization of S-adenosyl-L-methionine:benzoic acid carboxyl methyltransferase, the enzyme responsible for biosynthesis of the volatile ester methylbenzoate in snapdragon flowers. *Plant Phys.* 2001, **125**, 1-9.

61. POTT, M.B., PICHERSKY, E., PIECHULLA, B., Evening specific oscillations of scent emission, SAMT enzyme activity, and SAMT mRNA in flowers of *Stephanotis floribunda, J. Plant Physiol.*, 2002, **159**, 925-934.

62. RAGUSO, R.A., LEVIN, R.A., FOOSE, S.E., HOLMBERG, M.W., MCDADE, L.A., Fragrance chemistry, nocturnal rhythms and pollination "syndromes" in *Nicotiana. Phytochemistry*, 2003, **63**, 265-284.

63. THOMSON, D.J., Spectrum estimation and harmonic analysis. *Proc. IEEE,* 1982. **70**,1055-1091.

64. NIKIAS, C., PETROPOLU, A., Higher-order Spectra Analysis. Prentice-Hall, Upper Saddle River, NJ. 1996.

65. RAGUSO, R.A., Olfactory landscapes and deceptive pollination. *in*: Insect Pheromone Biochemistry and Molecular Biology: The Biosynthesis and Detection of Pheromones and Plant Volatiles. (G.L. Blomquist and R.G. Vogt, eds,), Elsevier, Amsterdam, Netherlands. 2003.

66. MCELFRESH, J.S., MILLAR, J.G., Geographic variation in the pheromone system of the saturniid moth, *Hemileuca eglanterina. Ecol.*, 2001, **82**, 3505-3518.

67. FAEGRI, K., VAN DER PIJL.,L., The Principles of Pollination Ecology. 3rd ed. Pergamom Press, Oxford, UK. 1979.

68. RAGUSO, R.A., WILLIS, M.A., Synergy between visual and olfactory cues in nectar feeding by naive hawkmoths, *Manduca sexta. Anim. Behav.*, 2002, **64**, 685-695.

69. SCHIESTL, F.P., AYASSE, M., Post-pollination emission of a repellent compound in a sexually deceptive orchid: a new mechanism for maximising reproductive success? *Oecol.*, 2001, **126**, 531-534.

70. NEGRE, F., KISH, C.M., BOATRIGHT, J., UNDERWOOD, B., SHIBUYA, K., WAGNER, C., CLARK, D.G., DUDAREVA, N., Regulation of methylbenzoate emission after pollination in snapdragon and petunia flowers. *Plant Cell*, 2003, **15**, 2992-3006.

71. KNUDSEN, J.T., TOLLSTEN, L., Trends in floral scent chemistry in pollination syndromes - floral scent composition in moth-pollinated taxa. *Bot. J. Linn. Soc.*, 1993, **113**, 263-284.

72. BOATRIGHT, J., NEGRE, F., CHEN, X.L., KISH, C.M., WOOD, B., PEEL, G., ORLOVA, I., GANG, D, RHODES, D., DUDAREVA, N., Understanding in vivo benzenoid metabolism in Petunia petal tissue *Plant Physiol.*, 2004, **135**, 1993-2011.

73. DOBSON, H.E.M., BERGSTROM, G., The ecology and evolution of pollen odors. *Plant Syst. Evol.*, 2000, **222**, 63-87.

74. VAINSTEIN A., LEWINSOHN E., PICHERSKY E., WEISS, D., Floral fragrance: New inroads into an old commodity. *Plant Physiol.*, 2001, **127**, 1383-1389.

75. DOBSON, H.E.M., ARROYO, J., BERGSTROM, G., GROTH, I., Interspecific variation in floral fragrances within the genus *Narcissus* (Amaryllidaceae). *Biochem. Syst. Ecol.*, 1997, **25**, 685-706.

76. MACTAVISH, H.S., MENARY, R.C., Volatiles in different floral organs, and effect of floral characteristics on yield of extract from *Boronia megastigma* (Nees) *Annal. Bot.*, 1997, **80**, 305-311.

77. GRISON, L, EDWARDS, A.A., HOSSAERT-MCKEY, M., Interspecies variation in floral fragrances emitted by tropical *Ficus species. Phytochemistry*, 1999, **52**, 1293-1299.

78. ANDERSSON, S., NILSSON, L.A, GROTH, I., BERGSTROM, G., Floral scents in butterfly-pollinated plants: possible convergence in chemical composition, *Bot. J. Linn. Soc.*, 2002, **140**, 129-153.

79. KUNIN, W.E., Sex and the single mustard: population density and pollinator behavior effects on seed-set. *Ecology*, 1993, **74**, 2145-2160.

80. KUNIN, W., IWASA, Y., Pollinator foraging strategies in mixed floral arrays: Density effects and floral constancy. *Theor. Pop. Biol.*, 1996, **49**, 232-263.

81. WRIGHT, G.A., LUTMERDING, A., DUDAREVA, N, SMITH, B.H., Intensity and the ratios of compounds in the scent of snapdragon flowers affect scent discrimination by honey bees (*Apis mellifera*). *J. Comp. Phys. A*, 2005, in press.

82. KIM, H.J., KIM, K., KIM, N.S., LEE, D.S., Determination of floral fragrances of *Rosa hybrida* using solidphase trapping-solvent extraction and gas chromatography–mass spectrometry. *J. Chrom. A*, 2000, **902**, 389–404.

83. KIM, N.S., LEE, D.S., Comparison of different extraction methods for the analysis of fragrances from *Lavandula* species by gas chromatography-mass spectrometry. *J. Chrom. A*, 2002, **982**, 31-47.

84. AZUMA, H., TOYOTA, M., ASAKAWA, Y., Intraspecific variation of floral scent chemistry in *Magnolia kobus* DC. (Magnoliaceae),. *J. Plant Res.*, 2001, **114**, 411-422.

85. KNUDSEN, J.T., Variation in floral scent composition within and between populations of *Geonoma macrostachys* (Arecaceae) in the western Amazon. *Amer. J. Bot.*, 2002, **89**, 1772-1778.

86. DUFA, M., HOSSAERT-MCKEY, M., ANSTETT, M.C., Temporal and sexual variation of leaf-produced pollinator-attracting odours in the dwarf palm. *Oecologia*, 2004, **139**, 392-398.

87. DUDAREVA, N., MURFITT, L.M., MANN, C.J., GORENSTEIN, N., KOLOSOVA, N., KISH, C.M., BONHAM, C., WOOD, K., Developmental regulation of methyl benzoate biosynthesis and emission in snapdragon flowers. *Plant Cell*, 2000, **12**, 949-961.

88. ROBERTSON, G.W., GRIFFITHS, D.W., MACFARLANE SMITH, D. BUTCHER, R.D., The application of thermal desorption-gas chromatography-mass spectrometry to the analyses of flower volatiles from five varieties of oilseed rape (*Brassica napus* spp. *oleifera*). *Phytochem. Anal.*, 1993, **4**, 152-157.

89. PORTER, A.E.A., GRIFFITHS, D.W., ROBERTSON, G.W., SEXTON, R., Floral volatiles of the sweet pea, *Lathyrus odoratus*, *Phytochemistry*, 1999, **51**, 211-214.

90. AYASSE, M., SCHIESTL F.P., PAULUS, H.F., LOFSTEDT, C., HANSSON, B., IBARRA, F., FRANCKE, W., Evolution of reproductive strategies in the sexually deceptive orchid *Ophrys sphegodes*: how does flower specific variation of odor signals influence reproductive success?, *Evol.*, 2000, **54**, 1995-2006.

91. MATILE, P., ALTENBURGER, R., Rhythms of fragrance emission in flowers. *Planta*, 1988, **174**, 242-247.

92. SVENSSON, M.G.E., BENGTSSON, M., LOFQVIST, J., Individual variation and repeatability of sex pheromone emission of female turnip moths *Agrotis segetum*. *J. Chem. Ecol.*, 1997, **23**, 1833-1850.

93. GEMENO, C., LUTFALLAH, A.F., HAYNES, K.F., Pheromone blend variation and cross-attraction among populations of the black cutworm moth (Lepidoptera : Noctuidae) *Ann. Entomolog. Soc. Amer.*, 2000, **93**, 1322-1328.

94. SOKAL, R., ROHLF, F.J., Biometry. W.H. Freeman and Co., New York, NY. 1995.

95. RENCHER A.C., Interpretation of canonical discriminant functions, canonical variates, and principal components. *Am. Stat.*, 1992, **46**, 217-225.

96. JOHNSON, R.A., WICHERN, D.W., Applied Multivariate Statistical Analysis 4th ed. Prentice Hall, Upper Saddle River, NJ. 1998.

97. EISTHEN, H.L., Why are olfactory systems of different animals so similar? Brain *Behav. and Evol.*, 2002, **59**, 273-293.

98. HILDEBRAND, J.G., SHEPHERD G.M., Mechanisms of olfactory discrimination: converging evidence for common principles across phyla. *Annu. Rev. Neurosci.*, 1997, **20**, 595-631.

99. BUCK, L., AXEL, R., A novel multigene family may encode odorant receptors – A molecular-basis for odor recognition. *Cell*, 1991, **65** (1), 175-187.

100. VOSSHALL, L.B., AMREIN, H., MOROZOV, P.S., RZHETSKY, A., AND AXEL, R., A spatial map of olfactory receptor expression in the *Drosophila* antenna. *Cell*, 1999, **96**, 725-736.

101. MA, M.H., SHEPHERD, G.M., Functional mosaic organization of mouse olfactory receptor neurons. *Proc. Natl. Acad. Sci USA*, 2000, **97**, 12869-12874.

102. FIRESTEIN S., How the olfactory system makes sense of scents. *Nature*, 2001, **413**, 211-218.

103. MOMBAERTS, P., Targeting olfaction. *Curr. Opin. Neurobiol.*, 1996, **6**, 481-486.

104. MENZEL, R., Searching for the memory trace in a mini-brain, the honeybee. *Learn. Mem.*, 2001, **8**, 53-62.

105. FABER, T., MENZEL R., Visualizing mushroom body response to a conditioned odor in honeybees. *Naturwissenschaften*, 2001, **88**, 472-476.

106. JUSTUS, K.A., SCHOFIELD, S.W., MURLIS, J., CARDE, R.T., Flight behaviour of *Cadra cautella* males in rapidly pulsed pheromone plumes. *Phys. Entomol.*, 2002, **27**, 58-66.

107. BREED, M.D., GUZMAN-NOVOA, E., HUNT, G.J., Defensive behavior of honey bees: Organization, genetics, and comparisons with other bees. *Annu. Rev. Entomol.*, 2004, **49**, 271-298.

108. HUDSON R., From molecule to mind: the role of experience in shaping olfactory function. *J. Comp. Phys. A*, 1999, **185**, 297-304.

109. PAVLOV, I.P., Conditioned Reflexes. Oxford Univ. Press, Oxford, UK. 1927.

110. PEARCE, J.M., Similarity and discrimination: a selective review and a connectionist model. *Psychol. Rev.*, 1994, **101**, 587-607.

111. SHEPARD, R.N., Toward a universal law of generalization for psychological science. *Science*, 1987, **237**, 1317-1323.

112. WINSTON, M.L., The Biology of the Honey Bee. Harvard University Press. Cambridge, MA. 1987.

113. WRIGHT, G.A., SKINNER, B.D., SMITH, B.H., The ability of the honey bee, *Apis mellifera*, to detect and discriminate among the odors of varieties of canola flowers (*Brassica rapa* and *Brassica napus*) and snapdragon flowers (*Antirrhinum majus*). *J. Chem. Ecol.*, 2002, **28**, 721-740.

114. PRICE, E., DALLY, M., ERHARD, H., GERZEVSKE, M., KELLY, M., MOORE, N., SCHULTZE, A., TOPPER, C., Manupulating odor cues facilitates add-on fostering in sheep, *J. Anim. Sci.*, 1998, **76**, 961-964.

115. LEVY, F., KELLER, A., POINDRON, P., Olfactory regulation of maternal behavior in mammals. *Horm. Behav.*, 2004, **46**, 284-302.

116. GIURFA, M., The repellent scent-mark of the honeybee *Apis-mellifera-ligustica* and its role as communication cue during foraging. *Insectes Sociaux*, 1993, **40**, 59-67.

117. BOWDEN, R.M., WILLAMSON, S., BREED, M.D., Floral oils: their effect on nestmate recognition in the honeybee, *Apis mellifera. Insectes Sociaux*, 1998, **45**, 209-214.

118. MORITZ, R.F.A., NEUMANN, P., Differences in nestmate recognition for drones and workers in the honeybee, *Apis mellifera* (L.) *Anim. Behav.*, 2004, **67**, 681-688.

119. WOOD, M.J., RATNIEKS, F.L.W., Olfactory cues and *Vespula* wasp recognition by honey bee guards. *Apidol.*, 2004, **35**, 461-468.
120. WISE, P.M., CAIN, W.S.., Latency and accuracy of discriminations of odor quality between binary mixtures and their components. *Chem. Sens.*, 2000, **25**, 247-265.
121. LASKA, M., GALIZIA, C.G., GIURFA, M., MENZEL, R., Olfactory discrimination ability and odor structure-activity relationships in honeybees. *Chem. Sens.*, 1999, **22**, 457-465.
122. LASKA, M., TEUBNER, P. , Olfactory discrimination ability for homologous series of aliphatic alcohols and aldehydes. *Chem. Sens.*, 1999, **24**, 263-270.
123. LASKA, M., FREYER, D., Olfactory discrimination ability for aliphatic esters in squirrel monkeys and humans. *Chem. Sens.*, 1997, **22**, 457-465.
124. DALY, K.C., CHANDRA, S., DURTSCHI, M.L., SMITH, B.H., The generalization of an olfactory-based conditioned response reveals unique but overlapping odour representations in the moth *Manduca sexta. J..Exp. Biol,* 2001, **204**, 3085-3095.
125. LASKA, M., HUBENER, F., Olfactory discrimination ability for homologous series of aliphatic ketones and acetic esters. *Behav. Brain Res.*, 2001, **119**, 193-201.
126. LASKA, M,, HUDSON, R., Discriminating parts from the whole - determinants of odor mixture perception in squirrel-monkeys, *Saimiri-sciureus. J. Comp. Phys. A* , 1993, **173**, 249-256.
127. LASKA, M., HUDSON, R., Ability to discriminate between related odor mixtures. *Chem. Sens.*, 2002, **17**, 403-415.
128. GROSCH, W., Evaluation of key odorants of foods by dilution experiments, aroma models, and omission. *Chem. Sens.*, 2001, **26**, 533-545.
129. LAING, D.G., Perceptual odour interactions and objective mixture analyses. *Food Qual. Pref.*, 1994, **5**, 75-80.
130. CHANDRA, S.B.C., SMITH, B.H., An analysis of synthetic odor processing of odor mixtures in the honeybee, (*Apis mellifera*). *J. Exp. Biol.*, 1998, **201**, 3113-3121.
131. LAING, D.G., Perception of odor mixtures. *in*: Handbook of Olfaction and Gustation. (R.L. Doty, ed.) M. Dekker, New York, NY. pp.283-298.
132. MARR, D., Vision: A Computational Investigation into the Human Representation and Processing of Visual Information. W.H. Freeman, San Francisco, CA. 1982.
133. ZWAARDEMAKER, H., 1925. L'odorat. Doin, Paris. 1925.
134. BHAGAVAN, S., SMITH, B.H., Olfactory conditioning in the honeybee, *Apis mellifera*: the effects of odor intensity. *Physiol. Behav.,* 1996, **61**, 107-117.
135. PELZ, C., GERBER, B. MENZEL, R., Odorant intensity as a determinant for olfactory conditioning in the honeybee: Roles in discrimination, overshadowing, and memory consolidation. *J. Exp. Biol.*, 1997, **200**, 837-847.
136. WRIGHT, G.A., SMITH, B.H., Different thresholds for detection and discrimination of odors in the honey bee (*Apis mellifera*). *Chem. Sens.*, 2004b, **29**, 127-135
137. LASKA, M., SEIBT, A., Olfactory sensitivity for aliphatic alcohols in squirrel monkeys and pigtail macaques. *J. Exp. Biol.*, 2002, **205**, 1633-1643.
138. CLELAND, T.A. AND NARLA, V.A., Intensity modulation of olfactory acuity. *Behav. Neurosci.*, 2003, **117**, 1434-1440.

139. LAING, D.G., LEGHA, P.K., JINKS, A.L., HUTCHINSON, I. Relationship between molecular structure, concentration and odor qualities of oxygenated aliphatic molecules. *Chem. Sens.*, 2003, **28**, 57-69.

140. ARCTANDER, S., Perfume and Flavor Chemicals (Aroma Chemicals). Stefan's Arctander's Publications, Las Vegas. 1969.

141. GROSS-ISSEROFF, R., LANCET, D., Concentration-dependent changes of perceived odor quality. *Chem. Sens.*, 1988, **13**, 191-204.

142. MARFAING, P., ROUAULT, J. LAFFORT, P., Effect of the concentration and nature of olfactory stimuli on the proboscis extension of conditioned honeybees (*Apis mellifera ligustica*). *J. Insect Phys.*, 1989, **35**, 949-955.

143. WISE, P.M., Olsson, M.J., Cain, W.S., Quantification of odor quality. *Chem. Sens.*, 2000, **25**, 429-443.

144. DITZEN, M., EVERS, J.F., GALIZIA, C.G., Odor similarity does not influence the time needed for odor processing. *Chem. Sens.*, 2003, **28**, 781-789.

145. DETHIER, V.G. The Hungry Fly: A Physiological Study of the Behavior Associated with Feeding. Harvard Univ. Press, Cambridge, MA. 1975.

146. ERBILGIN, N., POWELL, J.S., RAFFA K.F., Effect of varying monoterpene concentrations on the response of *Ips pini* (Coleoptera : Scolytidae) to its aggregation pheromone: implications for pest management and ecology of bark beetles. *Agric. Forest Entomol.*, 2003, **5**, 269-274.

147. FADAMIRO, H.Y., BAKER, T.C., *Helicoverpa zea* males (Lepidoptera : Noctuidae) respond to the intermittent fine structure of their sex pheromone plume and an antagonist in a flight tunnel. *Physiol. Entomol.*, 1997, **22**, 316-324.

148. FADAMIRO, H.Y., COSSE, A.A., BAKER, T.C., Fine-scale resolution of closely spaced pheromone and antagonist filaments by flying male Helicoverpa zea. *J. Comp. Physiol. A*, 1999, **185**, 131-141.

149. BAU, J., JUSTUS, K.A., CARDE, R.T., Antennal resolution of pulsed pheromone plumes in three moth species. *J. Insect Phys.*, 2002, **48**, 433-442.

150. ROSPARS, J.P., LANSKY, P., KRIVAN, V., Extracellular transduction events under pulsed stimulation in moth olfactory sensilla. *Chem. Sens.*, 2000, **28**, 509-522.

151. ROSPARS, J.P., LANSKY, P., Stochastic pulse stimulation in chemoreceptors and its properties. *Math. Biosci.*, 2004, **188**, 133-145.

152. TODD, J.L. BAKER, T.C., Function of peripheral olfactory organs. *in*: Insect Olfaction. (B.S. Hansson, ed,), Springer, Berlin, Germany. 1999.

153. KELLER, T.A., WEISSBURG, M.J., Effects of odor flux and pulse rate on chemosensory tracking in turbulent odor plumes by the blue crab, *Callinectes sapidus. Biol. Bull.*, 2004, **207**, 44-55.

154. KAISSLING K.E., Flux detectors versus concentration detectors: Two types of chemoreceptors. *Chem. Sens.*, 1998, **23**, 99-111.

155. VICKERS, N.J., Defining a synthetic pheromone blend attractive to male *Heliothis subflexa* under wind tunnel conditions. *J. Chem. Ecol.*, 2002, **28**, 1255-1267.

156. FERVEUR, J.F., COBB, M., BOUKELLA, H., JALLON, J.M., World-wide variation in *Drosophila melanogaster* sex pheromone: Behavioural effects, genetic bases and potential evolutionary consequences. *Genetica*, 1996, **97**, 73-80.

157. RAFFA, K.F., Mixed messages across multiple trophic levels: the ecology of bark beetle chemical communication systems. *Chemoecology*, 2001, **11**, 49-65.

158. SYMONDS, M.R.E., ELGAR, M.A., The mode of pheromone evolution: Evidence from bark beetles. *Proc. Roy. Soc. B*, 2004, **271**, 839-846.

159. OSADA, K., YAMAZAKI, K., CURRAN, M., BARD, J., SMITH, B.P.C., BEAUCHAMP, G.K., The scent of age. *Proc. Roy. Soc. B*, 2003, **270**, 929-933.

160. RIBBANDS, C.R., The foraging method of individual honey bees. *J. Anim. Ecol.*, 1949, **18**, 47-66.

161. SMITH B.H., COBEY, S., The olfactory memory of the honeybee *Apis-mellifera* .2. Blocking between odorants in binary-mixtures. *J. Exp. Biol.*, 1994, **195**, 91-108.

162. COUVILLON, P.A, ARAKAKI, L., BITTERMAN M.E., Intramodal blocking in honeybees. *Anim. Learn. Behav.*, 1997, **25**, 277-282.

163. HOSLER, J.S., SMITH, B.H., Blocking and the detection of odor components in blends. *J. Exp. Biol.*, 2000, **203**, 2797-2806.

164. GIANNARIS, E.L., CLELAND, T.A, LINSTER, C., Intramodal blocking between olfactory stimuli in rats. Physiol. Behav., 2002, **75**, 717-722.

165. WRIGHT, G.A., SMITH, B.H., Variation in complex olfactory stimuli and its influence on odour recognition. Proc. Roy. Soc. B, 2004a, **271**, 147-152.

166. ROBERTS, S.C., GOSLING, L.M., Genetic similarity and quality interact in mate choice decisions by female mice. *Nature Gen.*, 2003, **35**, 103-106.

Chapter Nine

STRUCTURE AND FUNCTION OF INSECT ODORANT AND PHEROMONE-BINDING PROTEINS (Obps AND Pbps) AND CHEMOSENSORY-SPECIFIC PROTEINS (CSPs)

N. S. Honson, Y. Gong, E. Plettner*

Simon Fraser University
Dept. of Chemistry
8888 University Drive
Burnaby, B.C., CANADA V5A 1S6

Author for correspondence, email: plettner@sfu.ca

INTRODUCTION

Chemical communication in insects involves the detection of specific chemical signals, often present in very low doses against a diverse background of other chemicals. Chemical signals can be classified into three major groups, based on the source and detection of these chemicals. Species-specific semiochemicals used for conspecific chemical communication are known as pheromones. Semiochemicals used for interspecific communication are allelochemicals.[1] Compounds that are part of the general chemical landscape are grouped as general odorants and tastants. Insects use several chemosensory systems to perceive pheromones, allelochemicals, and general signals. For example, volatile pheromones are detected by specialized olfactory neurons in long sensory hairs on the antennae (*sensilla trichodea*),[2,3] while general odorants are detected by a distinct population of hairs, some with a different morphology.[4,5] These hairs can contain specialist or broadly tuned generalist neurons.[4,6] Finally, gustatory structures of various types on the antennae, legs, mouthparts, and ovipositor enable insects to detect non-volatile chemical cues from nearby surfaces, food and oviposition sites.[7,8] Sensory hairs are hollow cuticular structures, which encase the sensory neuron(s). The neurons are bathed and protected by an extracellular fluid, the sensillar lymph.[9,10] Some sensilla, such as sensilla *trichodea* and *basichonica*, have a single compartment, and others,[9,10] such as grooved peg and *coeloconic* sensilla, have two compartments filled with lymph.[11,12] The sensillar lymph is rich in various secreted proteins, the most abundant of which are in the class of odorant-binding proteins (OBPs) or in the class of chemosensory-specific proteins (CSPs).[8-11,13-14] Here, we explore what is known from recent studies of the structure and function of both classes of binding proteins. We also present possible roles of these proteins in the olfaction process.

PHEROMONE-BINDING PROTEINS AND ODORANT-BINDING PROTEINS (PBPS AND OBPS)

General Characteristics

Proteins that bind odorants or pheromones in insects belong to the OBP family. This family can be divided into three major classes: pheromone-binding proteins (PBPs),[15-18] general odorant-binding proteins (GOBP1 and GOBP2),[19] and the antennal binding protein X.[16,20,21] All of these binding proteins have the capability of binding low-molecular weight compounds (see below).

In chemosensory structures, such as sensory hairs, initial molecular recognition may take place at two levels. First, pheromone or odorant is bound and possibly transported by a PBP or OBP. These binding proteins are abundant (~ 10 mM) in the sensillar lymph.[15,22] The biological function of odorant binding by PBPs and OBPs is not clear, but it has been proposed that the binding proteins are essential for olfaction,[15,23] possibly because many pheromones and odorants are hydrophobic and would not efficiently traverse the aqueous barrier to the neuron.[24] However, solubilization of hydrophobic odorants cannot be the only role of OBPs and PBPs, because some compounds such as 3-hydroxy-butan-2-one (emitted by the male cockroach *Leucophaea maderae*),[25] and odorants such as ethanol (emitted from ripe fruit) are water soluble,[23] yet OBPs or PBPs are required for their detection. At the second molecular recognition level, pheromone interacts with a G-protein coupled transmembrane receptor located on the dendritic membrane. There are two possible mechanisms for receptor stimulation: (1) the odorant or pheromone may dissociate from the binding protein when the complex is near the membrane and bind to the olfactory receptor, or (2) the binding protein-odorant or pheromone complex may interact with the receptor. Many observations suggest that OBPs and PBPs play an active role in odor recognition. First, even water soluble odorants require LUSH (an OBP) to elicit an olfactory response.[23] The LUSH protein was found to co-crystallize with small alcohols[26] and to bind a variety of ligands (Table 9.1).[26,27] A similar picture emerges from electrophysiological recordings of receptor cells from the giant silkmoth *Antheraea polyphemus*, in which direct perfusion of the sensory hair with various PBP/odorant complexes elicits different response patterns.[28] The *A. polyphemus* PBPs are known to selectively bind different odorants[29] (Table 9.2) and are thought to have different conformations with various odorants.[30] Our mechanistic studies suggest that high salt concentration near the dendritic membranes may counterbalance the effect of low pH and prevent release of pheromone from PBPs,[31] which may mean that the OBP or PBP complex interacts with the receptor.[22] Thus, the events prior to membrane receptor stimulation are still poorly understood. Stimulation of the G-protein coupled transmembrane receptors[32,33] results in the production of inositol 1,4,5-triphosphate (IP_3), which triggers the opening of IP_3-gated sodium ion channels, allowing an influx of cations.[34] This ion influx in turn depolarizes the cell and initiates nerve impulses.[35,36]

Many insects express multiple OBPs or PBPs. The first evidence for the presence of two PBPs encoded by two different genes was in 1991 in the Chinese oak silkmoth, *Antheraea pernyi*[37] and in the gypsy moth *Lymantria dispar* which also has two PBPs.[38] Each of the gypsy moth PBPs was later shown to preferentially bind a different enantiomer of the pheromone (Table 9.2),[39] which suggests that these proteins are involved in some form of olfactory coding. Insect genome projects bring to light even more genes/insect species in that family (*e.g.* 51 genes for *Drosophila*[17] and 29 genes for *Anopheles*[18]). The presence of multiple OBPs and/or PBPs suggests that these proteins may be involved in odor recognition and possibly in transduction with the receptor.

Proteins in this family are highly acidic (pI ~ 5), and have molecular weights between 13 and 17 kDa (110-150 amino acids in length). There is considerable amino acid sequence similarity between PBPs and OBPs of the same moth species (50-60%), but little similarity with OBPs from *Drosophila melanogaster* and other non-lepidopteran species. OBPs and PBPs have six highly conserved cysteine residues, which form three interlocked disulfide bridges (with the pattern: first with third, second with fifth, and fourth with sixth)[40-42] (Honson and Plettner, unpublished). A few OBPs and PBPs have extra cysteines, for example, *Antheraea pernyi* PBPs Aper-1 and Aper-2 have an extra cysteine near the N-terminus[37] and related proteins such as the "plus-C" OBPs in *Drosophila*.[43] Various binding proteins in this family (often found in non-olfactory contexts such as hemolymph) have only four of the six conserved cysteine residues (they are missing the second and fifth conserved pair). In terms of their fold, all these proteins are comprised of six compact α-helices, which delineate a binding pocket in the middle (Fig. 9.1A and 9.1B).

Most ligands bound by PBPs and OBPs contain long, aliphatic hydrocarbon chains. Some have a polar functionality at one end of the chain such as (6*E*, 11*Z*) hexadeca-6,11-dienyl-1-acetate and the corresponding aldehyde which comprise the pheromonal blend of *A. polyphemus*,[15] or (10*E*,12*Z*) hexadec-10,12-dien-1-ol (bombykol), the first chemically characterized pheromone, from the silkworm moth *Bombyx mori*.[44] *L. dispar* has an epoxide moiety present in the most active identified pheromone blend component (7*R*, 8*S*) 7,8-epoxy-2-methyl-octadecane ((+) disparlure).[45-48] Furthermore, there are many widely distributed hydrocarbons, esters, and amides known to elicit responses in insects and known to bind to OBPs and PBPs (Tables 9.1 and 9.2). Functional groups and the conformation of semiochemicals may be important in selectively binding PBPs and OBPs (see below).[49]

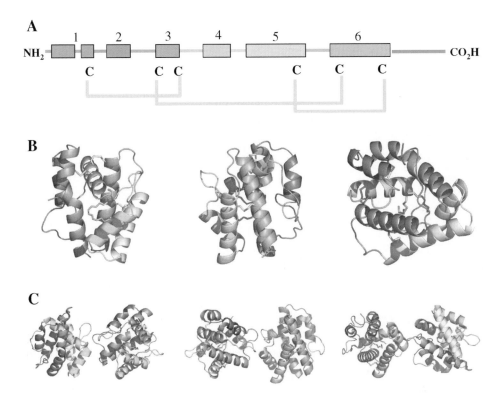

Figure 9.1: Overall fold and disulfide pairing of BmorPBP.[50] **A)** Secondary structure of BmorPBP. **B)** Side, top, and bottom views of BmorPBP with disulfide linkages and bombykol represented in stick form. **C)** Side, top, and bottom views of the BmorPBP asymmetric dimer. Images were prepared using PyMOL (Delano, W.L. The PyMOL Molecular Graphics System. (2004) DeLano Scientific, San Carlos, CA, USA. http://www.pymol.org).

Table 9.1: Binding data for OBPs

Species	OBP	Accession number	Ligand	K_d/ μM	Assay	Concentration/μM			pH	Ref
						OBP	Ligand	Reporter		
Apis mellifera L.	ASP2	AF166497	2-isobutyl-3-methoxypyrazine (IBMP)	K_{d1}: 0.24 K_{d2}: 0.21 K_{d3}: 233	isothermal titration calorimetry	20-30	0.74-12	n/a	7.0	98
			isopentyl acetate (isoamyl acetate)	K_{d1}: 2.2 K_{d2}: 3.6 K_{d3}: 11.8						
			1,3,3-trimethyl-2-oxa-bicyclo(2.2.2)octane (eucalyptol)	K_{d1}: 3.5 K_{d2}: 3.1 K_{d3}: 1.9						
			2-heptanone	K_{d1}: 7.1 K_{d2}: 8.3 K_{d3}: 4.6						
Drosophila melanogaster	LUSH	AF001621*	N-phenyl-1-naphthylamine (1-NPN)	1.5	1-NPN fluorescence	2	n/a	1-16	7.4	27
			dibutyl phthalate	5.1	1-NPN competitive fluorescence		1-16	2		
			diphenyl phthalate	4.4						
			di-(2-ethylhexyl) phthalate	10.7						
Locusta migratoria	Lmig OBP1	AF542076	N-phenyl-1-naphthylamine (1-NPN)	1.67	1-NPN fluorescence	2	n/a	1-20	7.4	99
			(9Z) octadec-9-enamide (oleoamide)	4.8	1-NPN competitive fluorescence		1-20	2		
			α-amylcinnamaldehyde	5.1						
Polistes dominulus	OBP-1	AY297026	N-phenyl-1-naphthylamine (1-NPN)	2.1	1-NPN fluorescence	2	n/a	1-20	7.4	67
			nonanoic amide (pelargonic amide)	11.0	1-NPN competitive fluorescence		2-20	2		
			tetradecanoic amide (myristic amide)	7.1						
			(9Z) octadec-9-enamide (oleic amide)	0.71						
			dodecanol	0.29						
			tetradecanol	2.8						
			octadecanol	11.0						
			dodecanoic acid (lauric acid)	7.9						
			octadecanoic acid (stearic acid)	2.8						
			methyl dodecanoate (methyl laurate)	0.89						

* The structure has been solved for this PBP. Please refer to Table 9.3.

Structure of Insect PBPs and OBPs

The disulfide bridge pairing for *Bombyx mori* PBP (BmorPBP) and *B. mori* GOBP-2 was elucidated by using a combination of chemical and enzymatic cleavage reactions.[40,41] The proteins proved difficult to digest, probably because they are compact, tightly folded, and reinforced with three disulfide bridges. Leal *et al.*[41] employed endoproteinase Lys-C, chymotrypsin, as well as cyanogen bromide to digest BmorPBP, while Scaloni *et al.*[40] utilized endoproteinase Lys-C and/or trypsin, and cyanogen bromide to digest BmorPBP and GOBP-2. Cysteine pairing results were identical for BmorPBP and GOBP-2 (Cys19-Cys54, Cys50-Cys108, and Cys97-Cys117). Since then, the honey bee PBP1 (AmelASP1)[42] and the gypsy moth PBPs (PBP1 and PBP2) have been found to have the same disulfide pairing by peptidic digestion (Honson and Plettner, unpublished obs.).

The first three-dimensional structure of an insect OBP was the X-ray structure of BmorPBP.[50] Since then, six more reports describing insect OBP structures determined either by X-ray diffraction or NMR studies have been published (Tables 9.3 and 9.4). The six compact α-helices proven in X-ray and NMR solved structures of both PBPs and OBPs confirm peptide mapping results, revealing three interlocking disulfide bridges. The disulfide bridges of BmorPBP are Cys19-Cys54, which connects α-helices 1 and 3, Cys50-Cys108 connecting helices 3 and 6, and finally Cys97-Cys117 reinforcing α-helices 5 and 6.[50] The heart-shaped binding pocket of the BmorPBP-bombykol complex is formed by four anti-parallel helices (α-helices 1, 4, 5, and 6) which converge to a narrow point (Fig. 9.1B). This cavity is enclosed by helix 3 and a flexible loop region between α-helix 3 and 4 (residues 60-69). The X-ray structure at physiological pH has revealed an unstructured C-terminal tail that resides on the outside of the protein surface (residues 125-137, residues 138-142 not resolved).[50] Interestingly, the NMR structure obtained at pH 4.5 showed the C-terminal tail conformation as α-helical, and the tail of this acidic form inserted into the hydrophobic pocket.[51] This binding-inactive form was termed "A". In a subsequent NMR structure with unliganded PBP at pH 6.5, the C-terminal tail is almost identical to the BmorPBP-bombykol complex, residing in various positions outside the binding cavity in an extended conformation.[52] This binding-active form was termed "B". Similarly, the C-terminus of *A. polyhemus* ApolPBP at pH 6.3 is also unstructured and on the outside of the protein.[53] However, the unstructured C-terminus of both *Apis mellifera* AmelASP1 at pH 5.5 and *D. melanogaster* LUSH at pH 4.6 and 6.5 folds into the core of the protein, forming one wall of the binding pocket, while helix 1 packs against the outside of the protein.[26,54] Interestingly, *L. maderae* LmaPBP lacks a C-

Table 9.2: Binding data for PBPs.

Species	PBP	Accession number	Ligand	K_d/ μM	Assay	Concentration/μM			pH	Ref
						PBP	Ligand	Reporter		
Antheraea pernyi	Aper-1	X96773	(6E, 11Z) hexadeca-6,11-dien-1-yl acetate	1.83	Vial adsorption assay	0.1	n/a	0.1-10	6.8	29
			(4E, 9Z) tetradeca-4,9-dien-1-yl acetate	29.4						
	Aper-2	X57562*	(6E, 11Z) hexadeca-6,11-dien-1-yl acetate	11.2	Vial adsorption assay	0.1	n/a	0.1-10	6.8	29
			(4E, 9Z) tetradeca-4,9-dien-1-yl acetate	3.75						
Antheraea polyphemus	ApolPBP	X17559*	(6E, 11Z) hexadeca-6,11-dien-1-yl acetate	0.64	Vial adsorption assay	0.1	n/a	0.1-10	6.8	29
				0.48	AMA competitive fluorescence	2	0-20	2	8.0	64
				0.80	Tryptophan fluorescence	2	0-8	n/a	6.8	30
			(4E, 9Z) tetradeca-4,9-dien-1-yl acetate	21	Vial adsorption assay	0.1	n/a	0.1-10	6.8	29
				0.51	AMA competitive fluorescence	2	0-20	2	8.0	64
				0.58	Tryptophan fluorescence	2	0-8	n/a	6.8	30
			(7R, 8S) 2-methyl-7,8-epoxyoctadecane	21.5	Mini column assay	1-3	n/a	0.02-0.40 nM	7.5	39
			1-aminoanthracene (AMA)	0.95	AMA fluorescence	2	n/a	0-20	8.0	64
			(6E, 11Z) hexadeca-6,11-dienal	0.50						
			(10E, 12Z) hexadec-10,12-dien-1-ol	0.54						
			(11Z) hexadec-11-enal	0.67						
			hexadecanoic acid (palmitic acid)	0.70						
			(9Z) hexadec-9-enoic acid (palmitoleic acid)	1.36						
			(9Z) octadec-9-enoic acid (oleic acid)	0.56						
Bombyx mori	BmorPBP	X94987*	(10E, 12Z) hexadec-10,12-dien-1-ol	1.1	Tryptophan fluorescence	1	0-10	n/a	8.0	64

* The structure has been solved for this PBP. Refer to Table 9.4.

Table 9.2: (continued).

Species	PBP	Accession number	Ligand	K_d/ μM	Assay	Concentration/μM			pH	Ref
Lymantria dispar	LdisPBP1	AF007867	(7R, 8S) 2-methyl-7,8-epoxyoctadecane	7.1	Mini column assay	1-3	n/a	0.02-0.40 nM	7.5	39
				0.2	Dansyl fluorescence	1	0-3	n/a	7.5	68
				0.1		3	0-3	n/a		
				0.3	Tryptophan fluorescence	1	0-2	n/a		
			(7S, 8R) 2-methyl-7,8-epoxyoctadecane	2.2	Mini column assay	1-3	n/a	0.02-0.40 nM	7.5	39
				2.8	Mini column assay	2	n/a	0.02-0.40 nM	7.5	68
				1.5	Competition assay with 2-methyl-*cis*-7,8-octadecene (20 μM)	1	n/a	0.02-0.40 nM		
			racemic-2-methyl-7,8-epoxyoctadecane	7.8	Mini column assay	1-3	n/a	0.02-0.40 nM	7.5	39
			(7R, 8S) 7,8-epoxyoctadecane	0.2	Dansyl fluorescence	1	0-3	n/a	7.5	68
				0.1		3	0-3	n/a		
			(7Z) 2-methyl-octadec-7-ene	4.7	Mini column assay	2	n/a	0.02-0.40 nM		
				0.5	Dansyl fluorescence	3	0-3	n/a		
				0.5	Tryptophan fluorescence	1	0-2	n/a		
			(7Z) octadec-7-ene	8.7	Mini column assay	2	n/a	0.02-0.40 nM		
				0.17	Dansyl fluorescence	3	0-3	n/a		
				0.3	Tryptophan fluorescence	1	0-2	n/a		
			(7R,8S) 2-methyl-7,8-aziridinyloctadecane	0.5	Dansyl fluorescence	3	0-3	n/a		
				0.2	Tryptophan fluorescence	1	0-2	n/a		
			(7S,8R) 2-methyl-7,8-aziridinyloctadecane	0.8	Dansyl fluorescence	3	0-3	n/a		
				0.4	Tryptophan fluorescence	1	0-2	n/a		

Table 9.2: (continued).

Species	PBP	Accession number	Ligand	K_d/ μM	Assay	Concentration/μM			pH	Ref
Leucophaea maderae	LmaPBP	AY116618*	8-Anilino-1-naphthalenesulfonic acid (ANS)	2.14	ANS fluorescence	1	n/a	0-13	8.0	65
			3-hydroxy-butan-2-one	3.8	ANS competitive fluorescence	1	0-90	4		
			butane-2,3-diol	2.5						
Lymantria dispar	LdisPBP2	AF007868	(7R, 8S) 2-methyl-7,8-epoxyoctadecane	1.8	Mini column assay	1-3	n/a	0.02-0.40 nM	7.5	39
			(7S, 8R) 2-methyl-7,8-epoxyoctadecane	3.2						
			racemic-2-methyl-7,8-epoxyoctadecane	4.9						
			(7Z) 2-methyl-octadec-7-ene	4.3	Mini column assay	2	n/a	0.02-0.40 nM	7.5	68
				0.5	Dansyl fluorescence	3	0-3	n/a		
			(7Z) octadec-7-ene	7.1	Mini column assay	2	n/a	0.02-0.40 nM		
				0.2	Dansyl fluorescence	3	0-3	n/a		
			(7R,8S) 2-methyl-7,8-aziridinyloctadecane	0.7						
			(7S,8R) 2-methyl-7,8-aziridinyloctadecane	0.7						
Mamestra brassicae	Mbra-1	AF051143	1-aminoanthracene (AMA)	6.0	Tryptophan fluorescence	1	0-10	n/a	8.0	64
				4.5	AMA fluorescence	1	n/a	0-10		
			hexadecanol (cetyl alcohol)	0.09	AMA competitive fluorescence	1	0-10	5		
			hexadecanoic acid (palmitic acid)	0.12						
			(Z9) hexadec-9-enoic acid (palmitoleic acid)	0.63						
			(11Z) hexadec-11-en-1-yl acetate	0.20						
			(11Z) hexadec-11-en-1-ol	0.17						
			(11Z) hexadec-11-enal	0.29						
			(10E, 12Z) hexadec-10,12-dien-1-ol	0.13						

* The structure has been solved for this PBP. Refer to Table 9.4.

terminal tail, and has a binding site that is much more open than other OBPs and PBPs that have been studied.[55] Possible roles of binding active and inactive forms of these proteins are discussed below.

Ligand Binding and Dissociation Mechanism

It is unclear how pheromone enters or exits the binding cavity. In particular, exiting the cavity is difficult, because the pheromone is completely encased in the binding pocket in most OBPs studied. It has been suggested that the low pH near a phospholipid membrane might trigger pheromone release.[56] However, high salt concentration near the membrane might counterbalance the pH effect, thereby preventing ligand release near the membrane.[31] All OBP and PBP structures reveal a pheromone binding pocket completely engulfed by α-helices except for a flexible loop region (varying from residues 55-75). Helix 1 exhibits conformational change depending on pH. At neutral pH, the segment is α-helical, while at acidic pH the N-terminal portion of the helix becomes disordered, and this region has been suggested as another entry/exit point for the pheromone.[51] These regions may function to allow ligand binding and perhaps ensure the ligand is not released until it is near the acidic membrane where olfactory receptors reside.

A mechanism of pheromone binding may involve PBPs functioning as dimers or multimers (see below). All solved NMR solution structures reveal PBPs in their monomeric form, while all X-ray structures (except for LUSH at pH 6.5) show PBPs and the one OBP forming asymmetric dimers (Fig. 9.1C). Interestingly, the dimeric interface spans the loop region, which may act as a flap, trapping the pheromone inside the binding cavity. In AmelASP1, the dimeric interaction may involve Arg19, His25, and Asp61 as ion pairs, and may also be pH dependent.[54] In BmorPBP, dimers form about a hydrophobic patch at Pro64 (which corresponds to the same position as Asp61 of AmelASP1) of one monomer, and Met131, Val133, and Lys38 of the other.[50] With such a high concentration of PBPs and OBPs in the sensillar lymph, it is possible that PBP and OBP monomers associate. Literature precedence suggests PBPs may function as dimers or higher order multimers. With *L. dispar* PBPs, dimers and higher-order multimers are routinely seen (Ling and Plettner, unpublished).[39] A pH-dependent monomer/dimer equilibrium of BmorPBP was demonstrated where both dimer and monomer co-eluted by gel filtration at neutral pH, while dissociation and conformational change of the dimer was exhibited at low pH.[57] Glutaraldehyde cross-linking experiments also show the presence of higher order multimers (Honson and Plettner, unpublished). Finally, a number of dimers have been detected by gel filtration and native PAGE[58,59] (Honson and Plettner, unpublished).

Table 9.3:   X-ray and NMR Structures of OBPs

Species	OBP	Ligand	Method	Resolution	pH	Unit *	Residues involved in binding / interesting features	Ref. {PDB code}
Drosophila melanogaster	LUSH	ethanol	X-ray	1.49 Å	4.6	D	-residues in alcohol binding cavity: Thr48, Ser52, Thr57, Val58, Phe64, Val106, Thr109, Ala110, Phe113, Trp123 -bottom of cavity contains nonpolar residues: Val106, Thr109, Ala110 -lip contains residues: Ser52, Thr57, Phe64, Phe113, Trp123 -alcohol OH H-bonds with Thr57 (H-bond donor) and Ser52 (H-bond acceptor); Ser52 H-bond donor to backbone carbonyl of Thr48 -C-terminal tail folds into core, forms part of alcohol binding pocket -helix 1 packs on outside of protein	26 {1OOF}
		propanol		1.45 Å			-propanol also contacts: Ala110, Phe113, Trp123	{1OOG}
		butanol		1.25 Å			-butanol also contacts: Ala55, Thr57, Leu76	{1OOH}
		butanol (not seen in solved structure)		2.04 Å	6.5	M	-small conformational changes at pH6.5 from pH4.6	{1OOI}

* M = monomer, D = dimer

Table 9.4: X-ray and NMR Structures of PBPs

Species	PBP	Ligand	Method	Resol-ution	pH	Unit*	Residues involved in binding / interesting features	Reference {PDB code}
Antheraea polymphemus	ApolPBP	n/a	NMR	n/a	6.3	M	-unstructured C-terminus (residues 126-142) -cavity contains mostly hydrophobic residues: Ile4, Met5, Asn7, Leu8, Ser9, Phe12, Met16, Leu33, Tyr34, Asn35, Phe36, Met43, Ala48, Ile 52, Asn53, Ala56, Thr57, Val61, Ala73, Lys74, Phe76, Ala77, Leu90, Ile94, Thr111, Ile112, Ala115, Phe118, Ile122 -Asn53 may form amide-carboxy H-bond with acetate ligands -Phe76 may provide CH/π interaction between methyl group of acetate ligands and phenyl ring -Trp37 outside lip of cavity; may interact with pheromone resulting in disposition of Trp37 into cavity	53, 100 {1QWV} {BMRB-5689}
Apis mellifera L.	AmelASP1	n-butyl-benzene-sulfonamide (additive in plastics)	X-ray	1.6 Å	5.5	D	-C-terminal tail folds inside cavity against one of its walls (residues Trp116, Phe117, Val118, Ile119) due to Pro113 -L-shaped cavity contains hydrophobic residues: Val9, Val13, Met49, Leu53, Ala55, Phe56, Leu 58, Leu74, Leu78, Met86, Tyr102 -lips of cavity: Trp4, Pro6, Pro7, Glu8, Leu12, Asp16, Ser57, Leu73, Pro75 -NH group of Leu119 H-bonds with SO group of ligand -CO group of Phe117 H-bonds with NH group of ligand -dimeric interaction involved Arg19, His24, Asp61 ion pairs, may be pH dependent	54, 101 {1R5R} 102 {BMRB-4940} (NMR, backbone sequential assignment only)

Table 9.4: (continued)

Species	PBP	Ligand	Method	Resolution	pH	Unit*	Residues involved in binding / interesting features	Reference {PDB code}
Bombyx mori	BmorPBP	(10E, 12Z) hexadec-10,12-dien-1-ol (bombykol)	X-ray	1.8 Å	8.2	D	-unstructured C-terminus tail (residues 125-137, residues 138-142 not solved) -residues found in cavity: Met5, Leu8, Ser9, Phe12, Phe36, Trp37, Ile52, Ser56, Met61, Leu62, Leu68, Phe76, Leu90, Ile91, Val94, Glu98, Thr111, Val114, Ala115, Phe118 -residues involved in the binding of bombykol: Leu8, Ser9, Phe12, Phe36, Trp37, Ser56, Met61, Phe76, Phe118; Ser 56 H-bonds with OH of bombykol, Phe12 and Phe118 π/π interactions with bombykol double bond -loop of residues 60-69 may be lid (loop between helices $\alpha 3$ and $\alpha 4$) -dimeric interaction involves hydrophobic patch at Pro64 (on loop) of one monomer and Met131, Val133, Lys38 of the other -pH-dependent conformational change may involve His69, His70, His95	[50] {1DQE}
		n/a	NMR	n/a	4.5	M	-C-terminal tail forms helix α_7 (residues 131-142), found inside hydrophobic core -residues which contact helix α_7: Ser9, Phe12, Ile52, Leu55, Ser56, Leu59, Leu62, Leu68, His70, Ala73, Phe76, Ala77, His80, Ala87, Leu90, Ile91, Val94, His95, Thr111, Phe118; following also interacts with bombykol in X-ray structure: Ser9, Phe12, Ile52, Ser56, Val94, Thr111, Phe118 -loop of residues 56-73 may be lid -histidine residues which may be involved in conformational change in bombykol-BmorPBP complex are more widely separated in this structure leading to less charge repulsion	[51] {1GM0}
		n/a	NMR	n/a	6.5	M	-unstructured C-terminal tail (residues 129-142) -cavity contains hydrophobic residues: Met5, Leu8, Phe12, Phe33, Tyr34, Phe36, Ile52, Met61, Leu62, Leu68, Ala73, Phe76, Ala77, Ala87, Leu90, Ile91, Val94, Trp110, Val114, Ala115, Phe118 -cavity contains four polar residues: Asp32, Thr48, Ser56, Glu98 -structure almost identical to bombykol-BmorPBP complex	[52] {1LS8}

Table 9.4: (continued)

Species	PBP	Ligand	Method	Resol-ution	pH	Unit *	Residues involved in binding / interesting features	Reference {PDB code}
Leucophaea maderae	LmaPBP	glycerol (from cryoprotectant)	X-ray	1.7 Å	8.5	D	-C-terminal tail is absent (last helix F consists of residues 103-117, LmaPBP is only 118 residues long) -binding cavity has residues: Leu36, Leu45, Ala46, Leu49, Leu54, Val89, Ile107, Phe110 (hydrophobic); Tyr5, Tyr75, Thr111 (polar, non-charged) -lip residues: Arg33, Asn34, Pro35, Lys85, Val114, Arg115; charged side chains may bind ligands entering cavity -residues contacting glycerol: Tyr5, Leu49, Tyr75, Lys85; electrostatic interactions	55 {1ORG} 103
		1-anilino-8-naphthalene sulfonate (ANS)		1.6 Å	8.5		-residues contacting ANS: Tyr5, Arg33, Leu45, Leu49, Val70, Met71, Leu74, Tyr75, Thr82, Lys85, Ala86, Val89, Thr111, Val114 -sulfate group of ANS is H-bond acceptor of Tyr75 OH group -ANS would clash with Met71 and Lys85 in apo protein; the following residues have moved in the ANS complex: Met71, Phe77, Lys85, Arg33	{1OW4}
		3-hydroxy-butan-2-one (H3B2)		1.7 Å	5.6		-binds both R and S enantiomers of H3B2 -residues contacting H3B2: Tyr5, Leu49, Leu54, Leu74, Tyr75, Lys85, Val89, Phe110, Thr111; electrostatic interactions -H3B2 interacts directly with Tyr5, Tyr75, Lys85, Phe110	{1P28}

* M = monomer, D = dimer

A model of ligand binding and dissociation for PBPs can be constructed from these experiments. When a hydrophobic pheromone enters the sensillar lymph, it first needs to pass through pore tubules on the surface of the antennae.[60] The molecule then interacts with the aqueous interface and most likely flees from this environment by adsorbing to the cuticular surface. The mechanism of pheromone desorption from this surface is unknown. At neutral pH, ligand enters the PBP binding cavity by approaching the flexible loop region, which moves to allow entry. This entry may be assisted by a number of residues which line the lip of the cavity.[26,54,55] Binding or a change in pH may lead to a conformational transition within the protein, resulting in closure of the entrance by the flexible loop region or multimerization of this binding active PBP form can occur, thereby trapping the ligand (see below). Conformational variation upon ligand binding has been observed for a number of insect PBPs, including ApolPBP[61] and BmorPBP.[56]

Ligand release from OBPs also is not well understood. A mechanism has been proposed, based on the pH-dependent behavior of PBPs. As the PBP-ligand complex nears the dendritic membrane, it likely enters a region of low pH and high ionic strength.[31] At low pH, α-helix 1 becomes disordered. The proposed pheromone entry point is now blocked because of a change in the loop between helices 4 and 5 near His 69 and His 70. The ligand must exit past the now flexible N-terminal peptide as the helix loses α-helical structure.[51] In this mechanism, different entrance and exit points for the ligand exist. In the case of AmelASP1 and LUSH, where the C-terminal tail forms one wall of the binding cavity while helix 1 packs on the outside,[26,54] the C-terminal tail exists as an extended polypeptide, and its position may be perturbed upon contact with the receptor or due to its presence in an acidic environment. However, the conformational changes in these studies were observed at low ionic strength. At the high salt concentrations likely present near the membrane, no pH-induced ligand dissociation has been observed.[31] Furthermore, measurement of ligand dissociation (see below) has revealed very long half-lives for PBP-pheromone complexes. There is thus the possibility that the ligand does not dissociate within the time frame of an olfactory signal (ms range).

Assessment of Ligand Binding

A number of binding assays have been employed to study the binding of pheromones and odorants. A widely used assay is a native PAGE experiment where isolated binding protein is incubated with a different test compound that has been radiolabeled. After the incubation period, the sample is loaded onto a native PAGE, electrophoresed, then either stained and quickly destained[38] or blotted onto a membrane.[39,62] The band corresponding to PBP is then excised and placed in a scintillation counter for counting radioactivity, or the blot is scanned with a TLC plate scanner. Although these gel assays have identified compounds which bind the proteins, a more quantitative assay is required for determination of binding constants. In 1995, Du and Prestwich[29] developed a vial adsorption binding assay where various concentrations of radiolabeled pheromone were incubated in a coated plastic vial with PBP. Bound PBP adsorbs to the vial surface while unbound pheromone remains in solution. A sample of the unbound pheromone is then counted for radioactivity. In this way, dissociation constants for pheromone or various ligand competitors can be measured.[29] This assay was refined by introducing a filtration step before measuring the radioactivity. The mini columns used contained a size exclusion gel matrix, which allowed the PBP-pheromone complex to be collected as filtrate, removing unbound pheromone in the solution.[39]

However, radiolabeled pheromone and pheromone analogs are difficult to synthesize,[63] thus, a relatively small number of radioligand is available. To allow testing of non-radiolabeled compounds, fluorescence binding assays were developed. The first involved binding of the fluorescence probe 1-aminoanthracene (AMA).[64] PBPs from certain moth species can bind AMA leading to a blue shift of the emission maximum and an increase in fluorescence intensity. To test ligand binding, the PBP-AMA complex is titrated with competitor, and the displacement of AMA is detected by an increase in fluorescence.[64] Of the PBPs tested, only ApolPBP and MbraPBP1 could bind AMA.[64] Since these initial studies, other probes have been tested and found to bind those PBPs that cannot bind AMA. 8-Anilino-1-naphthalene sulphonic acid (ANS) has been utilized in competitive fluorescence studies with LmaPBP,[65] while N-phenyl-1-naphthylamine (1-NPN) has been employed with LUSH,[27] the migratory locust *Locusta migratoria* LmigOBP1,[66] and the social wasp *Polistes dominulus* OBP-1.[67] However, there are two problems with the use of a competition assay: 1) determination of the dissociation constant is indirect since displacement of the reporter is being measured, and 2) it is not known whether these probes have an effect on competitor ligand binding. For example, LdisPBP1 and PBP2 have exhibited positive or negative blend effects with (-) and (+) disparlure when a pheromonal blend component is present (see below).[68]

To circumvent these problems, our group has covalently attached the fluorescent probe dansyl chloride to LdisPBP1 and PBP2 by selective reduction of one disulfide bridge and attachment of one dansyl group as a thiosulfonate. An increase in fluorescence emission occurs upon ligand binding as ligand is titrated into the PBP solution. However, the modified protein has a short lifetime (a few days), because the disulfide tends to reform, resulting in cleavage of the probe, or alternatively hydrolysis of the probe occurs.[68]

The intrinsic fluorescence of tryptophan also changes slightly with a variety of ligands.[30,68] Most ligands bind with a decrease in fluorescence, however, certain ones cause a small increase, suggesting a different conformation of the bound protein. The advantages of this assay are direct measurement of ligand binding, and no chemical modification of the protein is necessary. The disadvantage is a small relative change in fluorescence.

The above assays are summarized in Tables 9.1 and 9.2 for OBPs and PBPs where binding dissociation constants are available. Interestingly, the dissociation constants differ by an order of magnitude, depending on whether the assay is a vial adsorption/mini column assay, or a fluorescence assay. We believe this discrepancy may be the result of a dose effect attributed to the different ligand:PBP ratios used in binding assays (see below).

Ligand Recognition by OBPs and PBPs

Highly conserved residues that may contribute to ligand binding include Trp37, Ile52, Val61, Phe76, Ala115, and Phe118. Trp37 is present in most PBPs as part of the binding cavity. It is found to directly interact with bombykol in complex with BmorPBP,[50,69] while it is positioned outside the lip of the ApolPBP cavity.[53] Here, it is speculated, the pheromone may interact with this residue, resulting in movement of Trp37 into the binding pocket.[53] Positions 52 and 61 in ApolPBP and the corresponding position in other PBPs, as well as Thr48 in the OBP LUSH, are bulky residues that may help in defining the shape of the hydrophobic binding pocket. In BmorPBP, residue 61 is a methionine and acts as a hydrogen bond acceptor with bombykol along with Ser56.[69] Residue 61 in BmorPBP may also be part of the flexible lid region.[50] Phe76 (ApolPBP) is also present in the binding cavity of ApolPBP and BmorPBP,[50,53] allowing CH/π interactions with the methyl groups of its ligands.[69] Another highly conserved residue is Ala115, two residues downstream from the sixth cysteine. This residue is found to be part of the hydrophobic pocket in many of the PBPs with solved structures and with the LUSH OBP, suggesting that a small hydrophobic residue is needed at this position to support binding of long hydrocarbon chains. This alanine is replaced by the polar residue Thr111 in LmaPBP (Fig. 9.2E).[55] This is interesting since the pheromone blend of *L. maderae* consists of hydrophilic compounds such as 3-hydroxy-butan-2-one (H3B2) and short chain carboxylic acids,[25] and Thr111 directly contacts H3B2 *via* electrostatic interactions.[55] Finally, Phe118 is found in the binding cavity of ApolPBP, BmorPBP, LmaPBP (present as Val114), and the LUSH OBP (present as residue Phe113, this residue is not conserved in other OBPs), and has been shown within the crystal structure of BmorPBP to exhibit π/π interactions between the double bond of bombykol with Phe118 and Phe12 (Fig. 9.2B).[69] A summary of the specific residues involved in binding is given in Tables 9.3 and 9.4 for all solved OBP and PBP structures.

Although high sequence similarity gives us an idea of the evolutionary importance of certain conserved residues, other more variable residues provide information on the binding specificities within these proteins. For example, we have found a unique hydrogen bonding network in LdisPBP1 and PBP2 *via* homology modeling against the solved structure of BmorPBP.[31] The homology model of LdisPBP1 binding with its preferred enantiomer (-) disparlure shows the epoxide acting as an H-bond acceptor, pointing directly towards Thr9, which in turn acts as a H-bond acceptor with Trp37. Finally, the backbone carbonyl of this tryptophan operates as a donor with the protonated His124 (Fig. 9.2C). This same histidine has shown precedence as a likely candidate in ligand binding in a pH-dependent manner.[31] Conversely, the slightly weaker binder (+) disparlure has its epoxide oriented away from Thr9, and no H-bonding occurs. LdisPBP2 exhibits no H-bonding with its preferred enantiomer (+) disparlure.[31] Here, threonine is replaced by alanine, and neither enantiomer of disparlure forms an H-bond. However, the epoxide of (-) disparlure points toward the aromatic ring of Phe36, which may result in an electrostatic repulsion. The epoxide of (+) disparlure points away from this ring, resulting in stabilization of this enantiomeric form.

Binding of the OBP LUSH with ethanol, propanol, and butanol also involves an H-bonding network where Thr57 acts as an H-bond donor, and Ser52 acts as an H-bond acceptor with the alcohol hydroxyl group. Ser52, in turn, donates a hydrogen to the backbone carbonyl of Thr48 (Fig. 9.2F).[26] Besides direct interactions with the molecule, H-bonding networks reinforce protein structure, and perturbation of these networks may lead to different protein conformers. However, the binding pocket itself is quite rigid when compared to the flexible pheromone. Klusak *et al.*[69] have studied the binding of BmorPBP with bombykol using high-level *ab initio* methods. The shape of the cavity, which is critical for selectivity, is defined by amino acid residues that show no direct interaction with bombykol, yet these residues are not conserved and, therefore, are distinct for the BmorPBP binding cavity. In conclusion, specific hydrophobic and hydrogen bonding interactions account for the overall stability of binding.[69]

Conformational changes of PBPs have been demonstrated many times. BmorPBP undergoes a conformational transition between pH 5 and 6 when monitored by circular dichroism (CD), intrinsic fluorescence, and ANS fluorescence, which indicates a more flexible protein at lower pH.[56] OBPs from the large black chafer *Holotrichia parallela* and the yellowish elongate chafer *Heptophylla picea* each possess two OBPs that run as two separate bands on native gels, but that have been sequenced and identified as conformational isomers.[70] Other examples include the conformational changes induced by binding of pheromonal compounds to ApolPBP and PBP2, as demonstrated by CD and second derivative UV-difference spectroscopy.[61] The most dramatic changes in PBP structure have been

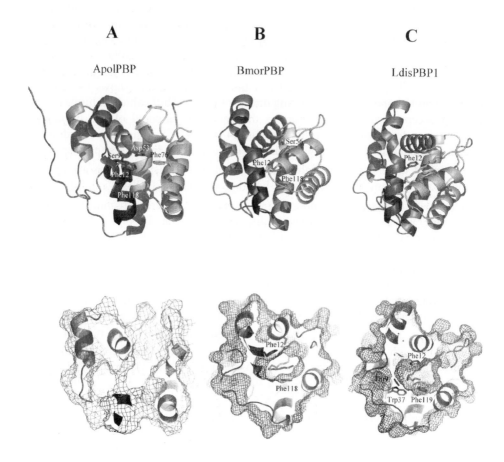

Fig. 9.2: Ligand binding and overall fold of OBPs with high sequence identity (A, B, and C) compared with those of lower sequence similarity (D, E, an F): A) Non-liganded ApolPBP,[53] B) BmorPBP complexed with bombykol,[50] C) Homology model of LdisPBP1 threaded onto the BmorPBP solved structure with (-) disparlure docked.

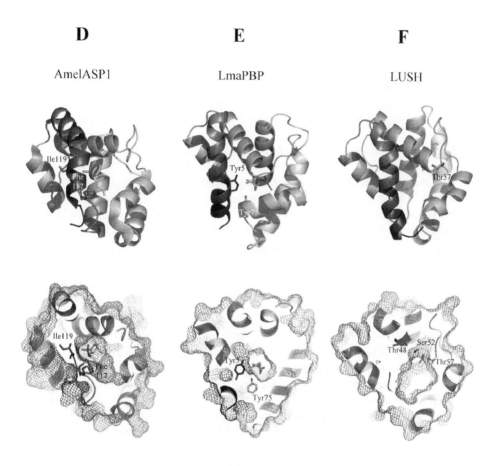

Fig. 9.2: (continued). D) AmelASP1 complexed with n-butyl-benzene-sulfonamide (a plastics additive),[54] E) LmaPBP with H3B2 bound,[55] F) LUSH complexed with ethanol.[26] The top row of structures reveals the overall fold of the protein, while the bottom row is a cutaway view of the molecule with the protein surface and binding cavity shown as mesh. Images were prepared using PyMOL (Delano, W.L. 2004).

demonstrated with the solved structures of BmorPBP (see A and B forms). Subtle conformational changes occur during ligand binding as demonstrated by tryptophan and dansyl fluorescence assays. For example, binding of a variety of pheromone analogs to LdisPBP1 caused a decrease in tryptophan fluorescence, but binding of (+) disparlure caused an increase.[68] These conformational changes may introduce another level of pheromone discrimination, not only assisting ligand binding but perhaps enabling the receptor to recognize a specific PBP-pheromone complex.

Link Between Structure and Function within Sensory Hairs

Binding assays with odorant-binding proteins have shed light on four effects, related to function: 1) concentration effect, 2) biphasic dissociation kinetics, 3) ligand (dose effect), and 4) blend effect. When binding was studied at different concentrations of PBP, the affinity increased abruptly above 1.5-2 μM of binding protein.[39] Scatchard analysis[68] and published crystal structures (Tables 9.3 and 9.4) of liganded PBPs reveal that only one ligand molecule can bind/binding site, so cooperativity is not possible. Furthermore, only one binding site/protein molecule has been detected in various physical studies, so allostery is not possible. This concentration effect has been attributed to 1) increased desorption of the hydrophobic pheromone ligand from the walls of the assay vial,[31] and 2) more than one population of binding sites, each with a different affinity (see below).[68]

Kinetics of ligand dissociation reveals an exponential decay pattern that fits two phases: a rapid one and a slow one (Fig. 9.3). The decay curves were obtained by pre-equilibrating PBP with the ligand, removing the unbound ligand (by gel filtration) and equilibrating the isolated PBP-ligand complex in fresh buffer. Samples were then probed for remaining PBP-ligand complex by gel filtration assay.[39] If there were only one population of binding sites, then one would expect a simple exponential decay in such an experiment. The observation of two phases of decay supports the idea that different populations of binding sites with different affinities are present in a PBP sample. Interestingly, (+) disparlure (the preferred ligand of PBP2, Table 9.2), associated and dissociated more rapidly overall than the less preferred ligand (-) disparlure. Thus, it is possible that ligand association and dissociation kinetics may encode more than the equilibrium binding constants usually measured.

Binding assays done at different ligand:PBP ratios (= doses) gave different equilibrium binding constants. Radioassays, which are done with low ligand doses, gave weaker binding constants than fluorescence assays, which are done with high

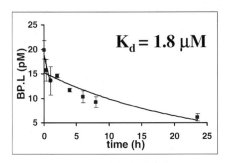

$$L + BP \xrightleftharpoons[k_{off}]{k_{on}} L.BP \qquad K_d = \frac{[L][BP]}{[L.BP]} = \frac{k_{off}}{k_{on}}$$

PBP2, (+) disparlure

$k_{off\,1} = 0.36\ h^{-1}\ t_{1/2} = 1.9\ h$

$k_{off\,2} = 0.04\ h^{-1}\ t_{1/2} = 17\ h$

$k_{on\,1} = 55.6\ M^{-1}s^{-1}$

PBP2, (−) disparlure

$k_{off\,1} = 0.12\ h^{-1}\ t_{1/2} = 5.8\ h$

$k_{off\,2} = 0.02\ h^{-1}\ t_{1/2} = 35\ h$

$k_{on\,1} = 10.6\ M^{-1}s^{-1}$

Fig. 9.3: Kinetics of ligand dissociation for LdisPBP2. Data were obtained by equilibrating isolated binding-protein-ligand complex (BP.L) in fresh buffer and separating protein-bound ligand from dissociated ligand at various time points (see text). Ligands tested were (+) disparlure ((7R, 8S) *cis*-7,8-epoxy-2-methyloctadecane) and the enantiomer, (−) disparlure. The dissociation is exponential and biphasic, with a rapid phase ($k_{off\,1}$) and a slow phase ($k_{off\,2}$). The association rates given were calculated from the measured dissociation constants, K_d, and the rapid off rate ($k_{off\,1}$). The half-life for each rate dissociation is given as $t_{1/2}$ (calculated from the initial value measured and the individual exponential decay).

ligand doses (Tables 9.1 and 9.2).[68] If there were one population of binding sites, the equilibrium binding constant should be independent of the ligand dose. This dose effect is also consistent with more than one population of binding sites.

In an effort to build a structure-activity profile for the *L. dispar* PBPs, we attempted to do competition assays, using a radiolabeled ligand as the reporter and a series of non-radiolabeled compounds as test ligands. In such assays, one expects no change in radioligand binding if the test ligand does not bind, and a decrease in radioligand binding if the test ligand binds and thereby displaces the radioactive reporter. To our surprise, binding of the radioligand was stronger in the presence of test ligands. This blend effect was especially pronounced with *Ldis*PBP1 and (-) disparlure as the test ligand.[68] Here too, the behavior of the PBP is consistent with more than one population of binding sites.

As mentioned earlier, pheromone-binding proteins have been shown to exist in monomeric, dimeric and higher-order multimeric form (Fig. 9.1C). In the crystal structures, where the PBP crystallized as a dimer (*B. mori* and *A. mellifera*, Table 9.4) the proteins aggregated in an asymmetric "head to tail" manner. In both cases, the binding site is buried within the protein, and the ligand is thought to enter/exit by movements near the interphase, which gives one binding cavity with no entry/exit site. Such a buried binding site would effectively trap a ligand, which would manifest itself as a higher apparent binding affinity.

In addition to multimers, PBPs exhibit two distinct folds of the monomeric form (the "A" and "B" forms). The "B" form itself has many conformers, which vary mainly in the orientation of the C-terminal portion. Interestingly, it has been proposed that different bound ligands will stabilize different conformers of the "B" form.[30] Furthermore, it has been proposed that the "A" form is preferentially monomeric and the "B" form aggregates.[57] Our recent experiments with fluorescently labeled PBPs suggest that a very slow equilibration process between at least two forms exists, and that addition of ligand stabilizes one form, while removal of ligand and/or low pH destabilizes that form (Gong and Plettner, unpublished).

With the above information, we can hypothesize how the four effects described above may arise. Two equilibrated forms of the protein (one binding inactive, the other binding active and able to multimerize) interact with a ligand. This ligand stabilizes the binding-active form and thereby shifts equilibrium towards that form. The ligand complex can then multimerize and thereby trap (scavenge) a portion of the ligand. This hypothesis could explain the concentration effect, because the higher the total protein concentration, the more likely multimerization of the binding-active form and the higher the apparent affinity for the ligand. The hypothesis could also explain the dose effect, because the more ligand there is relative to binding protein, the more the equilibrium between the binding inactive and active forms is shifted towards the active form. This, in turn would increase the probability of multimerization and of ligand scavenging. Furthermore, if the tendency of ligand-protein complex to multimerize depends on the ligand, then one can account for blend effects. A negative blend effect (where a combination of ligands exhibits a weaker equilibrium binding constant than the individual pure ligands) would result when the different ligand-protein complexes are not able to form mixed multimers and, therefore, *less* ligand trapping occurs. A positive blend effect (where a combination of ligands exhibits a stronger equilibrium binding constant than the individual ligands) would result when the ligand-protein complexes have a greater propensity to form mixed multimers and, therefore, *more* ligand trapping occurs.

A possible biological function of the dose effect could be to function as a signal gain control. The signal gain control is important in allowing insects to respond over a wide range of concentrations. In nature, insects need to follow a concentration gradient, which ranges over ~ 6 - 8 orders of magnitude[5] to find the source, and, in experimental situations (such as laboratory EAG trials or mating disruption attempts in the field), insects are exposed to doses up to ~10 orders of magnitude above their response threshold.[71,72] Interestingly, some species of insect are not saturated by these high (unnatural) doses of their sex pheromone. Such species are not amenable to pheromone-based mating disruption. A 10 order of magnitude detection range is highly remarkable, when compared to typical laboratory sensors, which can detect only ~ 3 - 4 orders of magnitude in concentration without saturating. A possible function of the dose effect described above is protecting the receptors at high doses of odorant. At low doses, negligible scavenging occurs, but as the dose increases, binding protein multimerization increases, and thus ligand trapping (scavenging) also increases.

A potential role for the blend effect described above would be 1) the filtering of relevant vs. irrelevant blends of odorants, and 2) the enhancement of certain blends that signal highly favorable conditions. For example, in EAG studies of isolated hairs of *H. virescens*, the sensilla responding to 11Z-hexadecenal (a pheromone component) showed significantly higher neuronal activity when the pheromone was presented together with linalool or 3Z-hexanol (plant odorants) than when the pheromone was presented by itself.[73] The sensory hair studied did not respond to linalool by itself, and the synergy was not due to a second silent neuron in the hair. Similar observations have been made with palm weevils, which respond much more strongly to a mixture of their pheromone and host plant odorants than to pheromone alone.[74] Some olfactory neurons were subsequently identified that are activated only by blends of pheromone and plant odorants.[5] The measurements in these studies were done exclusively on the peripheral part of the olfactory system, so the signal enhancements cannot be due to signal integration and analysis at the glomerular or post-glomerular level in the brain (see Mustaparta, this volume). A biological role for such signal filtering could be that, for example, a calling female moth located on a good food source might be more attractive to a male than a female located on a poor food source. Since plants differ widely in their volatile emissions, depending on species, sub-species, and health (see chapters by Schmidt, *et al.*; Martin and Bohlmann; Raffa, *et al.*; this volume), such olfactory signal processing may cause insects to focus on better resources.

A final question is why insects have multiple OBPs, often well separated in OBP phylogenies. In some cases, OBPs from one insect overlap somewhat in their binding selectivities. For example, in the gypsy moth, both PBP1 and PBP2 bind both disparlure enantiomers with different affinities: PBP2 prefers to bind (+) disparlure over the (-) enantiomer and PBP1 is of opposite preference.[39] In general, PBP1 discriminates more between different ligands, as determined from equilibrium binding constants. Many other examples of selectivity can be found in Tables 9.1 and 9.2. We hypothesize that having different binding proteins that bind a given odorant with different affinities may contribute to the wide range of concentrations insects can respond to. Different OBPs have been shown to express in distinct sub-populations of sensilla in several insects.[75] Furthermore, different olfactory neurons have also been shown to be randomly distributed within certain patches of sensory hairs.[76] Each neuron has a well-defined window of odorant selectivity. For example, in the gypsy moth, some pheromone neurons only respond to (+) disparlure while others only respond to (-) disparlure.[47,77,78] If there is a mosaic of sensory hairs containing various combinations of x neurons and y OBPs, then for each neuron type there will be y types of sensory hairs. The neuron within these hairs will respond to the same odorant, but in a different concentration window. Much like an experimenter needs to recalibrate an instrument when switching from a low-concentration to a high-concentration regime (for example, going from splitless to split gas chromatography), the insect may also "recalibrate" its olfactory system by having a population of sensory hairs that responds well within a low concentration window and another population of sensory hairs that responds well within a high concentration window. Once again, reports from laboratories that do EAG on isolated sensilla suggest that there are populations of sensilla which respond to the same odorant, but which differ in their dose responses.[5] (H. Mustaparta, pers. comm.)

CHEMOSENSORY-SPECIFIC PROTEINS (CSPS)

General Characteristics

Chemosensory-specific proteins (CSPs), which are also referred to as olfactory specific-D-like (OS-D-like) proteins or sensory appendage proteins (SAPs) are a newly discovered class of small soluble proteins whose molecular weights are generally around 13 kDa. They are relatively acidic with a pI of 5-6[79,80] and have high conformational stability.[81]   Since its first member, OS-D,[82] or A-10,[75] was reported to be expressed in the antennae of *Drosophila melanogaster*, this class of protein has drawn more and more interest. Until now, CSPs have been isolated from at least 20 species across 6 insect orders,[83] such as *Cactoblastis cactorum, Mamestra brassicae,*[8,84] *Periplaneta americana, P. fuliginosa, Blattella germanica,*[85] *Schistocerca gregaria,*[86] and *Eurycantha calcarata.*[87]

CSPs share no sequence homology with either PBPs or GOBPs. They are better conserved than OBPs across evolution, sharing more than 40-60% identity.[8,86,87]   They are shorter (110-115 amino acids) than PBPs or OBPs, and contain only 4 conserved cysteines forming two non-interlocked disulfide bridges in the pattern of CysI-CysII and CysIII-CysIV, *i.e.* Cys29-Cys38 and Cys57-Cys60 in the CSPs of *S. gregaria* and Cys29-Cys36 and Cys55-Cys58 in CSPs from *M. brassicae* (MbraCSPA6) (Fig. 9.4A).[81,86,88]   According to previous NMR analysis of MbraCSPA6 and SgreCSP4 (from *S. gregaria*), CSPs are mainly folded into α-helices.[81,89]   The recently resolved X-ray structure of MbraCSPA6 has revealed a novel V-shaped helical fold with six α-helices connected by α-α loops (Fig. 9.4B).[90]

CSPs have a large tissue distribution compared with OBPs and PBPs. They have been isolated from the ejaculatory bulb in *Drosophila,*[91] the antennae, tarsi, and mouth apparatus of *S. gregaria,*[86] the antennae and pheromone glands of *M. brassicae,*[92] the wing extracts and legs of *L. migratoria,*[66,79] the mouth organs, tarsi and cuticle of *E. calcarat,*[87] and the antennae of many other species such as *C. morosus,*[11] *B. mori,*[80] and *P. dominulus.*[67] Due to their localization in tissue, several roles involved in chemical communication and perception have been proposed. In *C. cactorum,* a CSP was found in the labial palp involved in CO_2 detection.[84]   Binding assays between *S. gregaria* CSPs and sodium bicarbonate or glucose, however, were not effective.[86]   The presence of the sex pheromone and a CSP in the ejaculatory bulb of *Drosophila* suggested the idea that such proteins could be a carrier for hydrophobic molecules.[91]   This view was supported by later experiments demonstrating reversible binding of the *Drosophila* pheromone vaccenyl acetate to the MbraCSPA6.[62]   P10, a small protein (Mw: 10 kDa) sharing around 50% sequence identity with other CSPs was isolated in the regenerated legs of the American cockroach *P. americana,* and is proposed to have a role in limb regeneration.[93]

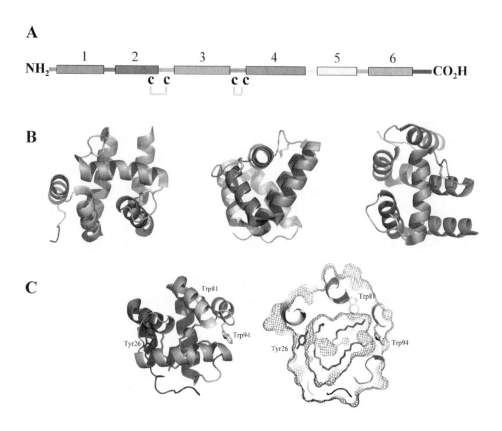

Fig. 9.4: Overall fold, disulfide pairing, and ligand binding of MbraCSPA6.[94] A) Secondary structure of MbraCSPA6. B) Side, top, and bottom views of MbraSCPA6 with disulfide linkages represented in stick form. C) Ligand binding and overall fold of MbraSCPA6 complexed with three molecules of 12-bromo-dodecanol. Interesting residues and bound ligands are shown as stick representations. Images were prepared using PyMOL (Dealno, W.L. 2004).

Structure of Insect CSPs

According to preliminary NMR data, CSPs are constituted mainly by α-helical domains. Only one CSP structure (MbraCSPA6) has been resolved so far (Fig. 9.4). MbraCSPA6 consists of six amphiphatic helices with dimensions of 25 x 30 x 32 Å.[90] Helices A and B together with helices D and E form two V-shaped structures with an opening angle of ~60°, which defines two parallel planes 12 Å apart. Helix C is perpendicular to the two planes and positioned between the four ends of the two V-shaped structures. Helix F is parallel to the V planes and packed to helices D and E outside the core of the protein. Because crystals were obtained under two different conditions, there were two varying crystal forms, form 1 and form 2. The narrow hydrophobic channel is defined by helices A and D, extending 14 Å within the core of monomer A of form 2. Six water molecules were visible in the channel. In monomer B, the Tyr26 side chain rotated by 100° to the protein surface, resulting in a shortened channel. The elongation of the channel by side chain rotation for binding long alkyl chains is suggested and confirmed by modeling studies.[90]

Based on fluorescence assays that indicated MbraCSPA6 binds halogenated alkyl alcohol or acids,[90] MbraCSPA6 was co-crystallized with 12-bromo-dodecanol (BrC12:OH). The solved structure of the BrC12:OH/MbraCSPA6 complex indicated a drastic conformational change upon complexation, after which helix C is pushed outwards by ~5Å and is split into two shorter helices at residue 47.[94] Since there are only two disulfide bridges conserved in CSPs, one would expect CSPs to be much more flexible when compared with OBPs and PBPs.

Three amino acid positions at 26, 81, 94 are indicated as highly conserved as aromatic residues after comparison of a total of 70 CSPs.[83] Tyr26 and Trp94 act as gates to the hydrophobic pocket in MbraCSPA6.[88,90] Trp81 in ASP3c may interact with ligand.[95] So, residues involved in ligand binding may be functionally conserved. Also, the comparison indicated three highly conserved motifs, the exposed parts of which may be involved in protein regulation and/or interactions. They are 1) N-terminal YTTKYDN(V/I)(N/D)(L/V)DEIL, (B) central DGKELKXX(I/L)PDAL, and (C) C-terminal KYDP.[83] These motifs were used to characterize several protein classes, revealing three classes of CSPs that have diverged from the conserved motifs. These classes share less sequence identity with MbraCSPA6, and were predicted to have a tertiary structure similar to MbraCSPA6 from homology modeling studies. For example, the homology model of the *Manduca sexta* SAP1 (MsexSAP1), which lacks both putative aromatic gate residues, maintains the general 3D structure.[83] Thus, the basic structure of MbraCSPA6 may be conserved within the whole CSP family.

Ligand Binding

Although CSPs' natural ligands have not yet been determined, the binding data indicate that they bind highly hydrophobic short to medium chain length (14-18 carbons) linear molecules similar to insect pheromones and fatty acids. It has been demonstrated that MbraCSPA6 from antennae is able to bind several components of pheromonal blend: *cis*-11-hexadecanyl acetate, *cis*-11-hexadecenol, hexadecanyl acetate, and *cis*-11-octadecenyl acetate (vaccenyl acetate).[8,62] CSPs in antennae and gland extracts of *M. brassicae* were able to bind the fatty acids, palmitic acid and oleic acid, but not progesterone. This may suggest that MbraCSPs prefer to bind fatty acids and/or molecules with a 16-18 carbon backbone.[92] Ban found that oleoamide was strongly associated with CSPs purified from the wings of *L. migratoria*, supporting that long-chain compounds are the natural ligands for CSPs (Table 9.5).[66]

Ligand binding assays, which were mainly based on fluorescence techniques, have been performed on MbraCSPA6, *S. gregaria* CSP4 (SgreCSP4), *A. mellifera* L. ASP3c (antennal-specific protein from honeybee, unlike ASP1 and ASP2, it is classified as a CSP based on the N-terminal homology),[96] *P. dominulus* CSP (PdomCSP), and *L. migratoria* CSPlm-II-10 (LmigCSPII-10), with dissociation constants in the μM range (Table 9.5).[66,67,90,95,97] From the binding data, CSPMbraA6 binds halogenated alkyl alcohol or acids such as 15-bromo-pentgadecanotic acid and 12-bromo-dodecanol.[90] SgreCSP4 failed to bind pheromones of the studied species, carboxylic acids and linear alcohols of 12, 14, and 18 carbon atoms, as well as derivatives of the above acids.[97] AmelASP3c can bind with 15-bromo-pentadecanoic acid (with a K_d of 0.65 uM), fatty acids, and the components of brood pheromone: methyl palmitate and methyl stearate. There is no binding of AmelASP3c with floral odorants or other components of honeybee pheromones.[95] The crystal structure of the MbraCSPA6 and 12-bromo-dodecanol complex revealed a 3:1 docking stoichiometry and expansion of the binding cavity upon ligand binding (Table 9.6). From fluorescence quenching experiments, a cooperative binding mechanism in successive steps to accommodate more than one ligand was confirmed.[94]

Table 9.5: Binding Data for CSPs

Species	CSP	Accession number	Ligand	K_d/ μM		Assay	Concentration / μM			pH	Ref
							CSP	Ligand	Re-porter		
Mamestra brassicae	MbraCSPA6	AF255918	12-bromo-dodecanol 15-bromo-pentadecanoic acid 9-bromo-octadecanoic acid (9-bromo-stearic acid)	0.90 1.60 0.35		Fluorescence	1	0.02-22.5	n/a	8.0	90
				K_{d1}	K_{d2}						94
			12-bromo-dodecanol 15-bromo-pentadecanoic acid 9-bromo-octadecanoic acid n-hexadecanoic acid (palmitic acid)	5.0 14 1.8 1.6	0.36 0.3 0.2 0.25						
Schistocerca gregaria	SgreCSP4	AF070964	N-phenyl-1-naphthylamine 2-amylcinnamaldehyde	4 9		Fluorescence	5 5	1-20 2-300	n/a 5	7.4	97
Apis mellifera L.	AmelASP3c	AF481963	15-bromo-pentadecanoic acid (+/-)-12-(9-anthroyloxy)octadecanoic acid n-tetradecanoic acid (myristic acid) n-hexadecanoic acid 1-octadecanotic acid (stearic acid) methyl hexadecanoate (methyl palmitate) methyl octadecanoate (methyl stearate)	0.65 0.57 1.64 0.51 0.80 1.02 1.23		Fluorescence	1	1-20 0.1-2 1-12	n/a n/a 1	7.5	95
Locusta migratoria	LmigCSPII-10	AY149658	N-phenyl-1-naphthylamine N-octadecyl-1-anthrylamine phenylacetonitrile N-decyl-1-anthrylamine 2-amylcinnamaldehyde phenyl-n-propyl ketone (butyrophenone)	6.2 7.1 6.4 4.3 4.3 12.1		Fluorescence	5	2-20	n/a 6	7.4	66

Table 9.5: (continued).

Species	CSP	Accession number	Ligand	K_d/ µM	Assay	Concentration / µM			pH	Ref
						CSP	Ligand	Re-porter		
Polistes dominulus	PdomCSP	AY297027	N-phenyl-1-naphthylamine	2.2	Fluorescence	2	1-20	n/a	7.4	67
			septanamide (pelargonic amide)	7.1						
			dodecanamide (lauric amide)	1.8						
			tetradecanamide (myristic amide)	0.36			1-30	4		
			hexadecanamide (palmitic amide)	0.36						
			(9E)octadec-9-enamide (elaidic amide)	0.53						
			(9Z)octadec-9-enamide (oleic amide)	0.71						
			dodecanol	0.89						
			tetradecanol	0.29						
			hexadecanol	0.36						
			octadecanol	3.2						
			octadecanoic acid (stearic acid)	0.64						
			methyl dodecanate (methyl laurate)	0.21						

Table 9.6: X-ray and NMR Structures of CSPs

Species	CSP	Ligand	Method	Resolution/ Å	pH	Unit *	Residues involved in binding/ interesting features	Ref. {PDB code}
Mamestra brassicae	MbraCSPA6	n/a	X-ray	Form1 2.80	8.4	M		{1KX8}
				Form2 1.65	5.5	D	*In monomer A* -residues contacting water: Arg68, Asn10, Asn6, Leu13, Asp9, His46, Glu62, Gly65, Ala66, Leu43, Val69, Leu47, Tyr26 -Tyr26 forms the bottom of the channel *In monomer B* -Tyr26 side chain is rotated by 100° around the χ1 angle	90 {1KX9}
		12-bromo-dodecanol	X-ray	Form3 1.39	6.5	D		{1N8V}
				Form4 1.80	6.5	M	-residues involved in conformational change: Leu43, Leu47, Ile51, Gln62, Ala66, Val69, Ile70, Leu84, Leu13, Tyr26, Tyr98 -Trp81 and Trp94 are close to ligands -residues forming cavity openings: Tyr98, Trp94, Gly51, Leu13, Asn61, Tyr4, His46	94 {1N8U}
		n/a	NMR	n/a	6.9	n/a	-residues forming hydrophobic channel mouths: Tyr26, Trp94, Tyr67, Tyr98 -Tyr26 and Trp94 are highly mobile	88,94 {1K19}

Cooperativity

Binding assays as well as crystallographic studies with CSPs have revealed that more than one ligand can be accommodated in the binding pocket. In the case studied (MbraCSPA6), up to three ligand molecules fit into one pocket.[94] The binding assay data also suggest that the binding of each ligand is positively cooperative, *i.e.*, the binding of the first ligand facilitates binding of the second, in turn facilitating binding of the third. Thus, at high ligand doses, the apparent affinity of the protein for the ligand is higher than at low doses. The biological significance of such a dose effect may be the same as with OBPs: to buffer and thereby protect receptors at high concentrations of ligand. Cooperativity with blends of ligands has not been investigated with these proteins, but remains an exciting possibility that should be explored.

SUMMARY

We have presented a comprehensive, up to date (September, 2004) review of the current understanding of odorant-binding protein (OBP) and chemosensory-specific protein (CSP) structure and function. Both types of protein are secreted and found in extracellular spaces, such as the lumen of chemosensory hairs. These proteins belong to individual superfamilies of insect proteins, each with a characteristic α-helical fold reinforced with disulfide bridges. Ligand binding has been studied in both families, and the following insights have arisen from these studies. 1) A variety of ligands is bound by each OBP or CSP, but these proteins also appear to exploit specific molecular interactions for ligand recognition. 2) Ligands are generally completely enclosed by these proteins, and it is not clear whether ligands are released within the time frame of olfaction. 3) For OBPs, the collective binding data indicate more than one binding-active form of the protein and, consequently, more than one type of binding site. 4) For CSPs, binding data and crystallography indicate cooperativity and multiple ligands binding to one binding site. 5) The functional significance of 3) and 4) may be concentration-dependent ligand binding and attenuation of very high doses, as well as early integration of blend information.

ACKNOWLEDMENTS

We thank the following agencies for support: NSERC (PGS-D to NH and E-RGPIN222923) and Research Corporation (R10519) and Simon Fraser University.

REFERENCES

1. NORDLUND, D.A., Semiochemicals: A review of the terminology, *in*: Semiochemicals. Their Role in Pest Control (D.A. Nordlund, R.L. Jones and W.J. Lewis, eds.), John Wiley & Sons, New York. 1981, pp. 13-28.
2. CHRISTENSEN, T.A., HEINBOCKEL, T., HILDEBRAND, J.G., Olfactory information processing in the brain: encoding chemical and temporal features of odors, *J. Neurobiol.*, 1996, **30**, 82-91.
3. GRANT, A.J., RIENDEAU, C.J., O-CONNELL, R.J., Spatial organization of olfactory receptor neurons on the antenna of the cabbage looper moth, *J. Comp. Physiol.*, 1998, **183**, 433-442.
4. HANSSON, B.S., LARSSON, M.C., LEAL, W.S., Green leaf volatile-detecting olfactory receptor neurones display very high sensitivity and specificity in a scarab beetle, *Physiol. Entomol.*, 1999, **24**, 121-126.
5. SAID, I., TAUBAN, D., RENOU, M., MORI, K., ROCHAT, D., Structure and function of antennal sensilla of the palm weevil *rhynchophorus palmarum* (coleoptera, curculionidae), *J. Insect Physiol.*, 2003, **49**, 857-872.
6. DEBRUYNE, M., CLYNE, P.J., CARLSON, J.R., Odor coding in a model olfactory organ: The *Drosophila* maxillary palp, *J. Neurosci.*, 1999, **19**, 4520-4532.
7. MATSUNAMI, H., AMREIN, H., Taste and pheromone perception in mammals and flies, *Genome Biol.*, 2003, **4**, Art. 220.
8. NAGNAN-LEMEILLOUR, P., CAIN, A.H., JACQUIN-JOLY, E., FRANCOIS, M.C., RAMACHANDRAN, S., MAIDA, R., STEINBRECHT, R.A., Chemosensory proteins from the proboscis of *Mamestra brassicae*, *Chem. Senses*, 2000, **25**, 541-553.
9. STEINBRECHT, R.A., LAUE, M., ZIEGELBERGER, G., Immunolocalization of pheromone-binding protein and general odorant-binding protein in olfactory sensilla of the silk moths *Antheraea* and *Bombyx*, *Cell Tissue Res.*, 1995, **282**, 203-217.
10. ZHANG, S.-G., MAIDA, R., STEINBRECHT, R.A., Immunolocalization of odorant-binding proteins in noctuid moths (insecta, lepidoptera), *Chem. Senses*, 2001, **26**, 885-896.
11. MONTEFORTI, G., ANGELI, S., PETACCHI, R., MINNOCCI, A., Ultrastructural characterization of antennal sensilla and immunocytochemical localization of a chemosensory protein in *Carausius morosus Bruenner* (Phasmida: *Phasmatidae*), *Arthropod Structure and Development*, 2002, **30**, 195-205.
12. DIEHL, P.A., VLIMANT, M., GUERENSTEIN, P., GUERIN, P.M., Ultrastructure and receptor cell responses of the antennal grooved peg sensilla of *Triatomainfestans* (Hmiptera: *Reduviidae*), *Arthropod Structure and Development*, 2003, **31**, 271-285.
13. MAIDA, R., STEINBRECHT, A., ZIEGELBERGER, G., PELOSI, P., The pheromone binding protein of *Bombyx mori*: Purification, characterization and

immunocytochemical localization, *Insect Biochem. Molec. Biol.*, 1993, **23**, 243-253.

14. PARK, S.-K., SHANBHAG, S.R., WANG, Q., HASAN, G., STEINBRECHT, R.A., PIKIELNY, C.W., Expression patterns of two putative odorant-binding proteins in the olfactory organis of *Drosophila melanogaster* have different implications for their functions, *Cell Tissue Res.*, 2000, **304**, 423-437.

15. VOGT, R.G., RIDDIFORD, L.M., Pheromone binding and inactivation by moth antennae, *Nature*, 1981, **293**, 161-163.

16. ROBERTSON, H.M., MARTOS, R., SEARS, C.R., TODRES, E.Z., WALDEN, K.K.O., NARDI, J.B., Diversity of odourant binding proteins revealed by an expressed sequence tag project on male *Manduca sexta* moth antennae, *Insect Mol. Biol.*, 1999, **8**, 501-518.

17. GRAHAM, L.A., DAVIES, P.L., The odorant-binding proteins of *Drosophila melanogaster*: Annotation and characterization of a divergent gene family, *Gene*, 2002, **292**, 43-55.

18. VOGT, R.G., Odorant binding protein homologues of the malaria mosquito *Anopheles gambiae*; possible orthologues of the OS-E and OS-F OBPs of *Drosophila melanogaster*, *J. Chem. Ecol.*, 2002, **28**, 2371-2376.

19. VOGT, R.G., RYBCZYNSKI, R., LERNER, M.R., Molecular cloning and sequencing of general-odorant binding proteins GOBP1 and GOBP2 from the tobacco hawk moth *Manduca sexta*: Comparisons with other insect OBPs and their signal peptides., *J. Neurosci.*, 1991, **11**, 2972-2984.

20. KRIEGER, J., VONNICKISCH-ROSENEGK, E., MAMELI, M., PELOSI, P., BREER, H., Binding proteins from the antennae of *Bombyx mori*, *Insect Biochem. Molec. Biol.*, 1996, **26**, 297-307.

21. VOGT, R.G., CALLAHAN, F.E., ROGERS, M.E., DICKENS, J.C., Cloning and expression of LAP, an adult specific odorant binding protein of the true bug *Lygus lineolaris* (hemiptera, heteroptera), *Chem. Senses*, 1999, **24**, 481-495.

22. PRESTWICH, G.D., Proteins that smell: Pheromone recognition and signal transduction, *Bioorg. Med. Chem.*, 1996, **4**, 505-513.

23. KIM, M.-S., REPP, A., SMITH, D.P., Lush odorant-binding protein mediates chemosensory responses to alcohols in *Drosophila melanogaster*, *Genetics*, 1998, **150**, 711-721.

24. VOGT, R.G., The molecular basis of pheromone reception: Its influence on behavior, *in*: Pheromone Biochemistry (G.D. Prestwich and G.J. Blomquist, eds.), Academic Press Inc., Orlando, Florida. 1987, pp. 385-431.

25. SIRUGUE, D., BONNARD, O., QUERE, J.L.L., FARINE, J.P., BROSSUT, R., 2-methylthiazolidine and 4-ethylguaiacol, male sex pheromone components of the cockroach *Nauphoeta cinerea* (dictyoptera, blaberidae): A reinvestigation, *J. Chem. Ecol.*, 1992, **18**, 2261-2276.

26. KRUSE, S.W., ZHAO, R., SMITH, D.P., JONES, D.N.M., Structure of a specific alcohol-binding site defined by the odorant binding protein lush from *Drosophila melanogaster*, *Nature Str. Biol.*, 2003, **10**, 694-700.

27. ZHOU, J.J., ZHANG, G.-A., HUANG, W., BIRKETT, M.A., FIELD, L.M., PICKETT, J.A., PELOSI, P., Revisiting the odorant-binding protein lush of *Drosophila melanogaster*: Evidence for odour recognition and discrimination, *FEBS Lett.*, 2004, **558**, 23-26.

28. POPHOF, B., Moth pheromone binding proteins contribute to the excitation of olfactory cells, *Naturwissenschaften*, 2002, **89**, 515-518.

29. DU, G., PRESTWICH, G.D., Protein structure encodes the ligand binding specificity in pheromone binding proteins, *Biochemistry (Mosc)*. 1995, **34**, 8726-8732.

30. BETTE, S., BREER, H., KRIEGER, J., Probing a pheromone binding protein of the silkmoth *Antheraea polyphemus* by endogenous tryptophan fluorescence, *Insect Biochem. Mol. Biol.*, 2002, **32**, 241-246.

31. KOWCUN, A., HONSON, N., PLETTNER, E., Olfaction in the gypsy moth, *Lymantria dispar*: Effect of pH, ionic strength and reductants on pheromone transport by pheromone-binding proteins, *J. Biol. Chem.*, 2001, **276**, 44770-44776.

32. CLYNE, P.J., WARR, C.G., FREEMAN, M.R., LESSING, D., KIM, J., CARLSON, J.R., A novel family of divergent seven-transmembrane proteins: Candidate odorant receptors in *Drosophila*, *Neuron*, 1999, **22**, 327-338.

33. VOSSHALL, L.B., WONG, A.M., AXEL, R., An olfactory sensory map in the fly brain, *Cell*, 2000, **102**, 147-159.

34. KRIEGER, J., BREER, H., Olfactory reception in invertebrates, *Science*, 1999, **286**, 720-723.

35. BREER, H., BOEKHOFF, I., STROTMANN, J., RAMING, K., TARELIUS, E., Molecular elements of olfactory signal transduction in insect antennae, *in*: Chemosensory Information Processing (D. Schild, ed.), Springer-Verlag, Berlin. 1990, pp. 77-86.

36. BREER, H., BOEKHOFF, I., TAREILUS, E., Rapid kinetics of second messenger formation in olfactory transduction, *Nature*, 1990, **345**, 65-68.

37. KRIEGER, J., RAMING, K., BREER, H., Cloning of genomic and complementary DNA encoding insect pheromone binding proteins: evidence for microdiversity, *Biochim. Biophys. Acta*, 1991, **1088**, 277-284.

38. VOGT, R.G., KOHNE, A.C., DUBNAU, J.T., PRESTWICH, G.D., Expression of pheromone binding proteins during antennal development in the gypsy moth, *Lymantria dispar*, *J. Neurosci.*, 1989, **9**, 3332-3346.

39. PLETTNER, E., LAZAR, J., PRESTWICH, E.G., PRESTWICH, G.D., Discrimination of pheromone enantiomers by two pheromone binding proteins from the gypsy moth *Lymnatria dispar*, *Biochemistry (Mosc)*. 2000, **39**, 8953-8962.

40. SCALONI, A., MONTI, M., ANGELI, S., PELOSI, P., Structural analysis and disulfide-bridge pairing of two odorant-binding proteins from *Bombyx mori*, *Biochem. Biophys. Res. Commun.*, 1999, **266**, 386-391.

41. LEAL, W.S., NIKONOVA, L., PENG, G., Disulfide structure of the pheromone binding protein from the silkworm moth, *Bombyx mori*, *FEBS Lett.*, 1999, **464**, 85-90.

42. BRIAND, L., NESPOULOUS, C., HUET, J.-C., PERNOLLET, J.-C., Disulfide pairing and secondary structure of asp1, an olfactory-binding protein from honeybee (*Apis mellifera* l.), *J. Peptide Res.*, 2001, **58**, 540-545.

43. ZHOU, J.-J., HUANG, W., ZHANG, G.-A., PICKETT, J.A., FIELD, L.M., "plus-c" odorant-binding protein genes in two *Drosophila* species and the malaria mosquito *Anopheles gambiae*, *Gene*, 2004, **327**, 117-129.

44. BUTENANDT, A., BECKMANN, R., STAMM, D., HECKER, E., Über den sexuallockstoff des seidenspinners *Bombyx mori*. Reindarstellung and constitution [on the sex attractant of the silkworm moth *Bombyx mori*. Isolation and structure], *Z. Naturforsch. B*, 1959, **14**, 283-284.

45. BIERL, B.A., BEROZA, M., COLLIER, C.W., Isolation, identification, and synthesis of the gypsy moth sex attractant, *J. Econ. Entomol.*, 1972, **65**, 659-664.

46. CARDE, R.T., DOANE, C.C., BAKER, T.C., IWAKI, S., MARUMO, S., Attractancy of optically active pheromone for male gypsy moths, *Environ. Entomol.*, 1977, **6**, 768-772.

47. GRANT, G.G., LANGEVIN, D., LISKA, J., KAPITOLA, P., CHONG, J.M., Olefin inhibitor of gypsy moth, *Lymantria dispar*, is a synergistic pheromone component of nun moth, *L. monacha*, *Naturwissenschaften*, 1996, **83**, 328-330.

48. GRIES, G., SCHAEFER, P.W., GRIES, R., LISKA, J., GOTOH, T., Reproductive character displacement in *Lymantria monacha* from northern Japan, *J. Chem. Ecol.*, 2001, **27**, 1163-1175.

49. PLETTNER, E., Insect pheromone olfaction: new targets for the design of species-selective pest control agents, *Curr. Med. Chem.*, 2002, **9**, 1075-1085.

50. SANDLER, B.H., NIKONOVA, L., LEAL, W.S., CLARDY, J., Sexual attraction in the silkworm moth: Sructure of the pheromone-binding-protein-bombykol complex, *Chemistry & Biology*, 2000, **7**, 143-151.

51. HORST, R., DAMBERGER, F., LUGINBUEHL, P., GUENTERT, P., PENG, G., NIKANOVA, L., LEAL, W.S., WUETHRICH, K., NMR structure reveals intramolecular regulation mechanism for pheromone binding and release, *Proc. Natl. Acad. Sci. USA*, 2001, **98**, 14374-14379.

52. LEE, D., DAMBERGER, F.F., PENG, G., HORST, R., GUNTERT, P., NIKANOVA, L., LEAL, W.S., WUTHRICH, K., NMR structure of the unliganded *Bombyx mori* pheromone-binding protein at physiological pH, *FEBS Lett.*, 2002, **531**, 314-318.

53. MOHANTY, S., ZUBKOV, S., GRONENBORN, A.M., The solution NMR structure of *A. polyphemus* PBP provides new insight into pheromone recognition by pheromone-binding proteins, *J. Mol. Biol.*, 2004, **337**, 443-451.

54. LARTIGUE, A., GRUEZ, A., BRIAND, L., BLON, F., BEZIRARD, V., WALSH, M., PERNOLLET, J.-C., TEGONI, M., CAMBILLAU, C., Sulfur single-wavelength anomalous diffraction crystal structure of a pheromone-binding protein from the honeybee *Apis mellifera* L., *J. Biol. Chem.*, 2004, **279**, 4459-4464.

55. LARTIGUE, A., GRUEZ, A., SPINELLI, S., RIVIERE, S., BROSSUT, R., TEGONI, M., CAMBILLAU, C., The crystal structure of a cockroach pheromone-binding protein suggests a new ligand binding and release mechanism, *J. Biol. Chem.*, 2003, **278**, 30213-30218.

56. WOJTASEK, H., LEAL, W.S., Conformational change in the pheromone-binding protein from *Bombyx mori* induced by ph and by interaction with membranes, *J. Biol. Chem.*, 1999, **274**, 30950-30956.

57. LEAL, W.S., Duality monomer-dimer of the pheromone-binding protein of *Bombyx mori*, *Biochem. Biophys. Res. Commun.*, 2000, **268**, 521-529.

58. CAMPANACCI, V., LONGHI, S., MEILLOUR, P.N.-L., CAMBILLAU, C., TEGONI, M., Recombinant pheromone binding protein 1 from *Mamestra brassicae* (MBraPBP1). Functional and structural characterization, *Eur. J. Biochem.*, 1999, **264**, 707-716.

59. DANTY, E., BIRAND, L., MICHARD-VANHEE, C., PEREZ, V., ARNOLD, G., GAUDEMER, O., HUET, D., HUET, J.-C., OUALI, C., MASSON, C., PERNOLLET, J.-C., Cloning and expression of a queen pheromone-binding protein in the honeybee: An olfactory-specific, developmentally regulated protein, *J. Neurosci.*, 1999, **19**, 7468-7475.

60. KEIL, T.A., Surface coats of pore tubules and olfactory sensory dendrites of a silkmoth revealed by cationic markers, *Tissue & Cell*, 1984, **16**, 705-717.

61. MOHL, C., BREER, H., KRIEGER, J., Species-specific pheromonal compounds induce distinct conformational changes of pheromone binding protein subtypes from *Antheraea polyphemus*, *Invert. Neurosci.*, 2002, **4**, 165-174.

62. BOHBOT, J., SOBRIO, F., LUCAS, P., NAGNAN-LEMEILLOUR, P., Functional characterization of a new class of odorant-binding proteins in the moth *Mamestra brassicae*, *Biochem. Biophys. Res. Commun.*, 1998, **253**, 489-494.

63. PRESTWICH, G.D., GRAHAM, S.M., KUO, J.-W., VOGT, R.G., Tritium-labeled enantiomers of disparlure. Synthesis and in vitro metabolism, *J. Am. Chem. Soc.*, 1989, **111**, 636-642.

64. CAMPANACCI, V., KRIEGER, J., BETTE, S., STURGIS, J.N., LARTIGUE, A., CAMBILLAU, C., BREER, H., TEGONI, M., Revisiting the specificity of *Mamestra brassicae* and *Antheraea polyphemus* pheromone-binding proteins with a fluorescence binding assay, *J. Biol. Chem.*, 2001, **276**, 20078-20084.

65. RIVIERE, S., LARTIGUE, A., QUENNEDY, B., CAMPANACCI, V., FARINE, J.-P., TEGONI, M., CAMBILLAU, C., BROSSUT, R., A pheromone-binding protein from the cockroach *Leucophaea maderae*: Cloning, expression and pheromone binding, *Biochem. J.*, 2003, **371**, 573-579.

66. BAN, L., SCALONI, A., BRANDAZZA, A., ANGELI, S., ZHANG, L., YAN, Y., PELOSI, P., Chemosensory proteins of *Locusta migratoria*, *Insect Mol. Biol.*, 2003, **12**, 125-134.

67. CALVELLO, M., GUERRA, N., BRANDAZZA, A., AMBROSIO, C.D., SCALONI, A., DANI, F.R., TURILLAZZI, S., PELOSI, P., Soluble proteins of chemical communication in the social wasp *Polistes dominulus*, *Cell. Mol. Life Sci.*, 2003, **60**, 1933-1943.

68. HONSON, N., JOHNSON, M.A., OLIVER, J.E., PRESTWICH, G.D., PLETTNER, E., Structure-activity studies with pheromone-binding proteins of the gypsy moth, *Lymantria dispar*, *Chem. Senses*, 2003, **28**, 479-489.

69. KLUSAK, V., HAVIAS, Z., RULISEK, L., VONDRASEK, J., SVATOS, A., Sexual attraction in the silkworm moth: Nature of binding of bombykol in pheromone binding protein - an ab initio study, *Chem. & Biol.*, 2003, **10**, 331-340.

70. DEYU, Z., LEAL, W.S., Conformational isomers of insect odorant-binding proteins, *Arch. Biochem. Biophys.*, 2002, **397**, 99-105.

71. KAISSLING, K.-E., PRIESNER, E., Die riechschwelle des seidenspinners, *Naturwissenschaften*, 1970, **57**, 23-28.

72. KAISSLING, K.-E., Control of insect behavior via chemoreceptor organs, *in*: Chemical Control of Insect Behavior. Theory and application (H.H. Shorey and J.J. McKelvey, eds.), Wiley, New York. 1977, pp. 45-65.

73. OCHIENG, S.A., PARK, K.C., BAKER, T.C., Host plant volatiles synergize responses of sex pheromone-specific olfactory receptor neurons in male *Helicoverpa zea*, *J. Comp. Physiol. A*, 2002, **188**, 325-333.

74. ROCHAT, D., MEILLOUR, P.N.-L., ESTEBAN-DURAN, J.R., MALOSSE, C., PERTHUS, B., MORIN, J.P., DESCOINS, C., Identification of pheromone synergists in american palm weevil, *Rhyncophorus palmarum*, and attraction of related *Dynamis borassi*, *J. Chem. Ecol.*, 2000, **26**, 155-187.

75. PIKIELNY, C.W., HASAN, G., ROUYER, F., ROSBACH, M., Members of a family of *Drosophila* putative odorant-binding proteins are expressed in different subsets of olfactory hairs, *Neuron*, 1994, **12**, 35-49.

76. VOSSHALL, L.B., AMREIN, H., MOROZOV, P.S., RZHETSKY, A., AXEL, R., A spatial map of olfactory receptor expression in the *Drosophila* antenna, *Cell*, 1999, **96**, 725-736.

77. SCHNEIDER, D., KAFKA, W.A., BEROZA, M., BIERL, B.A., Odor receptor responses of male gypsy and nun moths (lepidoptera, lymantriidae) to disparlure and its analogues, *J. Comp. Physiol. A*, 1977, **113**, 1-15.

78. HANSEN, K., Discrimination and production of disparlure enantiomers by the gypsy moth and the nun moth, *Physiol. Entomol.*, 1984, **9**, 9-18.

79. PICIMBON, J.-F., DIETRICH, K., BREER, H., KRIEGER, J., Chemosensory proteins of *Locusta migratoria* (Orthoptera: *Acrididae*), *Insect Biochem. Mol. Biol.*, 2000, **30**, 233-241.

80. PICIMBON, J.-F., DIETRICH, K., ANGELI, S., SCALONI, A., KRIEGER, J., BREER, H., PELOSI, P., Purification and molecular cloning of chemosensory proteins from *Bombyx mori*, *Arch. Bioch. Biophys.*, 2000, **44**, 120-129.

81. PICONE, D., CRESCENZI, O., ANGELI, S., MARCHESE, S., BRANDAZZA, A., FERRERA, L., PELOSI, P., SCALONI, A., Bacterial expression and conformational analysis of a chemosensory protein from *Schistocerca gregaria*, *Eur. J. Biochem.*, 2001, **268**, 4794-4801.

82. MCKENNA, M.P., HEKMAT-SCAFE, D.S., GAINES, P., CARLSON, J.R., Putative *Drosophila* pheromone-binding proteins expressed in a subregion of the olfactory system, *J. Biol. Chem.*, 1994, **269**, 16340-16347.

83. WANNER, K.W., WILLIS, L.G., THEILMANN, D.A., ISMAN, M.B., FENG, Q., PLETTNER, E., Analysis of the insect os-d-like gene family, *J. Chem. Ecol.*, 2004, **30**, 889-911.

84. MALESZKA, R., STANGE, G., Molecular cloning, by a novel approach, of a cDNA encoding a putative olfactory protein in the labial palps of the moth *Cactoblastis cactorum*, *Gene*, 1997, **202**, 39-43.

85. PICIMBON, J.-F., LEAL, W.S., Olfactory soluble proteins of cockroaches, *Insect Biochem Molec. Biol.*, 1999, **29**, 973-978.

86. ANGELI, S., CERON, F., SCALONI, A., MONTI, M., MONTEFORTI, G., MINNOCCI, A., PETACCHI, R., PELOSI, P., Purification, structural characterization, cloning and immunocytochemical localization of chemoreception proteins from *Schistocerca gregaria*, *Eur. J. Biochem.*, 1999, **262**, 745-754.

87. MARCHESE, S., ANGELI, S., ANDOLFO, A., SCALONI, A., BRANDAZZA, A., MAZZA, M., PICIMBON, J.-F., LEAL, W.S., PELOSI, P., Soluble proteins from chemosensory organs of *Eurycantha calcarata* (insects, phasmatodea), *Insect Biochem. Mol. Biol.*, 2000, **30**, 1091-1098.

88. MOSBACH, A., CAMPANACCI, V., LARTIGUE, A., TEGONI, M., CAMBILLAU, C., DARBON, H., Solution structure of a chemosensory protein from the moth *Mamestra brassicae*, *Biochem. J.*, 2003, **369**, 39-44.

89. CAMPANACCI, V., MOSBAH, A., BORNET, O., WECHSELBERGER, R., JACQUIN-JOLY, E., CAMBILLAU, C., DARBON, H., TEGONI, M., Chemosensory protein from the moth *Mamestra brassicae*. Expression and secondary structure from ^{1}H and ^{15}N NMR, *Eur. J. Biochem.*, 2001, **268**, 4731-4739.

90. LARTIGUE, A., CAMPANACCI, V., ROUSSEL, A., LARSSON, A.M., JONES, T.A., TEGONI, M., CAMBILLAU, C., X-ray structure and ligand binding study of a moth chemosensory protein, *J. Biol. Chem.*, 2002, **277**, 32094-32098.

91. DYANOV, H.M., DZITOEVA, S.G., Method for attachment of microscopic preparations on glass for *in situ* hybridization, PRINS and *in situ* PCR studies, *Biotechniques*, 1995, **18**, 822-824.

92. JACQUIN-JOLY, E., VOGT, R.G., FRANCOIS, M.-C., MEILLOUR, P.N.-L., Functional and expression pattern analysis of chemosensory proteins expressed in antennae and pheromonal gland of *Mamestra brassicae*, *Chem. Senses*, 2001, **26**, 833-844.

93. KITABAYASHI, A.N., ARAI, T., KUBO, T., NATORI, S., Molecular cloning of cDNA for p10, a novel protein that increases in the regenerating legs of *Periplaneta americana* (American cockroach), *Insect Biochem. Molec. Biol.*, 1998, **29**, 785-790.

94. CAMPANACCI, V., LARTIGUE, A., HALLBERG, B.M., JONES, T.A., GIUDICI-ORTICONI, M.-T., TEGONI, M., CAMBILLAU, C., Moth chemosensory protein exhibits drastic conformational changes and cooperativity on ligand binding, *Proc. Nat. Acad. Sci. USA*, 2003, **100**, 5069-5074.

95. BRIAND, L., SWASDIPAN, N., NESPOULOUS, C., BEZIRARD, V., BLON, F., HUET, J.-C., EBERT, P., PERNOLLET, J.-C., Characterization of a chemosensory protein (ASP3c) from honeybee (*Apis mellifera* L.) as a brood pheromone carrier, *Europ. J. Biochem.*, 2002, **269**, 4586-4596.

96. DANTY, E., ARNOLD, G., HUET, J.-C., HUET, D., MASSON, D., PERNOLLET, J.-C., Separation, characterization and sexual heterogeneity of multiple putative odorant-binding proteins in the honeybee *Apis mellifera* L. (Hymenoptera: *Apidea*), *Chem. Senses*, 1998, **23**, 83-91.

97. BAN, L., ZHANG, L., YAN, Y., PELOSI, P., Binding properties of a locust's chemosensory protein, *Bioch. Biophys. Res. Commun.*, 2002, **293**, 50-54.

98. BRIAND, L., NESPOULOUS, C., HUET, J.-C., TAKAHASHI, M., PERNOLLET, J.-C., Ligand binding and physico-chemical properties of ASP2, a recombinant odorant-binding protein from honeybee (*Apis mellifera* L.), *Eur. J. Biochem.*, 2001, **268**, 752-760.

99. BAN, L., SCALONI, A., D'AMBROSIO, C., ZHANG, L., YAN, Y., PELOSI, P., Biochemical characterization and bacterial expression of an odorant-binding protein from *Locusta migratoria*, *Cell. Mol. Life Sci.*, 2003, **60**, 390-400.

100. MOHANTY, S., ZUBKOV, S., CAMPOS-OLIVAS, R., Letter to the editor: ^{1}H, ^{13}C and ^{15}N backbone assignments of the pheromone binding protein from the silk moth *Antheraea polyphemus* (ApolPBP), *J. Biomol. NMR*, 2003, **27**, 393-394.

101. LARTIGUE, A., GRUEZ, A., BRIAND, L., PERNOLLET, J.-C., SPINELLI, S., TEGONI, M., CAMBILLAU, C., Optimization of crystals form nanodrops: Crystallization and preliminary crystallographic study of a pheromone-binding protein from the honeybee *Apis mellifera* L., *Acta Crystallographica Section D: Biological Crystallography*, 2003, **D59**, 919-921.

102. BIRLIRAKIS, N., BRIAND, L., PERNOLLET, J.-C., GUITTET, E., ^{1}H, ^{13}C and ^{15}N chemical shift assignment of the honey bee pheromone carrier protein ASP1, *J. Biomol. NMR*, 2001, **20**, 183-184.

103. LARTIGUE, A., RIVIERE, S., BROSSUT, R., TEGONI, M., CAMBILLAU, C., Crystallization and preliminary crystallographic study of a pheromone-binding protein from the cockroach *Leucophaea maderae*, *Acta Cryst.*, 2003, **D59**, 916-918.

Chapter Ten

OLFACTION AND LEARNING IN MOTHS AND WEEVILS LIVING ON ANGIOSPERM AND GYMNOSPERM HOSTS

Hanna Mustaparta* and Marit Stranden

Department of Biology
Neuroscience Unit
Norwegian University of
Science and Technology
NO-7489 Trondheim

Author for correspondence, email: hanna.mustaparta@bio.ntnu.no

INTRODUCTION

The olfactory system of herbivorous insects is challenged by the large diversity of volatiles emitted by plants. The knowledge of how plants synthesize and regulate the production of compounds has increased during recent years, based on the identified genome of *Arabidopsis thaliana*[1] and the advanced methods of isolation and characterizations of plant genes. Interesting evolutionary aspects of the enzymes involved in the various pathways of biosynthesis have appeared in studies of different plant species[2]. Focus has also been on responses of plants to abiotic and biotic factors causing variability of the released blends.[3,4] Light, temperature, humidity, airborne or water mediated chemicals, as well as feeding by various herbivore larvae can induce the release of compounds, not only in increased amounts but also as a changed composition. In trying to determine how plant volatiles influence the behavior of insects, it is important to consider all emitted volatiles as potential odorants. Obviously, volatiles released by the host plants of insects should contain the most relevant odorants guiding the insects to a suitable source for feeding, mate finding, or egg-laying. Also, volatiles released by non-hosts may be important in giving clues about unsuitable plants. However, various species of plants produce hundreds of compounds of which many are common in both hosts and non-hosts.[5] Even plants of gymnosperms and angiosperms, have enzymes producing the same compounds, such as the different terpene synthases that are suggested to have undergone convergent evolution.[2,6] Also, considerable differences exist between the emitted blends of these plants, reflecting the specificity of the various groups of enzymes, which gives the chemical signature of the species and individual plants. Monoterpenes dominate in gymnosperms, whereas sesquiterpenes are typical for angiosperms.

To meet the challenge of detecting the large diversity of molecules, the olfactory system operates with a large number of receptor protein types, each type expressed in subsets of receptor neurons, as shown in vertebrates and the fruit fly *Drosophila melanogaster*.[7-11] The number of receptor protein types in mice and rats is in the range of 1000, whereas 60 types are present in *Drosophila*, about 40 of which are expressed on the antennae. One may ask whether each type of receptor protein and, thus, each subpopulation of receptor neurons is evolved for receiving information about a small or a large number of odorants, which would further extend the number of odorants detected by each organism. This question is a particularly interesting one to study in herbivorous insects, because of their intimate interactions and specializations on certain species of host plants.

Odorants reach the receptor proteins after binding to particular proteins in the lymph of the olfactory sensilla in insects.[12] Two major kinds, the pheromone binding proteins (PBPs) and the general odor binding proteins (GOBPs) are commonly present in separate sensilla, specified for pheromones and plant odorants, respectively[13] (cf. Plettner this volume). Although the binding proteins filter out the odorants to be transported to the dendrite membrane, the receptor proteins in the membrane determine the specificity of the olfactory receptor neurons.[11,14] The presence of receptor proteins also in the axons of neurons, has made it possible to determine their projections in the brain.[10,5-18] The axons of each subset of neurons terminate in one or two specific glomeruli of the primary olfactory center, the antennal lobe of insects, and the olfactory bulb of vertebrates. This principle, called "the logic of the sense of smell" implies that there is a certain relationship between the number of receptor neuron types and the number of glomeruli. Honeybees and parasitic wasps have 160-200 glomeruli,[19,20] whereas 60-67 are present in the antennal lobes of the moth species studied[21-25] (Skiri *et al.* unpublished data), and 43 in the fruitfly.[26]

In our studies of heliothine moths, considered as a monophyletic insect group,[27] and distantly related moth and weevil species, we are interested in how plant odor information is encoded in the receptor and central interneurons leading to behavioral responses. This includes identification of naturally produced plant odorants. By examining closely vs. distantly related species and polyphagous vs. oligophagous species, we compare the olfactory coding mechanisms across them. In addition to the innate ability to respond to odors, the insects can also learn to associate an odor with a reward, which subsequently influences the behavioral responses to the odor. Learning of odors is particularly well studied in the honeybee by the use of the proboscis extension responses in appetitive learning with sucrose rewards, as well as by other behavioral bioassays.[28-30] Also, moths, including heliothines, can learn odors, as demonstrated by the use of the proboscis extension response and by wind-tunnel experiments.[31-37] Results from field studies suggest that abundance of a plant species, as in monocultures, may increase the use of a particular species because of the conditioning to the odors taking place in the insects.[38] Another hypothesis concerning the abundant use of these plants globally is that some heliothine moths on different continents may have adapted to cultivated plants during many years in monocultures, which may have influenced their olfactory receptors. This applies to the large cotton pest insects, the American tobacco budworm moth *Heliothis virescens*, and the Eurasian cotton bollworm moth *Helicoverpa armigera*, two of the heliothine species included in our studies.

IDENTIFICATION OF BIOLOGICALLY RELEVANT PLANT ODORANTS AND RECEPTOR NEURON SPECIFICITY

For identifying the biologically relevant plant odorants used by insects, the most precise method is gas chromatography linked to electrophysiological recordings from single olfactory receptor neurons (GC-SCR).[39-42] This allows naturally produced plant odorants to be separated in the column of the gas chromatograph and tested directly on each neuron. By the use of two columns with different properties (*e.g.*, polarity and chirality), each neuron can be tested with the same mixture separated by specific compound sequences and also pure enantiomers.[43,44] This is important for the further analyses by gas chromatography linked to mass-spectrometry (GC-MS) in which the identity of the active compounds is determined. Retesting the same types of receptor neurons with authentic or synthetic samples follows to prove the identification. The results obtained with this method not only provide knowledge about biologically relevant odorants and in which plant species they are produced, but also which odorants the single receptor neurons respond to, termed the molecular receptive range of the neurons. This knowledge is important for selecting odorants that should be tested on insect behavior. In order to collect as many plant substances as possible for neuron testing, aeration ("head-space" technique) has been made from a variety of plant species, both hosts as well as non-hosts. Intact plants are used for obtaining naturally emitted volatiles and cut plant materials for collecting larger quantity of the compounds. The organic molecules are trapped on different adsorbents and eluted by solvents. These solutions are then used for testing the receptor neurons on the insect antennae.

FUNCTIONAL CHARACTERIZATIONS AND CLASSIFICATION OF RECEPTOR NEURONS

With this method, we have functionally characterized and classified a large number of plant odor receptor neurons in different insect species that use angiosperms and gymnosperms as hosts. The species include heliothine moths (*H. virescens, H. armigera,* and *Helicoverpa assulta*)[43-47] (Røstelien *et al.* unpublished data), the cabbage moth *Mamestra brassicae* (Ulland *et al.*unpublished data), weevils (*Hylobius abietis, Pissodes notatus,* and *Anthonomus rubi*),[42,48-51] Bichão *et al.* unpublished data), the eucalyptus woodborer *Phoracantha semipunctata*,[52] and the melaleuca weevil *Oxyops vitiosa* (Gregory S. Wheeler personal communication). From the most recent experiments that employ the same test protocols, the results have enabled a comparison of the receptor neuron specificity across these species. Comparison with other species for which different protocols have been used, is more limited.[53-55] In the GC-SCR studies, the typical response mode to plant odorants is excitation, shown for all recorded neurons. Inhibition has only appeared in the strawberry weevil to one compound in a neuron type that was excited by another odorant (Bichão *et al.* unpublished data). This is in contrast to recent results obtained in *Drosophila* where direct screening of various odorants has shown that both excitation and inhibition are common response modes.[56] Another difference from *Drosophila* is a narrower receptor neuron tuning recorded in the moths and the weevils. Neurons in these species have shown responses to only a few (1-9) compounds out of the hundreds of test constituents present in various plant blends. Typical is a high sensitivity of each neuron to a single compound, termed primary odorant, and weaker responses to the others (secondary odorants) that are structurally related molecules. The term secondary odorants does not imply that they are behaviorally unimportant. According to the compounds eliciting responses, the receptor neurons fall into distinct functional types, named after the primary odorants. Below are given a few examples of receptor neurons in moth and beetle species living on different host plants of angiosperms and gymnosperms (Table 10.1).

RECEPTOR NEURONS IN MOTHS AND WEEVILS LIVING ON ANGIOSPERMS

Germacrene D Receptor Neurons

One type of receptor neurons, tuned to the sesquiterpene germacrene D, frequently occurred in recordings from heliothine moths, and was the first type identified in these species[57] (Fig. 10.1). The response was elicited by mixtures obtained from many plant species, both hosts and non-hosts. The identification, carried out by GC-MS and nuclear magnetic resonance (NMR), was proved by retesting an isolated fraction of germacrene D from essential oil of cubeb pepper via the gas chromatograph. Since germacrene D consists of two enantiomers, a chiral column was installed in the gas chromatograph to test pure enantiomers on this neuron type.[58] All germacrene D neurons recorded in the two polyphagous *H. armigera* and *H. virescens* species, as well as in the oligophagous *H. assulta* species, belong to one functional type[59] (Table 10.1). They all show the same specificity for the various compounds, *i.e.*, (-)-germacrene D had the best effect, being 10-fold stronger than the (+)-enantiomer, followed by weak effects of a few other sesquiterpenes with related structures, *e.g.*, (-)-α-ylangene, (+)-β-ylangene, (+)-α-copaene, and β-copaene. Thus, the germacrene D receptor neurons did not display any difference either across the oligophagous and the polyphagous species, or across species living on different continents that are exposed to different plant species. Interestingly, the two germacrene D enantiomers are produced by two different chiral enzymes in the plants,[60-62] suggesting that they may play different roles and, therefore, are detected by different receptor neurons in herbivorous insects. However, since the heliothine moths have not evolved different receptor types for detecting the two optical configurations of germacrene D, discrimination between the two enantiomers may not be important in these species. The higher sensitivity to (-)-germacrene D than to the (+)-enantiomer correlates well with the finding that (-)-germacrene D dominates in higher plants. For instance, the host plant sunflower, contains exclusively (-)-germacrene D.[44] The presence of the (+)-enantiomer, as well as the secondary odorants, was found in plants that are not typical hosts of heliothine moths.[46]

Receptor neurone types	Moths				Weevils	
	H. virescens	*H. armigera*	*H. assulta*	*M. brassica*	*A. rubi*	*O. vitiosa*
(-)-Germacrene D receptor neuron type	**(-)-Germacrene D** (+)-Germacrene D (-)-α-Ylangene (+)-β-Ylangene (+)-α-Copaene β-Copaene	**(-)-Germacrene D** (+)-Germacrene D (-)-α-Ylangene (+)-β-Ylangene (+)-α-Copaene β-Copaene	**(-)-Germacrene D** (+)-Germacrene D (-)-α-Ylangene (+)-β-Ylangene (+)-α-Copaene β-Copaene		**(-)-Germacrene D** (+)-Germacrene D (-)-β-Caryophyllene α-Humulene β-Bourbonene 3-Pentanone 3-Hexanone	
(+)-Linalool receptor neuron type	**(+)-Linalool** (-)-Linalool Dihydrolinalool*	**(+)-Linalool** (-)-Linalool Dihydrolinalool*			**(+)-Linalool** (-)-Linalool (+)-Dihydrolinalool (-)-Dihydrolinalool	**(+)-Linalool** (-)-Linalool (+)-Dihydrolinalool (-)-Dihydrolinalool
(-)-Linalool receptor neuron type				**(-)-Linalool** (+)-Linalool (+)-Dihydrolinalool (-)-Dihydrolinalool	**(-)-Linalool** (+)-Linalool (+)-Dihydrolinalool (-)-Dihydrolinalool	
Methyl salicylate receptor				**Methyl salicylate** Methyl benzoate	**Methyl salicylate** Methyl benzoate Ethyl benzoate	
E-β-Ocimene receptor neuron type	***E*-β-Ocimene** β-Myrcene Z-β-Ocimene Dihydromyrcene *E*-DMNT	***E*-β-Ocimene** β-Myrcene Z-β-Ocimene Dihydromyrcene *E*-DMNT	***E*-β-Ocimene** β-Myrcene Z-β-Ocimene Dihydromyrcene *E*-DMNT		***E*-β-Ocimene** β-Myrcene Z-β-Ocimene Dihydromyrcene *E*-DMNT Citronellol Geraniol Geranial Neral Limonene γ-Terpinene β-Phellandrene	

Table 10.1: Primary and secondary odorants of five receptor neuron types identified by GC-SCR and GC-MS in four species of moth (*Heliothis virescens, Helicoverpa armigera, Helicoverpa assulta,* and *Mamestra brassica*)[43-47,63] (Ulland *et al.* unpublished data) and two weevils (*Anthonomus rubi* and *Oxyops vitiosa*)[51] (Bichão *et al.* unpublished data, Gregory S. Wheeler personal communication). Primary odorants marked with bold. *E*-DMNT: 3*E*-4,8-dimethyl-1,3,7-nonatriene, *: enantiomers not tested.

In a distantly related species, the strawberry weevil *A. rubi*, germacrene D receptor neurons have also been identified[51] (Fig. 10.1). Interestingly, they display similar enantioselectivity, by responding stronger to (-)-germacrene D than to the (+)-enantiomer. However, differences are found for the secondary odorants. Whereas the sesquiterpenes (-)-β-caryophyllene, α-humulene, and β-bourbonene as well as the small size molecules 3-pentanone and 3-hexanone elicited weak responses in the germacrene D neurons of *A. rubi*, they had no effect on the germacrene D neurons in heliothine moths (Table 10.1). Vice versa, the secondary odorants for the heliothine germacrene D receptor neurons had no effect on the *A. rubi* germacrene D neurons. Thus, in spite of similar enantioselectivity, the receptor neurons in these distantly related species have different molecular receptive ranges, suggesting either a convergent evolution of the receptors, or chance mutations of a common ancestral receptor type. Both in heliothine moths and in *A. rubi*, the ten carbon ring and the direction of the isopropyl group of the germacrene D molecule is important for binding to the receptor. The direction of the isopropyl groups makes the difference between the enantiomeric effect, which is 10-fold in heliothine moths and 10– 100-fold in *A. rubi* (Fig. 10.1C, D). With other molecular changes, the stimulatory effect becomes reduced, however, in a different manner in the receptors of the distantly related species. So far, no receptor neurons tuned to germacrene D have been found in the weevils *H. abietis* and *P. notatus* living on conifers, although the compound is present in a relative large amount in their host plants. Further studies may reveal whether receptor neurons tuned to germacrene D are present in other insect species living on conifers.

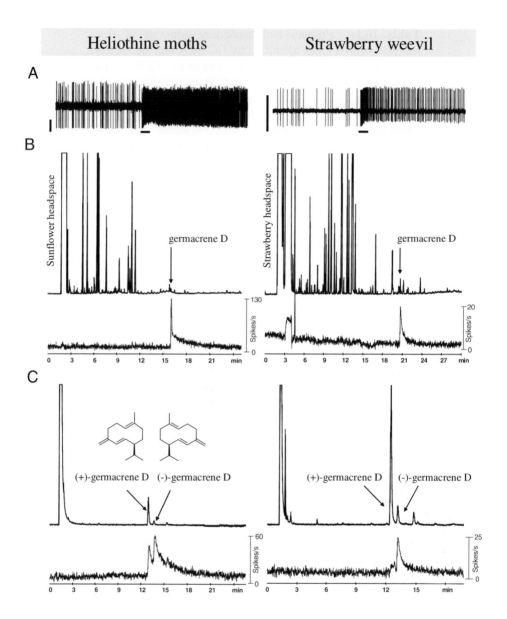

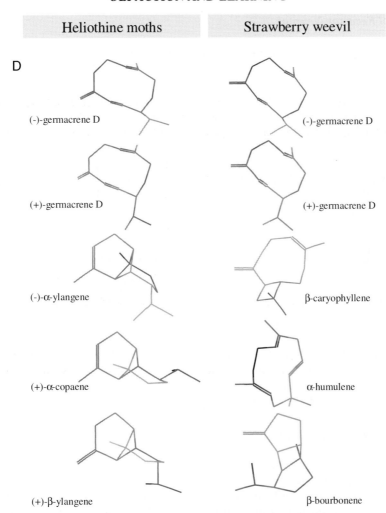

Heliothine moths	Strawberry weevil
D	
(-)-germacrene D	(-)-germacrene D
(+)-germacrene D	(+)-germacrene D
(-)-α-ylangene	β-caryophyllene
(+)-α-copaene	α-humulene
(+)-β-ylangene	β-bourbonene

Figure 10.1: The (-)-germacrene D receptor neuron type in heliothine moths and the strawberry weevil. **A.** Spike activity of a receptor neuron of *Heliothis virescens* (left) and *Anthonomus rubi* (right) recorded during direct stimulation of germacrene D (0.5 s, indicated by horizontal bars). Vertical bars are 0.5 mV. **B.** Gas chromatograms of headspace of hosts and simultaneously recorded neuron activity (spikes/s) of *H. virescens* (left) and *A. rubi* (right). Both neurons responded to a minor amount of (-)-germacrene D in the host. The neuron of *A. rubi* also responded to solvent and 3-pentanone. **C.** Gas chromatograms of germacrene D reference samples in two different concentrations [90% (+) and 10% (-)] and recorded activity of *H. virescens* (left) and *A. rubi* neuron (right). Higher sensitivity to (-)- than to (+)-germacrene D is shown by the strongest response to the smaller amount of this enantiomer. **D.** Molecular structures of the primary odorant (-)-germacrene D and some of the secondary odorants activating this receptor neuron type in *H. virescens* (left) and *A. rubi* (right).

Linalool Receptor Neurons

Receptor neurons responding primarily to the monoterpene linalool, a typical floral compound, have been identified in *H. virescens, M. brassicae, A. rubi*, and recently in *O. vitiosa*, living on angiosperms[63] (Røstelien *et al.* unpublished data, Bichão *et al.* unpublished data, Gregory S. Wheeler personal communication) (Table 10.1). The enantioselectivity of the neurons in these species have been tested by stimulation with linalool via a chiral gas chromatography column. All of them show the strongest response either to (+)- or to (-)-linalool, in both cases with a 10-fold stronger effect than the other enantiomer. In the strawberry weevil, both receptor neuron types have been demonstrated (Bichão *et al.* unpublished data), whereas only one type tuned to (-)-linalool has been found in *M. brassicae*[63] and to (+)-linalool in *H. virescens* and the weevil *O. vitiosa* (Røstelien *et al.* unpublished data , Gregory S. Wheeler personal communication). The enantioselectivity of the neurons in heliothine moths has only been tested by direct stimulation with enantiomeric samples, not 100% pure. All the linalool receptor neurons in these species respond secondarily to dihydrolinalool, a compound present in synthetic linalool but not yet identified in any of the samples collected from plants. Also, enantioselective responses to the two enantiomers of dihydrolinalool appear. Comparison of the other secondary odorants is under investigation. Linalool receptor neurons have also been reported in several other species living on angiosperms; moths,[41,64-67] beetles,[68,52,69] and the fruit fly.[70] So far, only a single receptor neuron tuned to linalool has been recorded in the weevil *H. abietis* living on conifers of which the headspace contained only a minor amount of the compound.[48] In the other weevil *P. notatus* with a conifer host, no linalool receptor neurons have been identified.

E-β-Ocimene Receptor Neurons

A relative large number of receptor neurons in heliothine moths respond strongly to *E*-β-ocimene, and more weakly to the structurally related compounds β-myrcene, *Z*-β-ocimene, dihydromyrcene, and *E*-DMNT (3*E*-4,8-dimethyl-1,3,7-nonatriene)[43,47] (Fig.10.2, Table 10.1). In all three species (*H. virescens, H. armigera, H. assulta*), the neurons show the same molecular receptive range and sensitivity to the five compounds. Also, in the weevil *A. rubi*, one neuron type responds strongly to *E*-β-ocimene and weakly to β-myrcene, *Z*-β-ocimene, dihydromyrcene, and *E*-DMNT (Fig.10.2).[51] However, in the weevil, these neurons have, in addition, several other secondary odorants (citronellol, geraniol, geranial, neral, limonene, γ-terpinene, and β-phellandrene), indicating a specificity of the receptors different from that in the heliothine moths. *E*-β-Ocimene receptor neurons have also been recorded in the eucalyptus woodborer *P. semipunctata*.[52] However, in the pine weevil species *H. abietis* and *P. notatus*, no receptor neurons responding to *E*-β-ocimene have been found despite numerous recordings. Instead, a receptor neuron type responding strongest to β-myrcene and more weakly to other related monoterpenes has been recorded in *H. abietis*.[48]

Methyl Salicylate Receptor Neurons

Receptor neurons detecting the typical plant stress compound methyl salicylate have been identified in the moth *M. brassicae* (Ulland *et al.* unpublished data) and in the strawberry weevil *A. rubi*,[51] but they are not found in heliothine moths or other weevils (Table 10.1). In both species, secondary responses to methyl benzoate were obtained. A difference between the specificity of the neuron type of these two species was demonstrated by a secondary response to ethyl benzoate, only occurring in the strawberry weevil. Interestingly, methyl salicylate and methyl benzoate are both produced by methylation of salicylic and benzoic acids, respectively, controlled by the same gene in *A. thaliana*.[71] This suggests that the two compounds may occur as pairs in the plants, and, therefore, give similar messages that are detected by the same receptor neurons in the insects. A different receptor neuron type identified in *H. armigera* tuned to methyl benzoate, with ethyl benzoate as a secondary odorant, did not respond to methyl salicylate, which was present in the tested plant material (Røstelien *et al.* unpublished data). Receptor neurons tuned to methyl salicylate have also been reported in other species on angiosperms, the fruit chaefer *Pachnoda marginata*[69] and the cabbage seed weevil *Ceuthonychus assimilis*.[41]

COMPARISON OF RECEPTOR NEURONS OF INSECT SPECIES LIVING ON GYMNO- AND ANGIOSPERM HOSTS

From the relatively large number of olfactory receptor neurons classified in moths and weevils, a major difference has appeared across species that use gymnosperm and angiosperm hosts. In the pine weevil *H. abietis* living on the Norway spruce (*Picea abies*), about 30 receptor neuron types have been found.[48] The majority of the recorded neurons respond to monoterpenes, and the largest group is tuned to bicyclic monoterpenes. Much fewer respond to sesquiterpenes, of which the primary odorant has only been identified for two out of the five neuron types, as α-copaene and a farnesene analogue, respectively. Also, in the weevil *P. notatus*, living on *Pinus pinaster*, the vast majority of receptor neurons are tuned to bicyclic monoterpenes, and fewer to monocyclic monoterpenes, and only one to a sesquiterpene.[50] In comparison, a relative larger number of the receptor neurons recorded in heliothine moths and the weevil *A. rubi* living on angiosperms, are tuned to sesquiterpenes. Thus, in the species presented here, a correlation appears between the primary odorants of the majority of receptor neurons and the principal

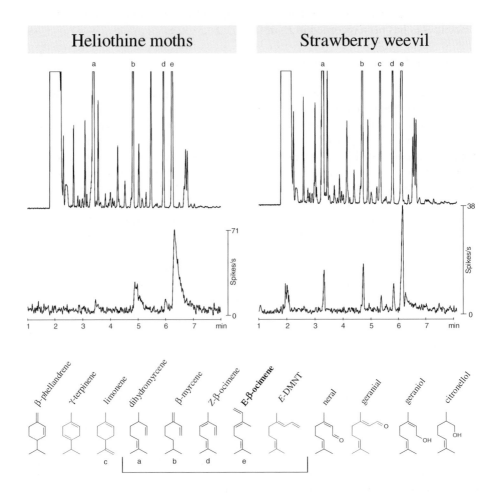

Figure 10.2: The *E*-β-ocimene receptor neuron type in heliothine moths and the strawberry weevil. Gas chromatograms of a standard sample and simultaneously recorded neuron activity (spikes/s) of a receptor neuron of *Heliothis virescens* (left) and of *Anthonomus rubi* (right). Both neurons responded to dihydromyrcene, β-myrcene, *Z*-β- and *E*-β-ocimene. The neuron of *A. rubi* also responded to the solvent and limonene. Below: The molecular structures of odorants activating the *E*-β-ocimene receptor neuron type in the strawberry weevil. The bracket indicates the molecular receptive range of the *E*-β-ocimene receptor neuron type in *H. virescens*.

compounds in the two groups of host plants. In addition, receptor neuron types with the same primary but different secondary odorants are present in the distantly related species of moths and weevils living on different host groups. Also, the two weevil species, *H. abietis* and *P. notatus*, considered as distantly related, are equipped with receptor neurons that are unique to each species, *i.e.*, having either different primary or secondary odorants. This indicates an independent evolution of these receptors, or possibly chance mutations of common ancestral receptor genes in the two weevil species during the adaptation to their conifer hosts. Another feature to compare among the species is possible overlap of the molecular receptive ranges of the receptor neurons, implying whether a compound is detected by one or more neuron types. In general, we have found little overlap of the molecular receptive ranges of the relative large number of neurons characterized[43-47] (Røstelien *et al.* unpublished data). This is particularly the case across chemical groups, like mono- and sesquiterpenes. However, within the same chemical group, like bicyclic monoterpenes, some overlap is found in weevils[48,50,51] (Bichão *et al.* unpublished data).

In summary, the identified odorants activating receptor neurons are constitutive in plants. Most of them, if not all, may be induced by various stress factors, as shown for many of the primary odorants presented here, *e.g.*, germacrene D, *E*-β-ocimene, and methyl salicylate, which are released in larger amounts in strawberry when infested with *A. rubi*.[51] Another aspect is that compounds produced in the same biosynthetic pathway, and thus occurring together in a given plant, may give similar messages to insects. This correlates with the findings that compounds activating the same receptor neurons are produced in the same pathway, such as methyl salicylate and methyl benzoate. Obviously, similar molecular features may by chance cause interaction with the same receptor protein, implying that secondary odorants have not played a significant role in forming the specificity of the neuron. Some compounds produced in the same major pathway activate different receptor neuron types, suggesting that they contribute with additional information in the recognition of a particular plant species or individual. Examples are the separate receptor neurons evolved in forest weevils that receive information about different bicyclic monoterpenes in conifers. In contrast, the enantiomers of germacrene D produced by different enzymes activate one and the same neuron type. Possibly, in this case, the relevant information is mediated only by (-)-germacrene D.

PROCESSING OF PLANT ODOR INFORMATION IN THE BRAIN

The presence of different receptor neuron types forms the basis for detection of and discrimination among odorants in the environment. In addition, important processing of the information takes place in the brain, resulting in the behavioral responses. The three olfactory areas of the insect brain are the antennal lobe, the mushroom bodies (important in learning and memory), and the lateral protocerebrum/lateral horn (premotoric area) (reviewed by[72]). In the antennal lobe, each subpopulation of receptor neurons converging in one or two specific glomeruli transmit the odor information to two major types of neurons, the local interneurons and the projection neurons innervating the particular glomeruli. The synaptic connections between them form a neuronal network for processing the information. In addition, modulatory neurons may influence the responses.[73] Local interneurons mediate interglomerular inhibition, causing enhanced contrast between activated and non-activated glomeruli, which seems important for odor discrimination. Projection neurons convey the odor information out of the antennal lobe via three antenno-cerebral tracts to higher orders of neurons in the protocerebrum, the Kenyon cells of the calyces of the mushroom bodies, and premotoric neurons in the lateral protocerebrum. The projection pattern of the receptor neuron types (responding mainly by excitation to plant odorants) indicates that the odor qualities are represented by activation of specific glomeruli. Thus, the quality of a plant odorant that is detected by a single type of receptor neurons (one subpopulation) is represented by activation of the one or two glomeruli in which these neurons converge. Obviously, an odorant activating more types elicits activation of more glomeruli. Thus, knowledge about the biologically relevant odorants and the degree of overlap of the molecular receptive ranges of the receptor neurons are important aspects for understanding how the biological relevant odor information is handled in the neuronal network of the antennal lobe of a species.

Processing of odor information in insects particularly has been elucidated for the pheromone system in male moths, based on the knowledge about the odorants and their behavioral effects.[74-77] Additionally, the olfactory system for pheromone information is simpler than the plant odor system, as expressed by the small number, usually 2-4 glomeruli dealing with pheromone information as compared to 60-67 glomeruli involved in plant odor information processing. Progress has also been made in studies of plant odor systems, by using intracellular recordings combined with staining of projection neurons or multiple recordings from populations of neurons.[78-87] The results have elucidated general characteristics of mechanisms involved in information processing, including spatial as well as temporal coding of odor qualities. In addition, optical recordings have been used successfully to map plant odor qualities to specific glomeruli and to study the correspondence between glomerular input and output, particularly in honeybees and in *D. melanogaster*.[88-91] However, knowledge about relevant plant odorants and the specificity of the receptor neurons is limited or lacking in many of the various species studied. The intention in our studies of heliothine moths is to reveal how relevant odor information mediated by the functionally characterized types of plant odor receptor neurons is transmitted and processed in the antennal lobe, and how these odor qualities are mapped to specific glomeruli. Obviously, this kind of study is challenging, both because of the large number of glomeruli and the technical difficulties involved particularly in intracellular recordings from the small neurons.

In studies of *H. virescens*, optical recordings have been performed during stimulation with a few of the identified primary and secondary plant odorants.[92,93] In these recordings, covering about 20% of the glomeruli in the anterior antennal lobe, each primary odorant elicited activity in specific glomeruli, according to what was expected. Additionally, three glomeruli were activated by two or three primary odorants, which is in contrast to the principle of each glomerulus receiving input from one subpopulation of receptor neurons. The results from optical recordings, indicating receptor neuron input to the glomeruli, have been compared with responses of projection neurons, representing the output. The projection neurons are characterized Physiologically and morphologically by intracellular recordings combined with staining with fluorescent dyes for visualization of the neuron by confocal laser scanning microscopy (CLSM), followed by 3-dimensional reconstruction in the software AMIRA[94] (Rø *et al.* unpublished data). These comparisons are aimed at finding out whether or not there is a correspondence of input and output of the odor quality information, and how different odor qualities are mapped to identified glomeruli in the atlases of the antennal lobes[24] (Skiri *et al.* unpublished data). So far, different projection neurons have been characterized by response properties, dendrite arborization in one or more glomeruli, axon in one of the three antenno-cerebral tracts, and projection pattern in the calyces of the mushroom bodies[94] (Rø *et al.* unpublished data). More data are needed in order to obtain an overall representation of odor quality of primary and secondary odorants in the glomeruli of the antennal lobe and to find out whether the different antenno-cerebral tracts in these moth species convey different biologically relevant information. Further important questions elucidated in other species concern synchronization of firing in populations of projection neurons[86,87] and how the information is transmitted to populations of Kenyon cells in the mushroom bodies and to the premotoric neurons in the lateral protocerebrum.[29,95]

BEHAVIORAL RESPONSES: INNATE AND MODIFIED BY OLFACTORY LEARNING

After knowing for which odorants the receptor neurons are tuned, behavioral studies should begin to gain information about the behavioral significance of single odorants and mixtures. In such time consuming experiments, one may first focus on the primary odorants that activate the largest groups of receptor neuron types. As mentioned above, most, if not all, neuron types are tuned to constitutive compounds that can also be induced. This indicates a variety of possible messages mediated by odorants. In heliothine moths, the behavioral effect has been tested for (-)-germacrene D, indicating attraction and increased oviposition of mated *H. virescens* females.[96] Behavioral experiments carried out in laboratories and in the field by other groups, have indirectly suggested attraction or inhibition by several compounds identified as primary odorants in the heliothine moths, discussed by Røstelien *et al.* (Røstelien *et al.* unpublished data). However, additional studies are needed to specifically determine the behavioral responses to more of the primary odorants and their mixtures. In the pine weevil, *H. abietis*, behavioral responses to some of the single, primary odorants have been tested, indicating attraction as well as inhibition, or no attraction[97] (Roten *et al.* unpublished data). Subsequently behavioral responses to mixtures containing the most attractive odorants have indicated a synergistic effect of the constituents, whereas other mixtures have indicated inhibition of attraction by two primary odorants. These studies continue to elucidate the biological significance of the primary odorants in these weevils.

In addition to the innate responses to odorants, changed behavior due to learning induced by previous experience must be considered. Important results on learning and memory and the underlying mechanisms come from studies of honeybees, which are based on experiments with conditioning of the proboscis extension response in appetitive learning.[29] The principle is that an odor stimulus (conditioned stimulus) followed by stimulation of the antennal taste sensilla with sucrose (unconditioned stimulus) also being the reward, makes the insect associate the odor with the reward. After one or more pairing trials, the insect extends the proboscis when stimulated with the odor alone, which is not the case in naive insects. Similar experiments carried out in heliothine moths have demonstrated their ability to learn odors, although not as fast as honeybees.[35,31] In order to trace the neuronal connections between the pathways mediating the unconditioned (taste) and the conditioned (olfactory) information, the projections of the gustatory receptor neurons have been traced. Two populations of gustatory receptor neurons are present, one located in the contact chemosensilla on the antennae (*sensilla chaeticae*) and another in *sensilla styloconicae* on the proboscis. By the use of fluorescent dyes for visualization in CLSM and reconstruction in AMIRA, the projections are found in two closely located finger-like areas in the suboesophageal ganglion.[98,99] One neuron has been found that might form the connection between the taste and the olfactory pathway (Rø *et al.* unpublished data). The morphology of this neuron resembles the neuron (VUMmx1) involved in associative learning in the honeybee, which is shown to have the ability when depolarized to substitute the sucrose reward in the learning performance.[100] Possibly, the neuron in the heliothine moths may have a similar function.

The identified primary and secondary odorants have been selected for testing the ability of heliothine moths and weevils to learn these odors. As mentioned above, in heliothine moths, we have used the proboscis extension response in appetitive learning, as in the honeybees.[31] Also, the hypothesis that odorants activating different receptor neurons can more easily be discriminated than those activating the same neurons, has been tested by differential conditioning. The data show that although the insect was able to distinguish between the primary odorant *E*-β-ocimene and β-myrcene activating the same neuron type, the two compounds were more easily confused than the two primary odorants tested. Further studies of olfactory learning in these species are being performed by classical conditioning of the proboscis extension response. Other groups carry out other interesting learning studies using wind-tunnel experiments, where heliothine moths that have been exposed to an odor in connection with feeding are tested for preference to the odorant.[32] In the pine weevil, *H. abietis,* olfactory learning has been studied by a multiple-choice bioassay (Roten *et al.* unpublished data). Here, the pine weevils that fed on different hosts of conifers were shown to prefer different enantiomers of an attractant when given equal choices. Interestingly, they preferred the enantiomer that dominates in the host they had experienced. The data obtained in the various learning studies, show that olfactory learning in addition to the innate responses to odors may play a significant role for host attraction and recognition in these herbivorous insect species.

SUMMARY

In this paper we have made comparisons between receptor neuron specificities in a few species of moths and weevils that use angiosperm and gymnosperm hosts. The recorded plant odor receptor neurons are narrowly tuned and classified according to one compound having strongest effect, defined as primary odorant, and to a few related compounds having weak effect, defined as secondary odorants. Neurons with the same primary odorant have been found in closely as well as distantly related species. Those of closely related heliothine moths have also the same secondary odorants, whereas the molecular receptive ranges differ across distantly related species. This suggests conservation of olfactory receptors in the related moths and convergent evolution or chance mutations of olfactory genes in the distantly related species during the adaptation to their host plants. Major differences between these particular species using gymnosperm and angiosperm hosts appear in the number of neurons tuned to mono- and sesquiterpenes, respectively. Obviously, data on more species are needed in order to find out whether the similarities and differences discussed apply to other species that use the two categories of hosts. Future studies *e.g.,* of moth living on conifers as well as more weevil species living on angiosperms, may shed more light on this topic. Knowledge about receptor neuron specificity is also important for further studies of how plant odor quality is handled in the central nervous system and how plant odorants influence the behavior of the insects, both as innate and as learned responses. Processing of the odor information and representation of odor qualities in the antennal lobe is elucidated by intracellular recordings combined with staining and optical recordings. Also the neuronal connection between the olfactory and taste pathways, *i.e.,* the pathways of the unconditioned and conditioned stimuli in appetitive learning, is search for by intracellular recordings and staining of neurons. The importance of olfactory learning also in weevils is shown by changed preference for odorants after experience with different host plants.

ACKNOWLEDGEMENTS

The Norwegian Research Council is acknowledged for the financial support to our research. We thank Dr. Anna-Karin Borg-Karlson and Ph.D. student Helena Bichão for contributions to the figures.

REFERENCES

1. ARABIDOPSIS GENOME INITIATIVE, Analysis of the genome sequence of the flowering plant *Arabidopsis thaliana*, *Nature*, 2000, **408**, 796-816.
2. BOHLMANN, J., MEYER-GAUEN, G., CROTEAU, R., Plant terpenoid synthases: Molecular biology and phylogenetic analysis, *Proc. Natl. Acad. Sci. USA*, 1998, **95**, 4126-4133.
3. PARÉ, P. W. TUMLINSON, J. H., Plant volatiles as a defense against insect herbivores, *Plant Physiol.*, 1999, **121**, 325-331.
4. DUDAREVA, N., PIRCHERSKY, E., GERSHENZON, J., Biochemistry of plant volatiles, *Plant Physiol.*, 2004, **135**, 1893-1902.
5. KNUDSEN, J. T., TOLLSTEN, L., BERGSTRÖM, L. G., Floras scents - a checklist of volatile compounds isolated by headspace techniques, *Phytochemistry*, 1993, **33**, 253-280.
6. MARTIN, D. M., FÄLDT, J., BOHLMANN, J., Functional characterization of nine norway spruce tps genes and evolution of gymnosperm terpene synthases of the TPS-d Subfamily, *Plant Physiol.*, 2004, **135**, 1908-1927.
7. BUCK, L., AXEL, R., A novel multigene family may encode odorant receptors: a molecular basis for odour recognition, *Cell*, 1991, **65**, 175-187.
8. MOMBAERTS, P., Molecular biology of odorant receptors in vertebrates, *Annu. Rev. Neurosci*, 1999, **22**, 487-509.
9. BUCK L.B., The search for odorant receptors, *Cell*, 2004, **S116**, S117-S119.
10. VOSSHALL, L. B., WONG, A. M., AXEL, R., An olfactory sensory map in the fly brain, *Cell*, 2000, **102**, 147-159.
11. HALLEM, E. A., HO, M. G., CARLSON, J. R., The molecular basis of odor coding in the *Drosophila* antenna, *Cell*, 2004, **117**, 965-979.
12. VOGT, R. G. and RIDDIFORD, L. M., Pheromone binding and inactivation by moth antennae, *Nature*, 1981, **293**, 161-163.
13. STEINBRECHT, R. A., LAUE, M., ZIEGELBERGER, G., Immunolocalization of pheromone-binding protein and general odorant-binding protein in olfactory sensilla of the silk moths *Antheraea* and *Bombyx*, *Cell Tissue Res.*, 1995, **282**, 203-217.
14. KRAUTWURST, D., YAU, K.-W., REED, R. R., Identification of ligands for olfactory receptors by functional expression of a receptor library, *Cell*, 1998, **95**, 917-926.

15. RESSLER, K. J., SULLIVAN, S. L., BUCK, L. B., Information coding in the olfactory system: Evidence for a stereotyped and highly organized epitope map in the olfactory bulb, *Cell*, 1994, **79**, 1245-1255.
16. VASSAR, R., CHAO, S. K., SITCHERAN, R., NUÑEZ, J. M., VOSSHALL, L. B., AXEL, R., Topographic organization of sensory projections to the olfactory bulb, *Cell*, 1994, **79**, 981-991.
17. MOMBAERTS, P., WANG, F., DULAC, C., CHAO, S. K., NEMES, A., MENDELSOHN, M., EDMONDSON, J., AXEL, R., Visualizing an olfactory sensory map, *Cell*, 1996, **87**, 675- 686.
18. GAO, Q., YUAN, B., CHESS, A., Convergent projections of *Drosophila* olfactory neurons to specific glomeruli in the antennal lobe, *Nature Neurosci.*, 2000, **3**, 780-785.
19. GALIZIA, C. G., MCILWRATH, S. L., MENZEL, R., A digital three-dimensional atlas of the honeybee antennal lobe based on optical sections acquired by confocal microscopy, *Cell Tissue Res.*, 1999, **295**, 383-394.
20. SMID, H. M., BLEEKER, M. A., VAN LOON, J. J. A., VET, L. E., Three-dimensional organization of the glomeruli in the antennal lobe of the parasitoid wasps *Cotesia glomerata* and *C. rubecula*, *Cell Tissue Res.*, 2003, **312**, 237-248.
21. ROSPARS, J. P., Invariance and sex-specific variations of the glomerular organization in the antennal lobes of a moth, *Mamestra brassicae* and a butterfly *Pieris brassicae* , *J. Comp. Neurol.*, 1983, **220**, 80-96.
22. ROSPARS, J. P., HILDEBRAND, J. G., Anatomical identification of glomeruli in the antennal lobes of the sphinx moth *Manduca sexta, Cell Tissue Res.*, 1992, **270**, 205-227.
23. ROSPARS, J. P., HILDEBRAND, J. G., Sexually dimorphic and isomorphic glomeruli in the antennal lobes of the sphinx moth *Manduca sexta, Chem. Senses*, 2000, **25**, 119-129.
24. BERG, B. G., GALIZIA, C. G., BRANDT, R., MUSTAPARTA, H., Digital atlases of the antennal lobe in two species of tobacco budworm moths, the oriental *Helicoverpa assulta* (male) and the American *Heliothis virescens* (male and female), *J. Comp. Neurol.*, 2002, **446**, 123-134.
25. GREINER, B., GADENNE, C., ANTON, S., Three-dimensional antennal lobe atlas of the male moth, *Agrotis ipsilon*: A tool to study structure-function correlation, *J. Comp. Neurol.*, 2004, **475**, 202-210.
26. LAISSUE, P. P., REITER, C., HIESINGER, P. R., HALTER, S., FISCHBACH, K. F., STOCKER, R. F., Three-dimensional reconstruction of the antennal lobe in *Drosophila melanogaster*, *J. Comp. Neurol.*, 1999, **405**, 543-552.
27. MATTHEWS, M., Heliothine Moths of Australia. A Guide to Pest Bollworms and Related Noctuid Groups, 1999, Coolingwood, Australia: CSIRO Publishing.
28. BITTERMAN, M. E., MENZEL, R., FIETZ, A., SCHÄFER, S., Classical conditioning of proboscis extension in honeybees (*Apis mellifera*), *J. Comp. Psychol.*, 1983, **97**, 107-119.
29. MENZEL, R., Searching for the memory trace in a mini-brain, the honeybee, *Learning & Memory*, 2001, **8**, 53-62.
30. MENZEL, R., MÜLLER, U, Learning and memory in honeybees: From behavior to neural substrates, Annu.Rev.Neurosci, 1996, **19**, 379-404.

31. SKIRI, H. T., STRANDEN, M., SANDOZ, J. C., MENZEL, R., MUSTAPARTA, H., Associative learning of plant odorants activating the same or different receptor neurones in the moth *Heliothis virescens*, *J.Exp. Biol.*, 2005, **208**, 787-796.

32. CUNNINGHAM, J. P, MOORE, C. J, ZALUCKI, M. P, WEST, S. A, Learning, odour preference and flower foraging in moths, *J.Exp.Biology*, 2004, **207**, 87-94.

33. DALY, K. C., SMITH, B. H., Associative olfactory learning in the moth *Manduca sexta*, *J. Exp. Biol.*, 2000, **203**, 2025-2038.

34. FAN, R.-J., ANDERSON, P., HANSSON, B. S., Behavioural analysis of olfactory conditioning in the moth *Spodoptera littoralis* (Boisd.) (Lepidoptera: Noctuidae), *J. Exp. Biol.*, 1997, **200**, 2969-2976.

35. HARTLIEB, E., Olfactory conditioning in the moth *Heliothis virescens*, *Naturwissenschaften*, 1996, **83**, 87-88.

36. HARTLIEB, E., ANDERSON, P., HANSSON, B. S., Appetitive learning of odours with different behavioural meaning in moths, *Physiol.ogy & Behavior*, 1999, **67**, 671-677.

37. HARTLIEB, E., HANSSON, B. S., Sex of food? Appetetive learning of sex odors in a male moth, *Naturwissenschaften*, 1999, **86**, 396-399.

38. CUNNINGHAM, J. P., ZALUCKI, M. P., WEST, S. A., Learning in *Helicoverpa armigera* (Lepdioptera: Noctuidae): A new look at the behaviour and control of a polyphagous pest, *Bull. Ent. Res.*, 1999, **89**, 201-207.

39. WADHAMS, L. J., Coupled gas chromatography - Single cell recording: a new Technique for use in the analysis of insect pheromones, *Z. Naturforsch.*, 1982, **37c**, 947-952.

40. TØMMERÅS, B. Å., MUSTAPARTA, H., Single cell responses to pheromones, host and non-host volatiles in the ambrosia beetle *Typodendron lineatum*, *Entomol. Exp. Appl.*, 1989, **52**, 141-148.

41. BLIGHT, M. M., PICKETT, J. A., WADHAMS, L. J., WOODCOCK, C. M., Antennal perception of oilseed rape, *Brassica napus* (Brassicaceae), volatiles by the cabbage seed weevil *Ceutorhynchus assimilis* (Coleoptera, Curculionidae), *J. Chem. Ecol.*, 1995, **21**, 1649-1664.

42. WIBE, A., MUSTAPARTA, H, Encoding of plant odours by receptor neurons in the pine weevil (*Hylobius abietis*) studied by linked gas chromatography-electrophysiology, *J. Comp. Physiol. A*, 1996, **179**, 331-344.

43. RØSTELIEN, T., BORG-KARLSON, A.-K., MUSTAPARTA, H., Selective receptor neurone responses to *E*-β-ocimene, β-myrcene, *E,E*-α-farnesene and *homo*-farnesene in the moth *Heliothis virescens*, identified by gas chromatography linked to electrophysiology, *J. Comp. Physiol.. A*, 2000, **186**, 833-847.

44. STRANDEN, M., BORG-KARLSON, A.-K., MUSTAPARTA, H., Receptor neuron discrimination of the germacrene D enantiomers in the moth *Helicoverpa armigera*, *Chem. Senses*, 2002, **27**, 143-152.

45. RØSTELIEN, T., BORG-KARLSON, A.-K., FÄLDT, J., JACOBSSON, U., MUSTAPARTA, H., The plant sesquiterpene germacrene D specifically activates a major type of antennal receptor neuron of the tobacco budworm moth *Heliothis virescens*, *Chem. Senses*, 2000, **25**, 141-148.

46. STRANDEN, M., LIBLIKAS, I., KÖNIG, W. A., ALMAAS, T. J., BORG-KARLSON, A.-K., MUSTAPARTA, H., (-)-Germacrene D receptor neurones in three species of heliothine moths: structure-activity relationships, *J. Comp. Physiol. A*, 2003, **189**, 563-577.

47. STRANDEN, M., RØSTELIEN, T., LIBLIKAS, I., ALMAAS, T. J., BORG-KARLSON, A.-K., MUSTAPARTA, H., Receptor neurones in three heliothine moths responding to floral and inducible plant volatiles, *Chemoecology*, 2003, **13**, 143-154.

48. WIBE, A, BORG-KARLSON, A.-K, NORIN, T, MUSTAPARTA, H, Identification of plant volatiles activating single receptor neurons in the pine weevil (*Hylobius abietis*), J. Comp. Physiol. A, 1997, 180, 585-595.

49. WIBE, A., BORG-KARLSON, A.-K., NORIN, T., MUSTAPARTA, H., Enantiomeric composition of monoterpene hydrocarbons in some conifers and receptor neuron discrimination of alpha-pinene and limonene enantiomers in the pine weevil, *Hylobius abietis, J. Chem. Ecol.*, 1998, **24**, 273-287.

50. BICHÃO, H., BORG-KARLSON, A.-K., ARAÚJO, J., MUSTAPARTA, H., Identification of plant odours activating receptor neurones in the weevil *Pissodes notatus* F. (Coleoptera, Curculionidae), *J. Comp. Physiol. A*, 2003, **189**, 203-212.

51. BICHÃO, H., BORG-KARLSON, A.-K., ARAÚJO, J., MUSTAPARTA, H., Five Types of Olfactory Receptor Neurons in the Strawberry Blossom Weevil Anthonomus rubi: Selective Responses to Inducible Host-plant Volatiles, *Chem. Senses*, 2005, **30**, 153-170.

52. BARATA, E. N, MUSTAPARTA, H, PICKETT, J. A, WADHAMS, L. J, ARAÚJO, J, Encoding of host and non-host plant odours by receptor neurones in the eucalyptus woodborer, *Phoracantha semipunctata* (Coleoptera: Cerambycidae), *J. Comp. Physiol. A*, 2002, 188, 121-133.

53. BLIGHT, M. M., PICKETT, J. A., WADHAMS, L. J., WOODCOCK, C. M., Antennal perception of oilseed rape, *Brassica napus* (Brassicaceae), volatiles by the cabbage seed weevil *Ceutorhynchus assimilis* (Coleoptera, Curculionidae), *J. Chem. Ecol.*, 1995, **21**, 1649-1664.

54. STENSMYR, M. C., LARSSON, M. C., BICE, S., HANSSON, B. S., Detection of fruit- and flower-emitted volatiles by olfactory receptor neurons in the polyphagous fruit chafer *Pachnoda marginata* (Coleoptera: Cetoniinae), *J. Comp. Physiol. A*, 2001, **187**, 509-519.

55. STENSMYR, M. C., GIORDANO, E., BALLIO, A., ANGIOY, A. M., HANSSON, B. S., Novel natural ligands for *Drosophila* olfactory receptor neurones, *J. Exp. Biol.*, 2003, **206**, 715-724.

56. HALLEM, E. A., HO, M. G., CARLSON, J. R., The Molecular Basis of Odor Coding in the *Drosophila* Antenna, *Cell*, 2004, **117**, 965-979.

57. RØSTELIEN, T., BORG-KARLSON, A.-K., FÄLDT, J., JACOBSSON, U., MUSTAPARTA, H., The plant sesquiterpene germacrene D specifically activates a major type of antennal receptor neuron of the tobacco budworm moth *Heliothis virescens*, *Chem. Senses*, 2000, **25**, 141-148.

58. STRANDEN, M, BORG-KARLSON, A.-K, MUSTAPARTA, H, Receptor neuron discrimination of the germacrene D enantiomers in the moth *Helicoverpa armigera*, *Chem. Senses*, 2002, **27**, 143-152.

59. STRANDEN, M., LIBLIKAS, I., KÖNIG, W. A., ALMAAS, T. J., BORG-KARLSON, A.-K., MUSTAPARTA, H., (-)-Germacrene D receptor neurones in three species of heliothine moths: structure-activity relationships, *J. Comp. Physiol. A*, 2003, **189**, 563-577.

60. SCHMIDT, C. O., BOUWMEESTER, H. J., DE KRAKER, J.-W., KÖNIG, W. A., Biosynthesis of (+)- and (-)-germacrene D in *Solidago canadensis*: isolation and characterzation of two enantioselective germacrene D synthases, *Angew. Chem. Int. Ed.*, 1998, **37**, 1400-1402.

61. SCHMIDT, C. O., BOUWMEESTER, H. J., FRANKE, S., KÖNIG, W. A., Mechanisms of the biosynthesis of sesquiterpene enantiomers (+)- and (-)-germacrene D in *Solidago canadensis*, *Chirality*, 1999, **11**, 353-362.

62. PROSSER, I., ALTUG, I. G., PHILLIPS, A. L., KÖNIG, W. A., BOUWMEESTER, H. J., BEALE, M. H., Enantiospecific (+)- and (-)-germacrene D synthases, cloned from goldenrod, reveal a functionally active variant of the universal isoprenoid-biosynthesis aspartate-rich motif, *Arch. Biochem. Biophys.*, 2004, **432**, 136-144.

63. ULLAND, S., BICHÃO, H., STRANDEN, M., RØSTELIEN, T., BORG-KARLSON, A.-K., MUSTAPARTA, H., Olfactory receptor neurones tuned to linalool in moths and weevils: structural-activity relationships, Conference Abstract, 2004, The 16th Congress of the European Chemoreception Research Organization, Dijon,

64. ANDERSON, P., HANSSON, B. S., LØFQVIST, J., Plant.odour-specific receptor neurones on the antennae of female and male *Spodoptera littoralis*, *Physiol. Entomol.*, 1995, **20**, 189-198.

65. ANDERSON, P., LARSSON, M., LØFQVIST, J., HANSSON, B. S., Plant odour receptor neurones on the antennae of the two moths *Spodoptera littoralis* and *Agrotis segetum*, *Entomol. Exp. Appl.*, 1996, **80**, 32-34.

66. JÖNSSON, M., ANDERSON, P., Electrophysiological response to herbivore-induced host plant volatiles in the moth *Spodoptera littoralis*, *Physiol. Entomol.*, 1999, **24**, 377-385.

67. SHIELDS, V. D. C., HILDEBRAND, J. G., Responses of a population of antennal olfactory receptor cells in the female moth *Manduca sexta* to plant-associated volatile organic compounds, *J. Comp. Physiol. A*, 2001, **186**, 1135-1151.

68. DICKENS, J. C., Specialized receptor neurons for pheromones and host plant odors in the boll weevil, *Anthonomus grandis* Boh. (Coleoptera: Curculionidae), *Chem. Senses*, 1990, **15**, 311-331.

69. STENSMYR, M. C., LARSSON, M. C., BICE, S., HANSSON, B. S., Detection of fruit- and flower-emitted volatiles by olfactory receptor neurons in the polyphagous fruit chafer *Pachnoda marginata* (Coleoptera: Cetoniinae), *J. Comp. Physiol. A*, 2001, **187**, 509-519.

70. DE BRUYNE, M., FOSTER, K., CARLSON, J. R., Odor coding in the *Drosophila* antenna, *Neuron*, 2001, **30**, 537-552.

71. CHEN, F., D'AURIA, J. C., THOLL, D., ROSS, J. R., GERSHENZON, J., NOEL, J. P., PICHERSKY, E., An *Arabidopsis thaliana* gene for methylsalicylate biosynthesis, identified by a biochemical genomics approach, has a role in defense, *The Plant Journal*, 2003, **36**, 577-588.

72. YACK, J. E., HOMBERG, U., Nervous System, *in* Handbook of Zoology. Lepidoptera, Moths and Butterflies, Volume 2: Morphology, Physiology and Development (N. P. Kristensen, ed.), Walter de Gruyter, Berlin, 2003, pp. 229-265.

73. KLOPPENBURG, P., FERNS, D., MERCER, A. R., Serotonin Enhances Central Olfactory Neuron Responses to Female Sex Pheromone in the Male Sphinx Moth *Manduca sexta*, *J. Neurosci.*, 1999, **19**, 8172-8181.

74. HEINBOCKEL, T., CHRISTENSEN, T. A., HILDEBRAND, J. G., Representation of binary pheromone blends by glomerulus-specific olfactory projection neurons, *J.Comp. Physiol. A*, 2004, **190**, 1023-1037.

75. VICKERS, N. J., CHRISTENSEN, T. A., BAKER, T. C., HILDEBRAND, J. G., Odour-plume dynamics influence the brain's olfactory code, *Nature*, 2001, **410**, 466-470.

76. HANSSON, B. S., CHRISTENSEN, T. A., Functional Characteristics of the Antennal Lobe, *in* Insect Olfaction (B. S. Hansson, ed.), Springer-Verlag, Berlin Heidelberg, 1999, pp. 126-162.

77. MUSTAPARTA, H., Encoding of plant odour information in insects: peripheral and central mechanisms, *Entomol. Exp. Appl.*, 2002, **104**, 1-13.

78. ANTON, S., HANSSON, B. S., Sex-pheromone and plant-associated odor processing in antennal lobe interneurons of male *Spodoptera littoralis* (Lepidoptera, Noctuidae), *J. Comp. Physiol. A*, 1995, **176**, 773-789.

79. STOPFER, M., BHAGAVAN, S., SMITH, B. H., LAURENT, G., Impaired odour discrimination on desynchronization of odour-encoding neural assemblies, *Nature*, 1997, **390**, 70-74.

80. ROCHE KING, J., CHRISTENSEN, T. A., HILDEBRAND, J. G., Response characteristics of an identified, sexually dimorphic olfactory glomerulus, *J. Neurosci.*, 2000, **20**, 2391-2399.

81. GREINER, B., GADENNE, C., ANTON, S., Central processing of plant volatiles in *Agrotis ipsilon* males is age-independent in contrast to sex pheromone processing, *Chem. Senses*, 2002, **27**, 45-48.

82. MASANTE-ROCA, I., GADENNE, C., ANTON, S., Plant odour processing in the antennal lobe of male and female grapevine moths, *Lobesia botrana* (Lepidoptera: Tortricidae), *J. Insect Physiol.*, 2002, **48**, 1111-1121.

83. MÜLLER, D., ABEL, R., BRANDT, R., ZÖCKLER, M., MENZEL, R., Differential parallel processing of olfactory information in the honeybee, *Apis mellifera* L, *J. Comp. Physiol. A*, 2002, **188**, 359-370.

84. REISENMAN, C. E., CHRISTENSEN, T. A., FRANCKE, W., HILDEBRAND, J. G., Enantioselectivity of projection neurons innervating identified olfactory glomeruli, *J. Neurosci.*, 2004, **24**, 2602-2611.

85. STOPFER, M. LAURENT, G., Short-term memory in olfactory network dynamics, *Nature*, 1999, **402**, 664-668.

86. LAURENT, G., Olfactory network dynamics and the coding of multidimensional signals, *Nature Rev.*, 2002, **3**, 884-895.

87. LEI, H., CHRISTENSEN, T. A., HILDEBRAND, J. G., Spatial and Temporal Organization of Ensemble Representations for Different Odor Classes in the Moth Antennal Lobe, *J. Neurosci.*, 2004, **24**, 11108-11119.

88. PELZ, D., ROESKE, C. C., GALIZIA, C. G., Functional response spectrum of genetically identified olfactory sensory neurons in the fruit fly *Drosophila melanogaster*, Conference Abstract, 8th European Symposium in insect taste and olfaction, 2003.

89. SACHSE, S., GALIZIA, C. G., Role of inhibition for temporal and spatial odor representation in olfactory output neurons: A calcium imaging study, *J. Neurophysiol.*, 2002, **87**, 1106-1117.

90. SACHSE, S., GALIZIA, C. G, The coding of odour-intensity in the honeybee antennal lobe: local computation optimizes odour representation, Eur. J. Neurosci, 2003, 18, 2119-2132.

91. GALIZIA, C. G., MENZEL, R., The role of glomeruli in the neural representation of odours: results from optical recording studies, *J. Insect Physiol.*, 2001, **47**, 115-130.

92. GALIZIA, C. G., SACHSE, S., MUSTAPARTA, H., Calcium responses to pheromones and plant odours in the antennal lobe of the male and female moth *Heliothis virescens, J. Comp. Physiol. A*, 2000, **186**, 1049-1063.

93. SKIRI, H. T., GALIZIA, C. G., MUSTAPARTA, H., Representation of primary plant odorants in the antennal lobe of the moth *Heliothis virescens*, using calcium imaging, *Chem. Senses*, 2004, **29**, 253-267.

94. MÜLLER, D., RØ, H., MUSTAPARTA, H., Olfaction in Heliothine moths: IV. Coding properties and dendrite arborizations in identified glomeruli of antennal lobe output neurones, *FENS abstr*, 2002, **1**, 470.

95. HEISENBERG, M., Mushroom body memoir: from maps to models, *Nature Rev. Neurosci.*, 2003, **4**, 266-275.

96. MOZURAITIS, R., STRANDEN, M., RAMIREZ, M. I., BORG-KARLSON, A.-K., MUSTAPARTA, H., (-)-Germacrene D increases attraction and oviposition by the tobacco budworm moth *Heliothis virescens, Chem. Senses*, 2002, **27**, 505-509.

97. ROTEN, Ø. O., Behavioural responses by the pine weevil *Hylobius abietis* to conifer produced odorants detected by the receptor neurones, 2003, Master Thesis at NTNU, Trondheim.

98. JØRGENSEN, K., ALMAAS, T. J., BJAALIE, J. G., MUSTAPARTA, H., Taste in Heliothine moths:I. Physiological characteristics of the antennal taste receptor neurones and the projection pattern in the suboesophageal ganglion, *FENS abstr.*, 2002, **1**, 470.

99. KVELLO, P., ALMAAS, T. J., MUSTAPARTA, H., BJAALIE, J. G., Taste in Heliothine moths:II. Morphology of the contact chemosensilla on the proboscis and the projection patterns of the associated receptor neurones in the primary taste centre of the CNS, *FENS abstr.*, 2002, **1**, 470.

100. HAMMER, M., An identified neuron mediates the unconditioned stimulus in associative olfactory learning in honeybees, *Nature*, 1993, **366**, 59-63.

INDEX